AF540308

ENCYCLOPAEDIA OF NANOTECHNOLOGY-IV

NANOMEDICINE

By

Dr. M.P. Arora
Dept. of Zoology
M.M.H. Post Graduate College
Ghaziabad
(U.P.)

DISCOVERY PUBLISHING HOUSE PVT. LTD.
NEW DELHI-110 002

First Published-2008

ISBN 978-81-8356-280-5

Published by

DISCOVERY PUBLISHING HOUSE PVT. LTD.
4831/24, Ansari Road, Prahlad Street,
Darya Ganj, New Delhi-110002 (India)
Phone: 23279245 • Fax: 91-11-23253475
E-mail: dphbooks@rediffmail.com
dphtemp@indiatimes.com

Printed at:
Sachin Printers, Delhi

Preface

Everyone in the modern world knows what technology is. But what is nanotechnology? Taken from the Greek, *nano* means "one billionth part of" a whole. In modern parlance, it means very, very small. In today's world of smaller cell phones and portable computers, miniaturization is the current technology. Nanotechnology is the next step after miniaturization. In tomorrow's world, nanotechnology will be the new common technology. It will affect everyone of the planet and many change civilization as we know it.

Nanotechnology is the molecular engineering. It will soon create effective machines as small as DNA. It describes the ideas and techniques that are creating or new domain of science and technology. Nanotechnology holds the promise of revolutionizing field ranging from medical science to optical communications, but requires a substantial investment in time and money to make the promise become a readily. The results could be nothing less than a new industrial revolution.

In an undertaking as new as nanotechnology, many different approaches are studied in order to ascertain the best way to fashion working devices. At present, techniques can be divided into one of two general categories. One technique, termed the top-down method, physically manipulates molecules into nanostructures. The other technique, termed the bottom-up method, coaxes atoms and molecules to assemble themselves into nanostructures.

The present title **"Nanomedicine"** has been designed for undergraduate and post-graduate students as well as those involved in basic research in biological, biochemical and biophysical sciences.

In the preparation of this book large number of books and research papers have been consulted. So no authenticity is claimed.

The author expresses his gratitude to Mr. Wasan and staff of M/s Discovery Publishing House for their whole hearted co-operation in the publication of this book.

The author tried hard to be accurate and upto date in statement and realises the impossibility of completely avoiding errors therefore, the author will greatly appreciate having his attention called to any questionable statement.

Author

Contents

1. Introduction **1—9**

Magnetization, Quantum Dot, Application of Quantum Dots, Quantum Dots in Life Sciences, Dendrimer Molecules, Nanoshells.

2. Nanomedicine **10—33**

Molecular Body, Body as Workyard, Body as Construction Site, Body as Battlefield, Medicine Today, Crude Methods, Limited Abilities, Nanotechnology in Medicine, Working Outside Tissues, Working Within Tissues, Eliminating Invaders, Herding Cells, Rebuilding Tissues, Working on Cells, Eliminating Viruses by Cell Surgery, Molecular Repairs, Healing Body and Limb, Correcting Chemistry, New Organs and Limbs, First Aid, Plague Insurance, Threat, Inadequate Abilities, Doing Better, Solving Hard Problems, Aging, Restoring Species.

3. Paths to Nanotechnology **34—44**

Colonization Metaphor, Mining Metaphor, Molecular Savvy, Profitability, Healthy, Active Research, Commercial Infrastructure, Systems Mentality, Continuous Path to Nanotechnology, Existing Fields and Industries, Micromechanics, Microelectronics and Computers, Biotechnology, Chemistry.

4. Prospect of Nanomedicine **45—142**

Current Medical Practice, Evolution of Scientific Medicine, Volitional Normative Model of Disease, Treatment Methodology, Evolution of Bedside Practice, Changing View of the Human Body, Nanomedical Perspective, Nanomedicine and Molecular

Nanotechnology, Nanomedicine: History of the Idea, Biotechnology and Molecular Nanotechnology, Naturophilia.

5. Molecular Manufacturing **143–203**

Top-Down Approaches to Nanotechnology, Bottom-up Pathways to Molecular Manufacturing, Biotechnology, Molecular and Supramolecular Chemistry, Scanning Probe Technology, Molecular Components and Molecular Assemblers, Molecular Assemblers.

6. Molecular Transport and Sortation **204–250**

Human Body Chemical Composition, Diffusion Transport, Brownian Motion, Passive Diffusive Intake, Active Diffusive Intake, Diffusion Cascade Sortation, Nanocentrifugal Sortation, Membrane Filtration, Simple Nanosieving, Dynamic Pore Sizing, Gated Channels, Receptor-Based Transport, Transporter Pumps, Sorting Rotors, Internal Transport Streams, Molecular Receptor Engineering, Physical Forces in Molecular Recognition, Ligand-Receptor Affinity, Ligand-Receptor Specificity, Ligand-Receptor Dynamics, Diamondoid Receptor Design, Minimum Feature Size and Positioning Accuracy, Receptor Configurations, Ligand-Receptor Mapping, Large Molecule Binding, Sorting, and Transport.

7. Nanosensor and Nanoscale Scanning **251–323**

Chemical and Molecular Nanosensors, Broadband Receptor Arrays, Narrowband Receptor Arrays, Counting Rotors, Chemical Assay, Chemical Nanosensor Theoretical Limits, Spatial Concentration Gradients, Temporal Concentration Gradients, Chemotactic Sensor Pads, Receptor Sensors, Displacement and Motion Sensors, Displacement Sensors, Velocity and Flow Rate Sensors, Acceleration Sensors, Angular Displacement, Force Nanosensors, Minimum Detectable Force, Nanogravimeters, Single-Proton Massometer, Isotope Discrimination, Pressure Sensing, Minimum Detectable Pressure, Spatial Pressure Gradients, Temporal Pressure Gradients, Ullage Sensors, Thermal Nanosensors, Minimum Detectable Temperature Change, Piston-Based Temperature Sensors, Thermal-Expansion Temperature Sensors, Mechanochemical Temperature Sensors, Spatial Thermal Gradients, Temporal Thermal Gradients, Electric and Magnetic Sensing, Electric Fields, Magnetic Fields, Optical Sensing, Particulate and High-Energy Radiation, Cellular Bioscanning, Cellular Topographics, Transcellular Acoustic Microscopy, Magnetic Resonance Cytotomography, Near-Field Optical Nanoimaging, Cell Volume

Sensing, Noninvasive Neuroelectric Monitoring, Cellular RF and Microwave Oscillations, Macrosensing, Acoustic Macrosensing, Proprioceptive Macrosensing, Electric/Magnetic Macrosensing, Optical Macrosensing, Neural Macrosensing, Other Macrosensing.

8. Nanosurgery **324—340**

Market Potential, Applications of Cosmetic Nanosurgery, Hair Colour, Skin Colour, Baldness, Unwanted Hair, Permanent Breath Freshener, Wrinkle Repair, Slender Now, Techniques and Strategies, Foiling the Immune System, Sensory Equipment, Intercommunications, Licensing, Economics, Distant Prospects, Nonsurgical Nanomachine-based Cosmetics, Total Makeover.

9. Age of Nanotechnology **341—349**

Clothing, Nanosuit, Little House on the Prairie, Topless Towers of Ilium, Android Robots, An Interlude behind Closed Doors, Sex.

10. Nanotechnology and Life **350—365**

Nanoparticles, Demron, Carbon Nanotubes, Micro and Nano-lithography, Nanomanipulation, Molecular Beam Epitaxy, Nanoelectronics, Nanomemories, Nanobatteries, Proton-exchange Membrane Fuel Cell, Self-assembly, Application of Nanotechnology, Life: Nature's Nanotechnology, Mother of All Machines, From Biology to Nanotechnology.

11. Improvements **366—378**

Elbow Grease, Programmable Animal.

Index **379—384**

1

INTRODUCTION

The 1966 film *Fantastic Voyage* treated moviegoers to a bold Vision of *nanotechnology* applied to medicine: through mysterious means, an intrepid team of doctors and their high-tech submarine were shrunk to minute size so that they could travel through the bloodstream of an injured patient and remove a life—threatening blood clot in his brain. In the past 35 years, great strides have been made in fabricating complex devices at ever smaller scales, leading some people to believe that such forms of medical intervention are possible and that tiny robots will soon be coursing through everyone's veins. Indeed, in some circles the idea is taken so seriously that worries have emerged about the dark side of such technology: Could self-replicating nanometer-scale automatons run amok and destroy the entire biological world?

It is shared by most investigators, such thoughts belong squarely in the realm of science fiction. Still, nanotechnology can potentially enhance *biomedical research tools*—for example, by providing new kinds of labels for experiments done to discover drugs or to reveal which sets of genes are active in cells under various conditions. *Nanoscale* devices of moreover, play a part in quick diagnostic screens and in genetic tests, such as those meant to determine a person's susceptibility to different disorders or to reveal which specific genes are mutated in a patient's cancer. Investigators are also studying them as improved contrast agents for *noninvasive imaging* and as *drug-delivery vehicles*. The emerging technologies may not be as photogenic as a platelet-size Raquel Welch blasting away at a clot with a laser beam, but they are every bit as dramatic because, in contrast, the benefits they offer to patients and researchers are real.

How exactly can nanotechnology do all these things? The answer hinges on one's definitions. All of biology is arguably a form of nanotechnology. After all, even the most complicated creature is made up of tiny cells, which themselves are constructed of *nanoscale building blocks*: proteins, lipids, nucleic acids and other complex biological molecules. But by convention the term "*nanotechnology* is usually restricted to artificial constructions made, say, from semiconductors, metals, plastic or glass. A few inorganic structures of nanometer scale—minute crystals for instance—have already been commercialized, notably as contrast agents.

Magnetization

Nature itself provides a beautiful illustration of the usefulness of such inorganic crystals in a biological context: humble magnetotactic (*magnetic-sensing*) bacteria. Such organisms, which live in bodies of water and their muddy bottoms, thrive only at one depth in the water or sediment. Above this position, oxygen is too abundant for their liking; below, too scarce. A bacterium that drifts away from the right evel must swim back, and so, like many of its cousins, the microbe wields a whip like tail for propulsion. But how does the buoyant cell tell up from down when gravity has essentially no effect on it?

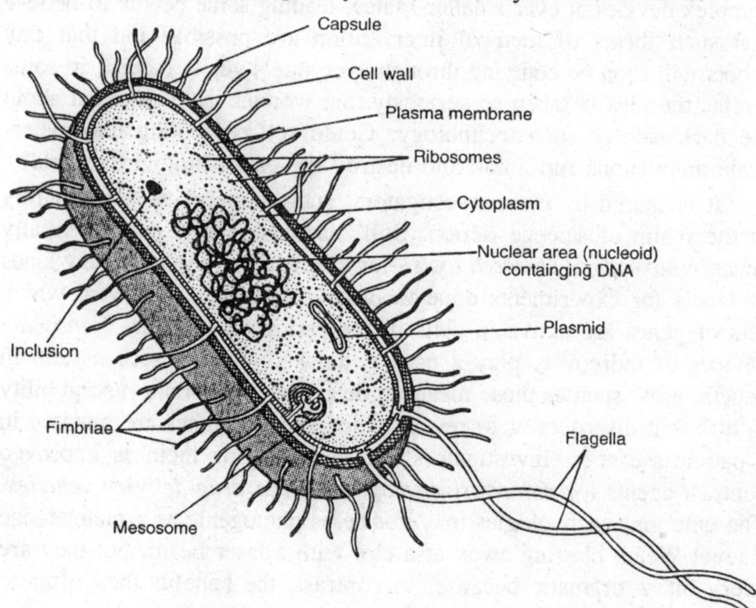

Fig. 1.1. Electron microscopic structure of a typical bacterial cell.

The answer is that this bacterium has fixed within it a chain of about 20 magnetic crystals that are each between 35 and 120 nanometers in diameter. Together these crystals constitute a *miniature compass*. Because the magnetic field of the earth is inclined in most places (it points not only north but also downward in the Northern Hemisphere and upward in the Southern), a *magnetotactic bacterium* can follow a magnetic field line up or down to its desired destination.

This compass is a marvel of natural *nanoscale engineering*. For one, it is made of the perfect material—either magnetite or grephite, both highly magnetic iron minerals. The use of multiple crystals is also no accident. At very small scales, the larger a magnetic particle is, the longer it stays magnetized. But if the particle becomes too large, it will spontaneously form two separate magnetic domains with oppositely directed *magnetizations*. Such a crystal has little overall magnetization and does not make for a very effective compass needle. By building its compass out of crystals that are of just the right size to exist as stable, single magnetic domains, the bacterium makes the best use of every bit of iron it lays down. Interestingly, when people design media for hard-disk storage, they follow exactly the same strategy, using magnetic *nanocrystals* that are of the proper size to be both stable and strong.

Artificial magnetic crystals of similar dimension might soon serve biomedical research in a novel way. Two groups, one in Germany and the other at my institution, the University of California at Berkeley, are exploring the use of magnetic nano particles to detect particular biological entities, such as microorganisms that cause disease.

Their method, like many of the techniques applied today, requires suitable antibodies, which bind to specific targets. The magnetic particles are affixed, as labels, to selected antibody molecules, which are then applied to the sample under study. To detect whether the antibodies have latched onto their target, an investigator applies a strong magnetic field (which temporarily magnetizes the particles) and then examines the specimen with a sensitive instrument capable of detecting the weak magnetic fields emanating from the probes. Labeled antibodies that have not docked to the sample tumble about so rapidly in solution that they give off no magnetic signal. Bound antibodies, however, are unable to rotate, and together their magnetic tags generate a readily detectable magnetic field.

Because the unbound probes produce no signal, this approach does away with the time-consuming washing steps usually required of such assays. The sensitivity demonstrated with this experimental technique is already better than with standard methods, and anticipated improvements in the apparatus should soon boost sensitivity by a factor of several hundred. Despite these advantages, the magnetic method probably will not completely replace the widespread practice of labeling probes with a fluorescent tag, typically an organic molecule that glows with a characteristic hue when it is energized by light of a particular colour. Colours are very useful in various diagnostic and research procedures, such as when more than one probe needs to be tracked.

The world of modern electronics is also full of light-emitting materials. Every CD player, for instance, reads the disc with light from a solid-state laser diode, which is made of an inorganic semiconductor. Imagine carving out a vanishingly small piece of that material, a scoop the size of a protein molecule. The result is a semiconductor *nanocrystal* or, in the talk of the trade, a "*quantum dot.*" Like nanoscale magnetic crystals, these minuscule dots have much to offer biomedical researchers.

Quantum Dot

As the name suggests, *quantum dots* owe their special properties to the weird rules of quantum mechanics, the same rules that restrict the electrons in atoms to certain discrete energy levels. An organic dye molecule absorbs only photons of light with just the right energy to lift its electrons from their quiescent state to one of the higher levels available to them. 'It is, the incident light must be exactly the right wavelength, or colour, to do the job. The molecule will subsequently emit a photon when the electron falls back to a lower energy level. This phenomenon is quite different from what happens in hulk semiconductors, which allow electrons to occupy two broad bands of energy. Such materials can absorb photons in a broad range of colour (all those that have enough energy to bridge the gap between these two bands), but they emit light only at one specific wavelength, corresponding to the hand-gap energy. Quantum dots are an intermediate case. Like bulk semiconductors, they absorb photons of all energies above the threshold of the band gap. But the wavelength of light a quantum dot emits—its colour depends very strongly on the dot's size. Hence, a single type of semiconductors dueling material can yield an entire family of distinctly coloured labels.

Application of Quantum Dots

Physicists first studied *quantum dots* in the 1970s, thinking that they might one day fashion new electronic or optical devices. Few of the pioneering investigators had any idea that these objects could help diagnose disease or discover new drugs. And none of them would have dreamed that the first real-world applications of *quantum dots* would be in biology and medicine. Making quantum dots that would function properly in biological systems did indeed require years of research, but they are now a reality.

The company leading the push to commercialize this technology, Quantum Dot Corporation, has licensed technique developed in my lab and at the Massachusetts Institute of Technology, Indiana University, Lawrence Berkeley National Laboratory and the University of Melbourne in Australia. Semiconductor nanocrystals have several advantages over conventional dye molecules. Small inorganic crystals can with stand significantly more cycles of excitation and light emission than can typical organic molecules, which soon decompose, And this stability allows investigators to track the goings-on in cells and tissues for longer intervals than can now be achieved. But the greatest benefit semiconductor nano crystals offer is less subtle—they come in more colours.

Quantum Dots in Life Sciences

Biological systems are very complex, and frequently several components must be observed simultaneously. Such tracking is difficult to achieve, because each organic dye must be excited with a different wavelength of light. But *quantum dots* make it possible to tag a variety of biological molecules, each with a crystal of a different size (and hence colour). And because all of' these crystals can be energized with a single light source, they an all be monitored at once.

This approach is being pursued actively, but *quantum dots* offer even more interesting possibilities. Imagine a small latex bead filled with a combination of quantum clots. The bead could, for instance, contain five different sizes of dots, or five colours, in a variety of concentrations. Alter the bead is illuminated, it will give off light, which when spread out by a prism will produce five distinct spectral lines with prescribed intensities—a spectral bar code, if you like. Such beads allow for an enormous number of distinct labels (billions, potentially), each of which could be attached, say, to DNA molecules composed of different sequences of genetic building blocks.

With these kinds of beads, technicians could easily compare the *genetic material* in a sample against a library of known *DNA* sequences, as might be done if an investigator wanted to find out which genes were active in certain cells or tissues. They would simply expose the sample to the full beaded library and read the spectral bar codes of the library DNAs that bind to sequences in the sample. Because binding takes place only when genetic sequences match closely (or more precisely, when one sequence complements the other), the results would immediately reveal the nature of the genetic material in the sample.

Semiconductor *quantum dots* should soon serve biomedical researchers in this way, but they are not the only nanostructures useful for optically sensing the genetic composition of biological specimens. Another example emerges from the work of Chad A. Mirkin and Robert L. Letsinger of Northwestern University, who recently developed an ingenious method to test for the presence of a specific genetic sequence in solution. Their scheme employs 13-nanometer gold particles studded with DNA.

The trick here is to use two sets of gold particles. The first set carries DNA that binds to one half of the target sequence; the second set carries DNA that binds to the other half. DNA with the complete target sequence readily attaches to both types of particles, linking them together. Because each particle has multiple DNA tentacles, bits of genetic material carrying the target sequence can glue many particles together. And when these gold specks aggregate, their optical properties shift markedly, changing the test solution from red to blue. Because the outcome of the test is easy to see without any instrumentation at all, such a system might be particularly useful for home DNA testing.

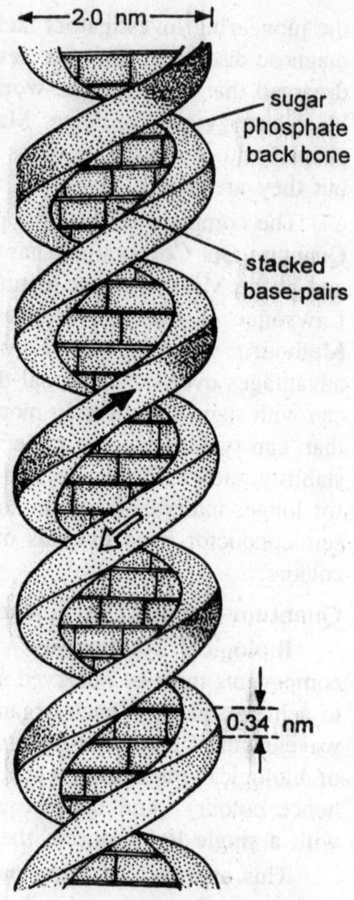

Fig. 1.2. Double stranded DNA.

No discussion of bio-nanotechnology would be complete with out at least a brief mention of one of the hottest instruments in science today—the *atomic force microscope*. Such devices probe materials in the same way an old-fashioned phonograph reads the grooves in a record: by dragging a sharp point over the surface and detecting the resulting deflections. The tip of an atomic force microscope is, however, much finer than a phonograph needle, so it can sense far smaller structures. Regrettably, fabricating tips that are both fine and sturdy for these microscopes has proved to the quite difficult.

The solution appeared in 1996, when workers at Rice University affixed a slender *carbon nanotube* to the tip of an atomic force microscope, making it possible to probe samples just a few nanometers in size. In 1998 Charles M. Lieber and his co-workers at Harvard University applied this approach to probing biomolecules, providing a very high resolution means to explore complex biological molecules and their interactions at the most basic level.

Last year James K. Gimzewski, at the IBM Zurich Research laboratory, showed with collaborators at IBM and the University of Basel that an array of micron-scale arms, or cantilevers, much like the ones employed in atomic force microscopes, could be used to screen samples for the presence of certain genetic sequences. They attached short strands of DNA to the tops of the cantilevers. When genetic material carrying a complementary sequence binds to the anchored strands, it induces a surface stress, which bends the (cantilever subtly—by just nonometers—but enough to be detected. By fabricating devices with many cantilevers and coating each with a different type of DNA, researchers should be able to test a biological sample rapidly for the presence of specific genetic sequences (as is now done routinely with gene chips) by nanomechanical means is without the need for labeling.

This example, like the others described earlier, illustrates that the connections between nanotechnology and the practice of medicine are often indirect, in that much of the new work promises only better research tools or aids to diagnosis. But in some cases, nano-objects being developed may themselves prove useful for therapy. One might, for instance, encapsulate drugs within nanometer-scale packages that control the medicines release in sophisticated ways.

Dendrimer Molecules

Consider a class of' artificial molecules called *organic dendrimers*. Two decades ago Donald A. Tomalia of the Michigan Molecular Institute in Midland fashioned the first of these intriguing structures. A *dendrimer molecule* branches successively from inside to outside.

Its shape resembles what one would get by taking many springs from a tree and poking them into a foam ball so that they shot out in every direction. Dendrimers are globular molecules about the size of a typical protein, but they do not come apart or unfold as easily as proteins do, because they are held together with stronger chemical bonds.

Like the lush canopies of mature trees, dendrimers contain voids. That is, they have an enormous amount of internal surface area. Interestingly, they can be tailored to have a range of different cavity sizes—spaces that are just perfect for holding therapeutic agents. Dendrimers can also be engineered to transport DNA into cells for gene therapy, and they might work more safely than the other leading method: genetically modified viruses.

Other types of nanostructure possess high surface area, and these too may prove useful for delivering drugs where they are needed. But *dendrimers* offer the greatest degree of control and flexibility. It may be possible to design dendrimers that spontaneously swell and liberate their contents only when the appropriate trigger molecules are present. This ability would allow a custom-made dendrimer to release its load of drugs in just the tissues or organs needing treatment.

Nanoshells

Other drug-delivery vehicles on the horizon include hollow *polymer capsules* under study by Helmuth Mohwald of the Max Planck Institute of Colloids and Interfaces in Golm, Germany. In response to certain signals, these capsules swell or compress to release drugs. Also intriguing are so-called *nanoshells*, recently invented by the researchers at Rice.

Nanoshells are extremely small beads of glass coated with gold. They can be fashioned to absorb light of almost any wave length, but *nanoshells* that capture energy in the near-infrared are of most interest because these wavelengths easily penetrate several centimeters of tissue. *Nanoshells* injected into the body can therefore be heated from the outside using a strong infrared source. Such a *nanoshell* could be made to deliver drug molecules at specific times by attaching it to a capsule made of a heat-sensitive polymer. The capsule would release its contents only when gentle heating of the attached nanoshell caused it to deform.

A more dramatic application envisioned for *nanoshells* is in cancer therapy. The idea is to link the gold-plated spheres to antibodies that bind specifically to tumour cells. Heating the *nanoshells* sufficiently would in theory destroy the cancerous cells, while leaving nearby tissue unharmed.

It is, of course, difficult to know for certain whether *nanoshells* will ultimately fulfill their promise. The same can be said for the myriad other minuscule devices being developed for medical use—among them, one-nanometer buckyballs made from just a few dozen carbon atoms. Yet it seems likely that some of the objects being investigated today will be serving doctors in the near future. Even more exciting is the prospect that physicians will make use of nanoscale building blocks to form larger structures, thereby mimicking the natural processes of biology. Such materials might eventually serve to repair damaged tissues. Research on these bold strategies is just beginning, but at least one enterprise already shows that the notion has merit: building scaffoldings on which to grow bone. Samuel I. Stupp of Northwestern is pioneering this approach using synthetic molecules that combine into fibers to which bone cells have a strong tendency to adhere.

What other marvels might the future hold? Although the means to achieve them are far from clear, sober nanotechnologists have stated some truly ambitious goals. One of the "*grand challenges*" of the National Nanotechnology Initiative is to find ways to detect cancerous tumours that are a mere few cells in size. Researchers also hope eventually to develop ways to regenerate not just bone or cartilage or skin but also more complex organs, using artificial scaffoldings that can guide the activity of seeded cells and can even direct the growth of a variety of cell types. Replacing hearts or kidneys or livers in this way might not match the fictional technology of *Fantastic Voyage*, but the thought that such medical therapies might actually become available in the not so distant future is still fantastically exciting.

2

NANOMEDICINE

Our bodies are filled with intricate, active molecular structures. When those structures are damaged, health suffers. Modern medicine can affect the workings of the body in many ways, but from a molecular viewpoint it remains crude indeed. *Molecular manufacturing* can construct a range of medical instruments and devices with far greater abilities. The body is an enormously complex world of *molecules*. With *nanotechnology* to help, we can learn to repair it.

MOLECULAR BODY

To understand what nanotechnology can do for medicine, we need a picture of the body from a molecular perspective. The human body can be seen as a workyard, construction site, and battleground for *molecular machines*. It works remarkably well, using systems so complex that medical science still doesn't understand many of them. Failures, though, are all too common.

Body as Workyard

Molecular machines do the daily work of the body. When we chew and swallow, muscles drive our motions. Muscle fibers contain bundles of molecular fibers that shorten by sliding past one another.

In the stomach and intestines, the molecular machines we call digestive *enzymes* break down the complex molecules in foods, forming smaller molecules for use as fuel or as building blocks. Molecular devices in the lining of the digestive tract carry useful molecules to the bloodstream.

Meanwhile, in the lungs, molecular storage devices called *hemoglobin* molecules pick up oxygen. Driven by molecular fibers, the

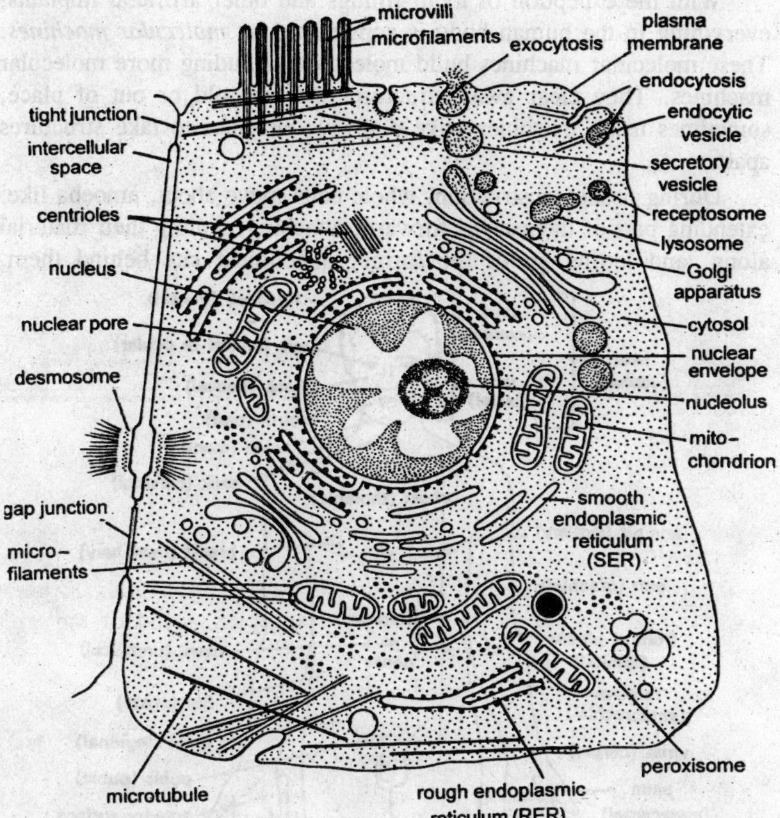

Fig. 2.1. Ultrastructure of a typical animal cells as seen in the electron microscope.

heart pumps blood laden with fuel and oxygen to *cells*. In the muscles, fuel and oxygen drive contraction based on sliding molecular fibers. In the brain, they drive the molecular pumps that charge nerve cells for action. In the liver, they drive molecular machines that build and break down a whole host of molecules. And so the story continues through all the work of the body. Yet each of these functions sometimes fails, whether through damage or inborn defect.

Body as Construction Site

In growing, healing, and renewing tissue, the body is a construction site. Cells take building materials from the bloodstream. Molecular machinery programmed by the cell's genes uses these materials to build biological structures: to lay down bone and collagen, to build whole new cells, to renew skin, and to heal wounds.

With the exception of tooth fillings and other artificial implants, everything in the human body is constructed by *molecular machines*. These molecular machines build molecules, including more molecular machines. They clear away structures that are old or out of place, sometimes using machinery like digestive enzymes to take structures apart.

During tissue construction, whole cells move about, amoeba like: extending part of themselves forward, attaching, pulling their material along, and letting go of the former attachment site behind them.

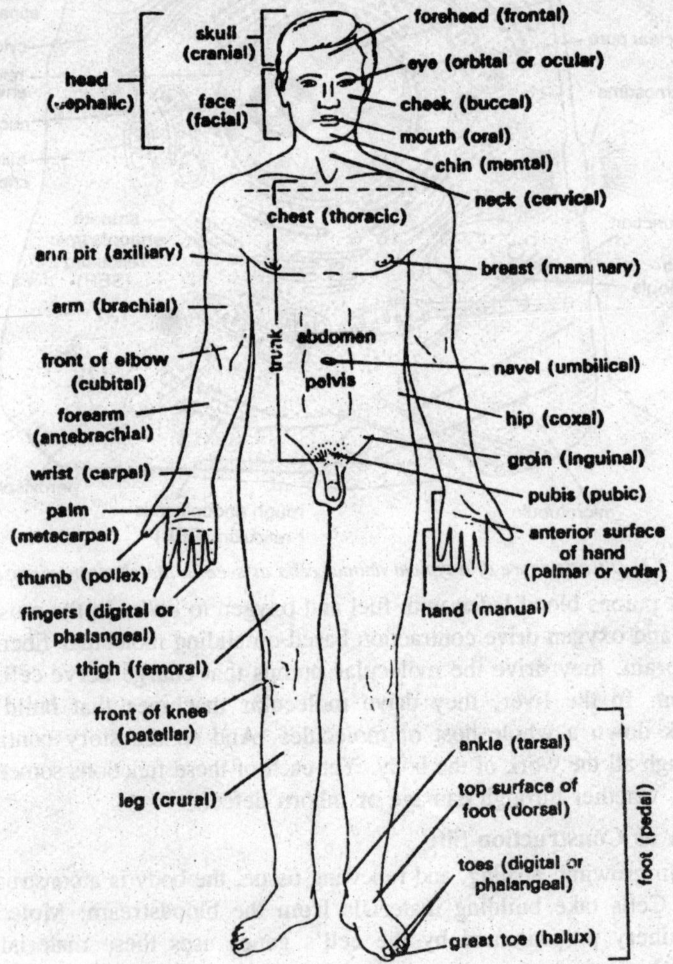

Fig. 2.2. Dorsal view showing the anatomical position.

Individual cells contain a dynamic pattern of molecules made of components that can break down but can also be replaced. Some molecular machines in the cell specialize in digesting molecules that show signs of damage, allowing them to be replaced by fresh molecules made according to genetic instructions. Components inside cells form their complex patterns by self-assembly, that is, by sticking to the proper partners.

Failures in construction increase as we age. Teeth wear and crack and aren't replaced; hair follicles stop working; skin sags and wrinkles. The eye's shape becomes more rigid, ruining close vision. Younger bodies can knit together broken bones quickly, making them stronger than before, but osteoporosis can make older bones so fragile that they break under minor stress.

Sometimes construction is botched from the beginning due to a missing or defective genetic code. In hemophilia, bleeding fails to stop due to the lack of blood clotting factor. Construction of muscle tissue is disrupted in 1 in 3,300 male births by *muscular dystrophy*, in which muscles are gradually replaced by scar tissue and fat; the molecule "*dystrophin*" is missing. *Sickle cell anemia* results from abnormal hemoglobin molecules.

Paraplegics and *quadriplegics* know that some parts of the body don't heal well. The spinal cord is an extreme—and extremely serious—case, but scarring and improper regrowth of tissues result from many accidents. If tissues always regrew properly, injury would do no permanent physical damage.

Body as Battlefield

Assaults from outside the body turn it into a battlefield where the aggressors sometimes get the upper hand. From parasitic worms to protozoa to fungi to *bacteria* to *viruses*, organisms of many kinds have learned to live by entering the body and using their molecular machinery to build more of themselves from the body's building blocks. To meet this onslaught, the body musters the defenses of the immune system—an armada of its own molecular machines. Your body's own amoebalike white blood cells patrol the bloodstream and move out into tissues, threading their way between other cells, searching for invaders.

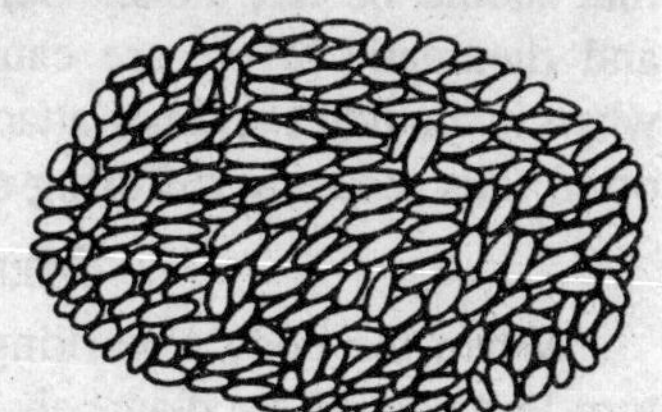

Fig. 2.3. Smallpox virus

How can the immune system distinguish the hundreds of kinds of cells that should be in the body from the invading cells and viruses that shouldn't? This has been the central question of the complex science of immunology. The answer, as yet only partially understood, involves a complex interplay of molecules that recognize other molecules by sticking to them in a selective fashion. These include free-floating antibodies—which are a bit like bumbling guided missiles—and similar molecules that are bound to the surface of white blood cells and other cells of the immune system, enabling them to recognize foreign surfaces on contact.

This system makes life possible, defending our bodies from the fate of meat left at room temperature. Still, it lets us down in two basic ways.

First, the immune system does not respond to all invaders, or responds inadequately. Malaria, tuberculosis, herpes, and AIDS all have their strategies for evading destruction. Cancer is a special case in which the invaders are altered cells of the body itself, sometimes successfully masquerading as healthy cells and escaping detection.

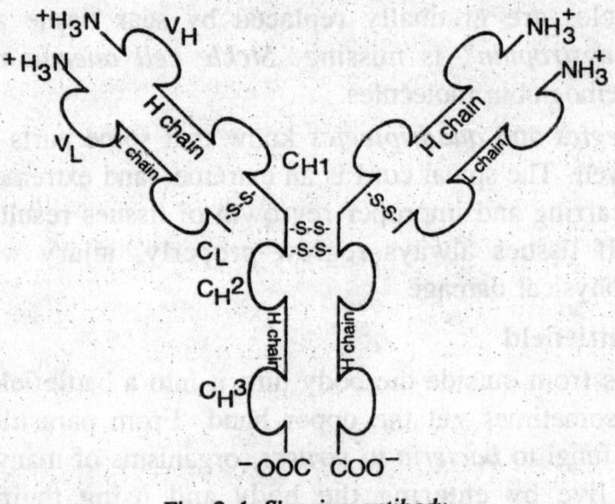

Fig. 2.4. Showing antibody

Second, the immune system sometimes overresponds, attacking cells that should be left alone. Certain kinds of arthritis, as well as lupus and rheumatic fever, are caused by this mistake. Between attacking when it shouldn't and not attacking when it should, the immune system often fails, causing suffering and death.

Medicine Today

When the body's working, building, and battling goes away, we turn to medicine for diagnosis and treatment. Today's methods, though, have obvious shortcomings.

Crude Methods

Diagnostic procedures vary widely, from asking a patient questions, through looking at X-ray shadows, through exploratory surgery and the microscopic and chemical analysis of materials from the body. Doctors can diagnose many ills, but others remain mysteries. Even a diagnosis does not imply understanding: doctors could diagnose infections before they knew about germs, and today can diagnose many syndromes with unknown causes. After years of experimentation and untold loss of life, they can even treat what they don't understand—a drug may help, though no one knows why.

Leaving aside such therapies as heating, massaging, irradiating, and so forth, the two main forms of treatment are surgery and drugs. From a molecular perspective, neither is sophisticated.

Surgery is a direct, manual approach to fixing the body, now practiced by highly trained specialists. Surgeons sew together torn tissues and skin to enable healing, cut out cancer, clear out clogged arteries, and even install pacemakers and replacement organs. It's direct, but it can be dangerous: anesthetics, infections, organ rejection, and missed cancer cells can all cause failure. Surgeons lack fine-scale control. The body works by means of molecular machines, most working inside cells. Surgeons can see neither molecules nor cells, and can repair neither.

Drug therapies affect the body at the molecular level. Some therapies—like insulin for diabetics—provide materials the body lacks. Most—like antibiotics for infections—introduce materials no human body produces. A drug consists of small molecules; in our simulated molecular world, many would fit in the palm of your hand. These molecules are dumped into the body (sometimes directed to a particular region by a needle or the like), where they mix and wander through blood and tissue. They typically bump into other molecules of all sorts in all places, but only stick to and affect molecules of certain kinds.

Antibiotics like penicillin are selective poisons. They stick to molecular machines in bacteria and jam them, thus fighting infection. Viruses are a harder case because they are simpler and have fewer vulnerable molecular machines. Worms, fungi, and protozoa are also difficult, because their molecular machines are more like those found in the human body, and hence harder to jam selectively. Cancer is the most difficult of all. Cancerous growths consist of human cells, and attempts to poison the cancer cells typically poison the rest of the patient as well.

Other drug molecules bind to molecules in the human body and modify their behaviour. Some decrease the secretion of stomach acid, others stimulate the kidneys, many affect the molecular dynamics of the brain. Designing drug molecules to bind to specific targets is a growth industry today, and provides one of the many short-term payoffs that is spurring developments in molecular engineering.

Limited Abilities

Current medicine is limited both by its understanding and by its tools. In many ways, it is still more an art than a science. Mark Pearson of Du Pont points out, "In some areas, medicine has become much more scientific, and in others not much at all. We're still short of what I would consider a reasonable scientific level. Many people don't realize that we just don't know fundamentally how things work. It's like having an automobile, and hoping that by taking things apart, we'll understand something of how they operate. We know there's an engine in the front and we know it's under the hood, we have an idea that it's big and heavy, but we don't really see the rings that allow pistons to slide in the block. We don't even understand that controlled explosions are responsible for providing the energy that drives the machine."

Better tools could provide both better knowledge and better ways to apply that knowledge for healing. Today's surgery can rearrange blood vessels, but is far too coarse to rearrange or repair cells. Today's drug therapies can target some specific molecules, but only some, and only on the basis of type. Doctors today can't affect molecules in one cell while leaving identical molecules in a neighbouring cell untouched because medicine today cannot apply surgical control to the molecular level.

Nanotechnology in Medicine

Developments in *nanotechnology* will result in improved *medical sensors*. As protein chemist Bill DeGrado notes, "Probably the first use you may see would be in diagnostics: being able to take a tiny amount of blood from somebody, just a pinprick, and diagnose for a hundred different things. Biological systems are already able to do that, and we should be able to design molecules or assemblies of molecules that mimic the biological system."

In the longer term, though, the story of nanotechnology in medicine will be the story of extending surgical control to the molecular level. The easiest applications will be aids to the immune system, which selectively attack invaders outside tissues. More difficult applications will require that medical *nanomachines* mimic white blood cells by entering tissues to interact with their cells. Further applications will involve the complexities of molecular-level surgery on individual cells.

As we look at how to solve various problems, you'll notice that some that look difficult today will become easy, while others that might seem easier turn out to be more difficult. The seeming difficulty of treating disorders is always changing: Once polio was frequent and incurable, today it is easily prevented. *Syphilis* once caused steady physical decline leading to insanity and death; now it is cured with a shot.

Athlete's foot has never been seen as a great scourge, yet it remains hard to cure. Likewise with the common cold. This pattern will continue: Deadly diseases may be easily dealt with, while minor ills remain incurable, or vice versa. As we will see, a *mature nanotechnology-based medicine* will be able to deal with almost any physical problem, but the order of difficulty may be surprising. Nature cares nothing for our sense of appropriateness. Horribleness and difficulty just aren't the same thing.

Working Outside Tissues

One approach to *nanomedicine* would make use of microscopic mobile devices built using molecular-manufacturing equipment. These would resemble the *ecosystem protectors* and mobile cleanup machines discussed in the last chapter. Like them, they would either be biodegradable, self-collecting, or collected by something else once they were done working. Like them, they would be more difficult to develop than simple, fixed-location nanomachines, yet clearly feasible and useful. Development will start with the simpler applications, so let's begin by looking at what can be done without entering living tissues.

The *skin* is the body's largest organ, and its exposed position subjects it to a lot of abuse. This exposed position, though, also makes it easier to treat. Among the earlier applications of molecular manufacturing may be those popular, quasimedical products, *cosmetics*. A cream packed with nanomachines could do a better and more selective job of cleaning than any product can today. It could remove the right amount of dead skin, remove excess oils, add missing oils, apply the right amounts of natural moisturizing compounds, and even achieve the elusive goal of "deep pore cleaning" by actually reaching down into pores and cleaning them out. The cream could be a smart material with smooth-on, peel-off convenience.

The mouth, teeth, and gums are amazingly troublesome. Today, daily dental care is an endless cycle of brushing and flossing, of losing ground to tooth decay and gum disease as slowly as possible. A mouthwash full of smart nanomachines could do all that brushing and flossing do and more, and with far less effort—making it more likely to be used.

This mouthwash would identify and destroy pathogenic bacteria while allowing the harmless flora of the mouth to flourish in a healthy ecosystem. Further, the devices would identify particles of food, plaque, or tartar, and lift them from teeth to be rinsed away. Being suspended in liquid and able to swim about, devices would be able to reach surfaces beyond reach of toothbrush bristles or the fibers of floss. As short-lifetime medical nanodevices, they could be built to last only a few minutes in the body before falling apart into materials of the sort found in foods (such as fiber). With this sort of daily dental care from an early age, tooth decay and gum disease would likely never arise. If under way, they would be greatly lessened.

Going beyond this superficial treatment would involve moving among and modifying cells. Let's consider what can be done with this treatment inside the body, but outside the body's tissues. The bloodstream carries everything from nutrients to immune-system cells, with chemical signals and infectious organisms besides.

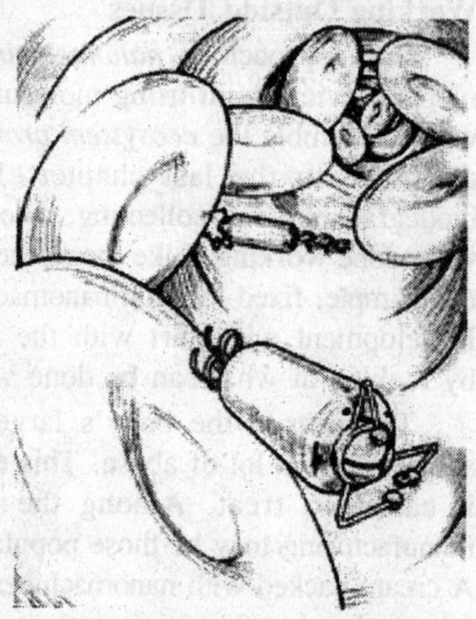
Fig. 2.5. Immune machines.

Here, it is useful to think in terms of medical nanomachines that resemble small submarines. Each of these is large enough to carry a *nanocomputer* as powerful as a mid-1980s mainframe, along with a huge database (a billion bytes), a complete set of instruments for identifying biological surfaces, and tools for clobbering viruses, bacteria, and other invaders. Immune cells, as we've seen, travel through the bloodstream checking surfaces for foreignness and—when working properly—attacking and eliminating what should not be there. These *immune machines* would do both more and less. With their onboard sensors and computers, they will be able to react to the same molecular signals that the immune system does, but with greater discrimination. Before being sent into the body on their search-and-destroy mission, they could be programmed with a set of characteristics that lets them clearly distinguish their targets from everything else. The body's immune system can respond only to invading organisms that had been encountered by that individual's body. Immune machines, however, could be programmed to respond to anything that had been encountered by world medicine.

Immune machines can be designed for use in the bloodstream or the digestive tract (the mouthwash described above used these abilities in hunting down harmful bacteria). They could float and circulate, as antibiotics do, while searching for intruders to neutralize. To escape being engulfed by white blood cells making their own patrols, immune machines could display standard molecules on their surface-molecules the body knows and trusts already—like a fellow police officer wearing a familiar uniform.

When an invader is identified, it can be punctured, letting its contents spill out and ending its effectiveness. If the contents were known to be hazardous by themselves, then the immune machine could hold on to it long enough to dismantle it more completely.

How will these devices know when it's time to depart? If the physician in charge is sure the task will be finished within, say, one day, the devices prescribed could be of a type designed to fall apart after twenty-four hours. If the treatment time needed is variable, the physician could monitor progress and stop action at the appropriate time by sending a specific molecule—aspirin perhaps, or something even safer—as a signal to stop work. The inactivated devices would then be cleared out along with other waste eliminated from the body.

Working Within Tissues

In most parts of the body, the finest blood vessels, capillaries, pass within a few cell diameters of every point. Certain white blood cells can leave these vessels to move among the neighbouring cells. *Immune machines* and similar devices, being even smaller, could do likewise. In some tissues, this will be easy, in some harder, but with careful design and testing, essentially any point of the body should become accessible for healing repairs.

Merely fighting organisms in the bloodstream would be a major advance, cutting their numbers and inhibiting their spread. Roving medical nanomachines, though, will be able to hunt down invaders throughout the body and eliminate them entirely.

Eliminating Invaders

Cancers are a prime example. The immune system recognizes and eliminates most potential cancers, but some get by. Physicians can recognize cancer cells by their appearance and by molecular markers, but they cannot always remove them all through surgery, and often cannot find a selective poison. *Immune machines*, however, will have no difficulty identifying cancer cells, and will ultimately be able to track them down and destroy them wherever they may be growing. Destroying every cancer cell will cure the cancer.

Bacteria, protozoa, worms, and other parasites have even more obvious molecular markers. Once identified, they could be destroyed, ridding the body of the disease they cause. Immune machines thus could deal with tuberculosis, strep throat, leprosy, malaria, amoebic dysentery, sleeping sickness, river blindness, hookworm, flukes, candida, valley fever, antibiotic-resistant bacteria, and even athlete's foot. All are caused by invading cells or larger organisms (such as worms). Health officials estimate that parasitic diseases, common in the Third World, affect more than one billion people. For many of these diseases, no satisfactory drug treatment exists. All can eventually be eliminated as threats to human health by a sufficiently advanced form of nanomedicine.

Herding Cells

Destroying invaders will be helpful, but injuries and structural problems pose other problems. Truly advanced medicine will be able to build up and restructure tissues. Here, medical nanodevices can stimulate and guide the body's own construction and repair mechanisms to restore healthy tissue.

What is healthy tissue? It consists of normal cells in normal patterns in a normal matrix all organized in a normal relationship to the surrounding tissues. Surgeons today (with their huge, crude tools) can fix some problems at the tissue level. A wound disrupts the healthy relationship between two different pieces of tissue, and surgical glues and sutures can partly remedy this problem by holding the tissues in a position that promotes healing. Likewise, coronary artery bypass surgery brings about a more healthy overall configuration of tissues—one that provides working plumbing to supply blood to the heart muscle. Surgeons cut and stitch, but then they must rely on the tissue to heal its wounds as best it can.

Healing establishes healthy relationships on a finer scale. Cells must divide, grow, migrate, and fill gaps. They must reorganize to form properly connected networks of fine blood vessels. And cells must lay down materials to form the structural, intercellular matrix—collagen to provide the proper shape and toughness, or mineral grains to provide rigidity, as in bone. Often, they lay down unwanted scar tissue instead, blocking proper healing.

With enough knowledge of how these processes work (and *nanoinstruments* can help gather that knowledge) and with good enough software to guide the process—a more difficult challenge—medical nanomachines will be able to guide this healing process. The problem here is to guide the motion and behaviour of a mob of active, living cells—a process that can be termed *cell herding*.

Cells respond to a host of signals from their environment: to chemicals in the surrounding fluids, to signal molecules on neighbouring cells, and to mechanical forces applied to them. Cell-herding devices would use these signals to spur cell division where it is needed and to discourage it where it is not. They would nudge cells to encourage them to migrate in appropriate directions, or would simply pick them up, move them along, and deliver them where needed, encouraging them to nestle into a proper relationship with their neighbours. Finally, they would stimulate cells to surround themselves with the proper intercellular-matrix materials. Or—like the owner of a small dog who, on a cold day, wraps the beast in a wool jacket—they would directly build the proper surrounding structures for the cell in its new location.

In this way, cooperating teams of cell-herding devices could guide the healing or restructuring of tissues, ensuring that their cells form healthy patterns and a healthy matrix and that those tissues have a healthy relationship to their surroundings. Where necessary, cells could even be adjusted internally, as we will discuss later.

Rebuilding Tissues

Again, skin provides easy examples and may be a natural place to start in practice. People often want hair where they have bare skin, and bare skin where they have hair. *Cell herding machines* could move or destroy hair follicle cells to eliminate an unwanted hair, or grow more of the needed cells and arrange them into a working follicle where a hair is desired. By adjusting the size of the follicle and the properties of some of the cells, hairs could be made coarser, or finer, or straighter, or curlier. All these changes would involve no pain, toxic chemicals, or stench. Cell-herding devices could move down into the living layers of skin, removing unwanted cells, stimulating the growth of new cells, narrowing unnaturally prominent blood vessels,

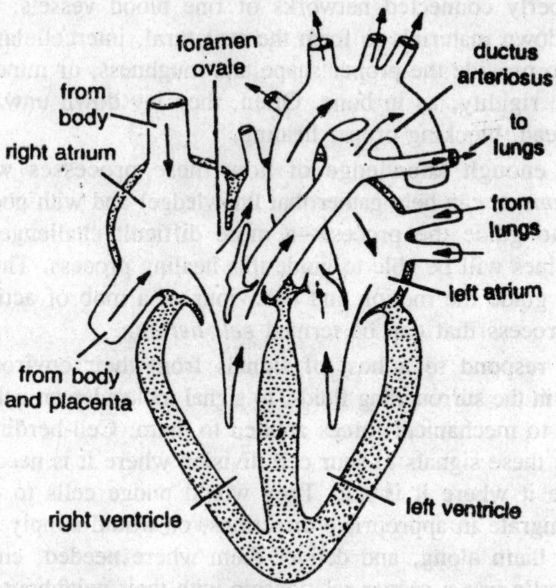

Fig. 2.6. Heart showing direction of blood flow.

insuring good circulation by guiding the growth of any needed normal blood vessels, and moving cells and fibers around so as to eliminate even deep wrinkles.

At the opposite end of the spectrum, cell herding will revolutionize treatment of life-threatening conditions. For example, the most common cause of heart disease is reduced or interrupted supply of blood to the heart muscle. In pumping oxygenated blood to the rest of the body, the heart diverts a portion for its own use though the coronary arteries. When these blood vessels become constricted, we speak of coronary-artery disease. When they are blocked, causing heart muscle tissue to die, we speak of someone "having a coronary," another term for heart attack.

Devices working in the bloodstream could nibble away at atherosclerotic deposits, widening the affected blood vessels. Cell herding devices could restore artery walls and artery linings to health, by ensuring that the right cells and supporting structures are in the right places. This would prevent most heart attacks.

But what if a heart attack has already destroyed muscle tissue, leaving the patient with a scarred, damaged, and poorly functioning heart? Once again, cell-herding devices could accomplish repairs, working their way into the scar tissue and removing it bit by bit, replacing it with fresh muscle fiber. If need be, this new fiber can be grown by applying a series of internal molecular stimuli to selected heart muscle cells to "remind" them of the instructions for growth that they used decades earlier during embryonic development.

Cell-herding capabilities should also be able to deal with the various forms of arthritis. Where this is due to attacks from the body's own immune system, the cells producing the damaging antibodies can be identified and eliminated. Then a cell-herding system would work inside the joint where it would remove diseased tissues, calcified spurs, and so forth, then rework patterns of cells and intercellular material to form a healthy, smoothly working, and pain-free joint. Clearly, learning to repair hearts and learning to repair joints will have some basic technologies in common, but much of the research and development will have to be devoted to specific tissues and specific circumstances. A similar process—but again, specially adapted to the circumstances at hand—could be used to strengthen and reshape bone, correcting osteoporosis.

In *dentistry*, this sort of process could be used to fill cavities, not with amalgam, but with natural dentin and enamel. Reversing the ravages of periodontal disease will someday be straightforward, with nanomedical devices to clean pockets, join tissues, and guide regrowth. Even missing teeth could be regrown, with enough control over cell behaviour.

Working on Cells

Moving through tissues without leaving a trail of disruption will require devices able to manipulate and direct the motions of cells, and to repair them. Much remains to be learned—and will be easy to learn with *nanoscale tools*—but today's knowledge of cells is enough for a start on the problem of how to do surgery on cells.

Cell biology is a booming field, even today. Cells can be made to live and grow in laboratory cultures if they are placed in a liquid with suitable nutrients, oxygen, and the rest. Even with today's crude techniques, much has been learned about how cells respond to different chemicals, to different neighbours, and even to being poked and cut with needles. Conducting a rough sort of surgery on individual cells has been routine for many years in scientific laboratories.

Today, researchers can inject new *DNA* into cells using a tiny needle; small punctures in a cell membrane automatically reseal. But both these techniques use tools that on a cellular scale are large and clumsy—like doing surgery with an ax or a wrecking ball, instead of a scalpel. *Nanoscale* tools will enable medical procedures involving delicate surgery on individual cells.

Eliminating Viruses by Cell Surgery

Some viral diseases will respond to treatments that destroy viruses in the nose and throat, or in the bloodstream. The flu and common cold are examples. Many others would be greatly improved by this, but not eliminated. All viruses work by injecting their genes into a cell and taking over its molecular machinery, using it to produce more viruses. This is part of what makes viral illnesses so hard to treat—most of the action is performed by the body's own molecular machines, which can't be interfered with on a wholesale basis. When the immune system deals with a viral illness, it both attacks free virus particles before they enter cells, and attacks infected cells before they can churn out too many more virus particles.

Some viruses, though, insert their genes among the genes of the cell, and lay low. The cell can seem entirely normal to the immune system, for months or years, until the viral genes are triggered into action and begin the infective process anew. This pattern is responsible for the persistence of herpes infections, and for the slow, deadly progress of AIDS.

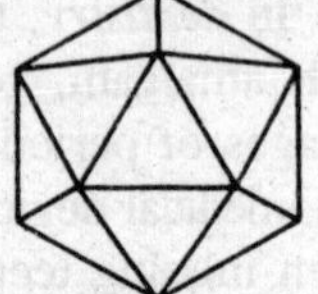

Fig. 2.7. Herpes simplex virus.

These viruses can be eliminated by molecular-level cellular surgery. The required devices could be small enough to fit entirely *within* the cell, if need be. Greg Fahy, who heads the *Organ Cryopreservation Project* at the American Red Cross's Jerome Holland Transplantation Laboratory, writes, "Calculations imply that molecular sensors, molecular computers, and molecular effectors can be combined into a device small enough to fit easily inside a single cell and powerful enough to repair molecular and structural defects (or to degrade foreign structures such as viruses and bacteria) as rapidly as they accumulate.... There is no reason such systems cannot be built and function as designed."

Equally well, a *cell surgery* device located outside a cell could reach through the membrane with long probes. At the ends of the probes would be tools and sensors along with, perhaps, a small auxiliary computer. These would be able to reach through multiple membranes, unpackage and uncoil DNA, read it, repackage it, and recoil it, "proofreading" the DNA by comparing the sequences in one cell to the sequences of other cells.

On reading the genetic sequence spelling out the message of the AIDS virus, a *molecular surgery* machine could be programmed to respond like an immune machine, destroying the cell. But it would seem to make more sense simply to cut out the AIDS virus genes themselves, and reconnect the ends as they were before infection. By doing this, and killing any viruses found in the cell, the procedure would restore the cell to health.

Molecular Repairs

Cells are made of billions of molecules, each built by *molecular machines*. These molecules self-assemble to form larger structures, many in dynamic patterns, perpetually disintegrating and reforming. Cell-surgery devices will be able to make molecules of sorts that may be lacking, while destroying molecules that are damaged or present in excess. They will be able not only to remove viral genes, but to repair chemical and radiation-caused damage to the cell's own genes. Advanced cell surgery devices would be able to repair cells almost regardless of their initial state of damage.

By activating and inactivating a cell's genes, they will be able to stimulate cell division and guide what types of cells are formed. This will be a great aid to cell herding and to healing tissues.

As surgeons today rely on the spontaneous, self-organizing ability of cells and tissues to join and heal the parts they manipulate, so cell-surgery devices will rely on the spontaneous self-organizing capabilities of molecules to join and "heal" the parts they put together. Healing of a surgical wound involves sweeping up dead cells, growing new cells, and a slow and genuinely painful process of tissue reorganization. In contrast, the joining of molecules is almost instantaneous and occurs on a scale far below that of the most sensitive pain receptor. "Healing" will not begin *after* the repair devices have done their work, as it does in conventional surgery: rather, when they complete their work, the tissue will have been healed.

Healing Body and Limb

The ability to herd cells and to perform *molecular repairs* and cell surgery will open new vistas for medicine. These abilities apply on a small scale, but their effects can be large scale.

Correcting Chemistry

In many diseases, the body as a whole suffers from misregulation of the signaling molecules that travel through its fluids. Many are rare: Cushing's disease, Grave's disease, Paget's disease, Addison's disease, Conn's syndrome, Prader-Labhart-Willi syndrome. Others are common: millions of older women suffer from osteoporosis, the weakening of bones that can accompany lowered estrogen levels.

Diabetes kills frequently enough to rank in the top ten causes of death in the United States; the number of individuals known to have it doubles every fifteen years. It is the leading cause of blindness in the United States, with other complications including kidney damage, cataracts, and cardiovascular damage. Today's *molecular medicine* tries to solve these troubles by supplying missing molecules: diabetics inject additional insulin. While helpful, this doesn't cure the disease or eliminate all symptoms. In an era of molecular surgery, physicians could choose instead to repair the defective organ, so it can regulate its own chemicals again, and to readjust the metabolic properties of other cells in the body to match. This would be a true healing, far better than today's partial fix.

Only now are researchers making progress on another frequent problem of metabolic regulation: obesity. Once this was thought to have one simple cause (consuming excess calories) and one main result (greater roundness than favoured by today's aesthetics), but both assumptions proved wrong. Obesity is a serious medical problem, increasing the risk of diabetes mellitus, osteoarthritis, degenerative diseases of the heart, arteries, and kidneys, and shortening life expectancy. And the supposed cause, simple overeating, has been shown to be incorrect—something dieters had always suspected, as they watched thinner colleagues gorge and yet gain no weight.

The ability to lay in stores of fat was a great benefit to people once upon a time, when food supplies were irregular, nomadism and marauding bands made food storage difficult and risky, and starvation was a common cause of death. Our bodies are still adapted to that world, and regulate fat reserves accordingly. This is why dieting often has perverse effects. The body, when starved, responds by attempting to build up greater reserves of fat at its next opportunity. The main effect of exercise in weight reduction isn't to burn up calories, but to signal the body to adapt itself for efficient mobility.

Obesity therefore seems to be a matter of chemical signals within the body, signals to store fat for famine or to become lean for motion. Nanomedicine will be able to regulate these signals in the bloodstream, and to adjust how individual cells respond to them in the body. The latter would even make possible the elusive "spot reduction program" to reshape the distribution of body fat.

Here, as with many potential applications of nanotechnology, the problem may be solved by other means first. Some problems, though, will almost surely require nanomedicine.

New Organs and Limbs

So far we've seen how *medical nanotechnology* would be used in the simpler applications outside tissues—such as in the blood—then inside tissues, and finally inside cells. Consider how these abilities will fit together for victims of automobile and motorcycle accidents.

Nanomanufactured medical devices will be of dramatic value to those who have suffered massive trauma. Take the case of a patient with a crushed or severed spinal cord high in the back or in the neck. The latest research gives hope that when such patients are treated promptly after the injury, paralysis may be at least partially avoidable, sometimes. But those whose injuries weren't treated—including virtually all of today's patients—remain paralyzed. While research continues on a variety of techniques for attempting to aid a spontaneous healing process, prospects for reversing this sort of damage using conventional medicine remain bleak.

With the techniques discussed above, it will become possible to remove scar tissue and to guide cell growth so as to produce healthy arrangements of the cells on a microscopic scale. With the right molecular-scale poking and prodding of the cell nucleus, even nerve cells of the sorts found in the brain and spinal cord can be induced to divide. Where nerve cells have been destroyed, there need be no shortage of replacements. These technologies will eventually enable medicine to heal damaged spinal cords, reversing paralysis.

The ability to guide cell growth and division and to direct the organization of tissues will be sufficient to regrow entire organs and limbs, not merely to repair what has been damaged. This will enable medicine to restore physical health despite the most grievous injuries.

If this seems hard to believe, recall that medical advances have shocked the world before now. To those in the past, the idea of cutting people open with knives *painlessly* would have seemed miraculous, but surgical anesthesia is now routine. Likewise with bacterial infections and antibiotics, with the eradication of smallpox, and the vaccine for polio: Each tamed a deadly terror, and each is now half-forgotten history. Our gut sense of what seems likely has little to do with what can and cannot be done by medical technology. It has more to do with our habitual fears, including the fear of vain hopes. Yet what amazes one generation seems obvious and even boring to the next. The first baby born after each breakthrough grows up wondering what all the excitement was about.

Besides, nanoscale medicine won't be a cure-all. Consider a fifty-year-old mentally retarded man, with a mind like a two-year-old's, or a woman with a brain tumour that has spread to the point that her personality has changed: How could they be "healed"? No healing of tissues could replace a missed lifetime of adult experience, nor can it replace lost information from a severely damaged brain. The best physicians could do would be to bring the patients to *some* physically healthy condition. One can wish for more, but sometimes it won't be possible.

First Aid

Throughout the centuries, medicine has been constrained to maintain functioning tissues, since once tissues stop functioning, they can't heal themselves. With molecular surgery to carry out the healing directly, medical priorities change drastically—function is no longer absolutely necessary. In fact, a physician able to use molecular surgery would prefer to operate on nonfunctioning, structurally stable tissue than on tissue that has been allowed to continue malfunctioning until its structure was lost.

Brain tumours are an example: They destroy the brain's structure, and with it the patient's skills, memories, and personality. Physicians in the future should be able to immediately interrupt this process, to stop the functioning of the brain to stabilize the patient for treatment.

Techniques available today can stop tissue function while preserving tissue structure. Greg Fahy, in his work on organ preservation at the American Red Cross, is developing a technique for vitrifying animal kidneys—making them into a low-temperature, crystal-free glass—with the goal of maintaining their structure such that, when brought back to room temperature, they can be transplanted. Some kidneys have been cooled to -30°C, warmed back up, and then functioned after transplantation.

A variety of other procedures can also stabilize tissues on a long-term basis. These procedures enable many cells—but not whole tissues—to survive and recover without help; advanced molecular repair and cell surgery will presumably tip the balance, enabling cells, tissues, and organs to recover and heal. When applied to stabilizing a whole patient, such a condition can be called *biostasis*. A patient in biostasis can be kept there indefinitely until the required medical help arrives. So in the future, the question "Can this patient be restored to health?"

will be answered "Yes, if the patient's brain is intact, and with it the patient's mind."

Sandra Lee Adamson of the National Space Society has her eyes on distant goals. Some have proposed that travel to the stars would take generations, preventing anyone on Earth from ever making the trip. But she notes that biostasis will "give hope to some fearless adventurers who will risk suspension and subsequent reanimation so they can see the stars for themselves."

Plague Insurance

Medical nanotechnologies promise to extend healthy life, but if history is any guide, they may also avert sudden massive death. The word *plague* is rarely heard today, except in relation to AIDS; it calls up visions of the Black Death of the Middle Ages, when one third of Europe died in 1346-50. A virulent influenza struck in 1918, half lost in the news of the First World War: how many of us realize that it killed at least 20 million? People often act as though plagues were gone for good, as if sanitation and antibiotics had vanquished them. But as doctors are forever telling their patients, antibiotics kill bacteria, but are useless for viruses. The flu, the common cold, herpes, and AIDS—none has a really effective treatment, because all are caused by viruses. In some African countries, as much as 10 percent of the population is estimated to be infected with the AIDS-causing HIV virus. Without a cure soon, the steep rise in deaths from AIDS still lies in the future. AIDS stands as a grim reminder that the great plagues of history are not behind us.

Threat

New diseases continue to appear today as they have throughout history. Today's population, far larger than that of any previous century, provides a huge, fertile territory for their spread.

Today's transportation systems can spread viruses from continent to continent in a single day. When ships sailed or churned their way across the seas, an infected passenger was likely to show full-blown disease before arrival, permitting quarantine. But few diseases can be guaranteed to show themselves in the hours of a single aircraft flight.

So far as is known, every species of organism, from bacterium to whale, is afflicted with viruses. Animal viruses sometimes "jump the species gap" to infect other animals, or people. Most scientists believe that the ancestors of the AIDS virus could, until recently, infect only certain African monkeys. Then these viruses made the interspecies jump. A similar jump occurred in the 1960s when scientists in West Germany, working with cells from monkeys in Uganda, suddenly fell ill. Dozens were infected, and several died of a disease that caused both blood clots and bleeding, caused by what is now named the Marburg virus. What if the Marburg virus had spread with a sneeze, like influenza or the common cold?

We think of human plagues as a health problem, but when they hit our fellow species, we tend to see them from an environmental perspective. In the late 1980s, over half the harbor-seal population in large parts of the North Sea suddenly died, leading many at first to blame pollution. The cause, though, appears to be a distemper virus that made the jump from dogs. Biologists worry that the virus could infect seal species around the world, since distemper virus can spread by aerosols—that is, by coughing—and seals live in close physical contact. So far its mortality rate has been 60 to 70 percent.

What of AIDS itself: Could it change and give rise to a form able to spread, say, as colds do? Nobel Laureate Howard M. Temin has said, "I think that we can very confidently say that this can't happen." Nobel Laureate Joshua Lederberg, president of Rockefeller University in New York City, replied, "I don't share your confidence about what can and cannot happen." He points out that "there is no reason a great plague could not happen again. We live in evolutionary competition with microbes—bacteria and viruses. There is no guarantee that we will be the survivors."

Inadequate Abilities

Bacterial diseases are mostly controllable today. Sanitation limits the ways in which plague can spread. These measures are just good enough to lull us into imagining the problem is solved.

Viruses are common, viruses mutate; some spread through the air, and some are deadly. Plagues show that fast-spreading diseases can be deadly, and effective antiviral drugs are still rare.

The only really effective treatments for viral diseases are preventive, not curative. They work either by preventing exposure, or by exposing the body beforehand to dead or harmless or fragmentary forms of the virus, to prepare the immune system for future exposure. As the long struggle for an AIDS vaccine shows, one cannot count on modern medicine to identify a new virus and produce an effective vaccine within a single month or year or even a single decade. But influenza epidemics spread fast, and Marburg II or AIDS II or something entirely new and deadly may do the same.

Doing Better

The deaths from the next great plague could have begun in a village last week, or could begin next year, or a year before we learn to deal with new viral illnesses promptly and effectively. With luck, the plague will wait until a year after.

Immune machines could be set to kill a new virus as soon as it is identified. The instruments nanotechnology brings will make viral identification easy. Some day, the means will be in place to defend human life against viral catastrophe.

From eliminating viruses to repairing individual cells, improving our control of the molecular world will improve health care. Immune machines working in the bloodstream seem about as complex as some engineering projects human beings have already completed—projects like large satellites. Other medical nanotechnologies seem to be of a higher order of complexity.

Solving Hard Problems

Somewhere in the progression from relatively simple immune devices to molecular surgery, we ve crossed the fuzzy line between systems that teams of clever biomedical engineers could design in a reasonable length of time and ones that might take decades or prove impossibly complex. Designing a nanomachine capable of entering a cell, reading its DNA, finding and removing a deadly viral DNA sequence, and then restoring the cell to normal would be a monumental job. Such tasks are advanced applications of nanotechnology, far beyond mere computers, manufacturing equipment, and half-witted "*smart materials*."

To succeed within a reasonable number of years, we may need to automate much of the engineering process, including software engineering. Today's best expert systems are nowhere near sophisticated enough. The software must be able to apply physical principles, engineering rules, and fast computation to generate and test new designs. Call it *automated engineering*.

Automated engineering will prove useful in advanced nanomedicine because of the sheer number of small problems to be solved. The human body contains hundreds of kinds of cells forming a huge number of tissues and organs. Taken as a whole (and ignoring the immune system), the body contains hundreds of thousands of different kinds of molecules. Performing complex molecular repairs on a damaged cell might require solving millions of separate, repetitive problems. The molecular machinery in cell surgery devices will need to be controlled by complex software, and it would be best to be able to delegate the task of writing that software to an automated system. Until then, or until a lot of more conventional design work gets done, nanomedicine will have to focus on simpler problems.

Aging

Where does aging fit in the spectrum of difficulty? The deterioration that comes with aging is increasingly recognized as a form of disease, one that weakens the body and makes it susceptible to a host of other diseases. Aging, in this view, is as natural as smallpox and bubonic plague, and more surely fatal. Unlike bubonic plague, however, aging results from internal malfunctions in the molecular machinery of the body, and a medical condition with so many different symptoms could be complex.

Surprisingly, substantial progress is being made with present techniques, without even a rudimentary ability to perform cell surgery in a medical context. Some researchers believe that aging is primarily the result of a fairly small number of regulatory processes, and many of these have already been shown to be alterable. If so, aging may be tackled successfully before even simple cell repair is available. But the human aging process is not well enough understood to enable a confident projection of this; for example, the number of regulatory processes is not yet known. A thorough solution may well require advanced nanotechnology-based medicine, but a thorough solution seems possible. The result would not be immortality, just much longer, healthier lives for those who want them.

Restoring Species

A challenging problem related to medicine (and to biostasis) is that of species restoration. Today, researchers are carefully preserving samples from species now becoming extinct. In some cases, all they have are tissue samples. For other species, they've been able to save germ cells in the hope that they will be able to implant fertilized eggs into related species and thus bring the (nearly?) extinct species back.

Each cell typically contains the organism's complete genetic information, but what can be done with this? Many researchers today collect samples for preservation thinking only of the implantation scenario: one that they know has already been made to work. Other researchers are taking a broader view: the Center for Genetic Resources and Heritage at the University of Queensland is a leader in the effort. Daryl Edmondson, coordinator of the gene library, explains that the center is unique because it will "actively collect data. Most other libraries simply collate their own collections." Director John Mattick describes it as a "genetic Louvre" and points out that if genes from today's endangered species aren't preserved, "subsequent generations will see we had the technology to keep [DNA] software and will ask why we didn't do it." With this information and the sorts of molecular repair and cell-surgery capabilities we have discussed, lost species can someday be returned to active life again as habitats are restored.

One such center isn't enough: the Queensland center focuses on Australian species (naturally enough) and has limited funds. Besides, anything so precious as the genetic information of an endangered species should be stored in many separate locations for safety. We need to take out an insurance policy on Earth's genetic diversity with a broader network of genetic libraries, concentrating special attention on gathering biological samples from the fast-disappearing rain forests. Scientific study can wait: the urgency of the situation calls for a vacuum-cleaner approach. The Foresight Institute is promoting this effort through its BioArchive Project.

3

PATHS TO NANOTECHNOLOGY

The framework for thinking about scientific and economic development is what we would like to call "*knowledge space.*" This is a richly multidimensional space where different areas correspond to different aspects of technology. There can be fertile regions that support many new companies, and many people working on those technologies. There can be barren regions where there is little economic gain to be had by working on the technologies in those regions. Distance metrics can be used for moving around in knowledge space, based upon how much money must be spent and how much time people must spend to uncover new knowledge and to develop new technologies. Different axes in this multidimensional knowledge space can represent levels of knowledge. The frontiers of this space expand with time as people do more research and development work.

COLONIZATION METAPHOR

It is difficult to think about a space with many dimensions, so we can put this concept more into human terms by thinking about a model in two or three dimensions. In two dimensions, it is possible to make a graph or a map, and this is somewhat suggestive of land or terrain. The process through time of new industries developing resembles expansion, pioneering, and settling as people migrate toward areas with more resources.

There may be very fertile valleys to settle—similar to the Ohio River valley. There may be places where there is nothing to support a company—similar to the desert or a high mountain range. As the pioneer industries push back the frontiers, behind them will come all the infrastructures and the support industries that can be compared in this model to the railroads, schools, and so on, that followed the pioneers to establish a real civilization.

Many people believe that somewhere "out there" is a sustainable fusion device that will give huge amounts of very cheap electric power. The problem is that it is taking billions and billions of dollars to make small amounts of progress toward this goal, and there are absolutely no profitable applications until we get all the way there. No one will pay money for a fusion reactor that consumes energy instead of producing energy. In this sort of industrial-development situation, there is no way to get there unless you have an extremely wealthy, extremely farsighted company, or a government that decides that developing this industry is important.

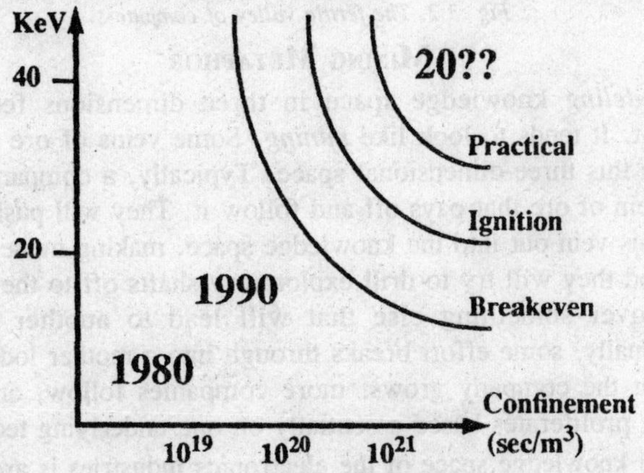

Fig. 3.1. The fusion desert.

An example of a fertile valley with which we are all familiar is *computer*. The rate of progress in this technology has been incredible. For a long while there was improvement in performance every two years, although recently this has slowed to every three years. The rate of progress here is not really constant. Nevertheless, we all fully expect that a couple of years from now, we will have inexpensive machines on our desktops with 100 MIPS and 100 MB of RAM. We can project the progress that we expect into the future. This area is so fertile that we see hundreds of companies, and numerous different computer architectures being tried. There have been many failures and that is a right. There is so much money to be made here that people will try things that do not work just on the chance that those things might work. Therefore, if there is a solution to a particular problem, someone will to find it.

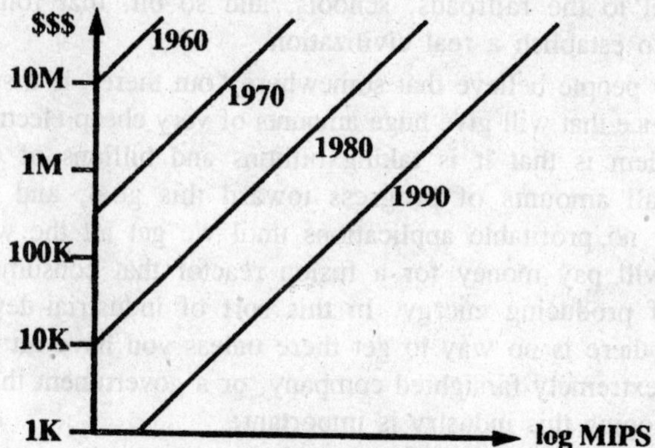

Fig. 3.2. The fertile valley of computers.

Mining Metaphor

Modeling knowledge space in three dimensions feels a little different. It tends to look like *mining*. Some veins of ore will be out there in this three-dimensional space. Typically, a company will find some vein of ore that pays off and follow it. They will push their way along this vein out into the knowledge space, making money all of the way, and they will try to drill exploratory shafts off to the side to try to discover something else that will lead to another rich vein. Occasionally, some effort breaks through into a mother lode that is so rich that the company grows, more companies follow, and a whole industry proliferates based essentially on one underlying technology.

The knowledge space of the electronics industries is around 1970. The known space at that time is represented by the area to the left and under the curve. Many technologies looked like they might be the next "*big thing*." Does anyone remember *Ovonics*? At that point in time, it was not yet clear which one of these technologies was going

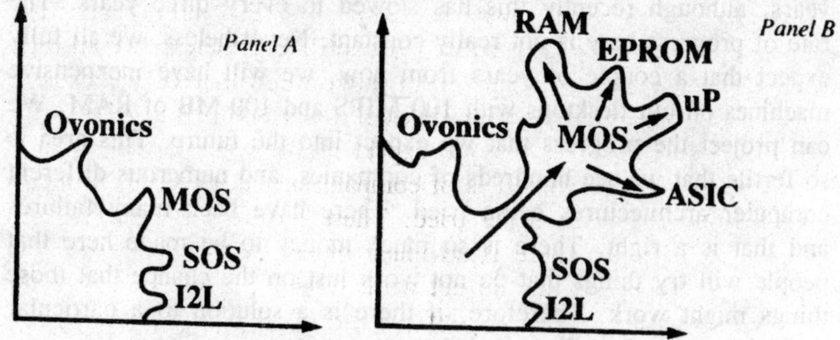

Fig. 3.3. The knowledge space of the electronics industry.

to pay off. So, companies were trying each of them. A little bit of progress occurred in several fields, but basically most of those technologies just withered because not enough useful things could be done with them, relative to what happened with MOS. So MOS proliferated into n-MOS, p-MOS, and c-MOS. Numerous kinds of devices could be built with each of those technologies. Rapid progress was made in all directions, and some of those little protrusions have since turned into major industries on their own.

Everyone who has thought about it believes that nanotechnology, is going to look something like the electronics industry. It will be an incredibly rich vein of ore, but it is deep in the mountain, very far away from where we are now. The real question now, therefore, is how are we going to get there in a roughly continuous manner that is going to pay off while we are moving from here to there?

What are the paths to nanotechnology? What properties of an industry make it conducive to the development of nanotechnology? Which one of those industries is most likely to contribute to a direct path to nanotechnology over the next couple of decades? To answer these questions, we must know what features of those industries will help them contribute to the development of molecular manufacturing.

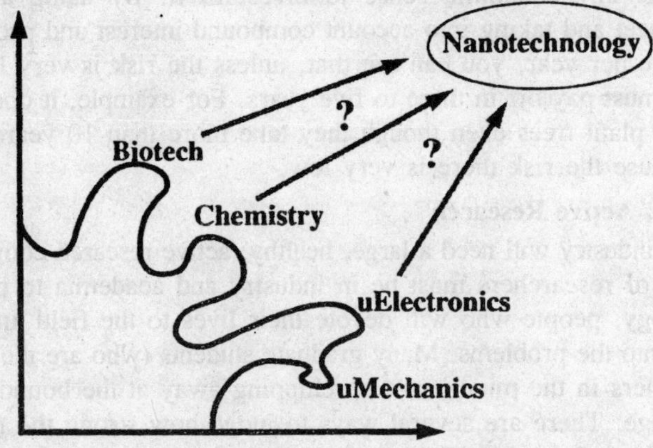

Fig. 3.4. Paths to nanotechnology.

Molecular Savvy

The first desirable property seems obvious. Nanotechnology will be based upon molecules, so at least some fraction of the practitioners in an industry that will contribute to developing nanotechnology must be comfortable thinking about and manipulating matter at the molecular level.

That does not necessarily mean everyone in that industry must think like that, any more than it is the case that everyone designing computers today must be very fluent in device physics. However, it does mean that at least one stratum of that industry must have that skill.

Profitability

The industry must be profitable. It cannot be going through a desert all of the time. Deserts are no fun. You must pack in all of your food and water. The load you carry is very heavy, and if you run out of supplies, you are in real trouble.

An industry that has profitable applications can make money now, and use that money to fund further research. In fact, the very profits that it generates will attract, through the normal function of greed, people who want to make more profits and who will push the state of the art forward. Because some risk is always involved in starting up new companies or going into new research projects, it is important that these ventures pay off in a relatively short time frame. If a research project is likely to pay off only in 10 years (if at all), and if there is any significant risk associated with that project, then it does not make any economic sense to invest in it. By using a simple spreadsheet and taking into account compound interest and probability of failure per year, you can see that, unless the risk is very low, the project must pay off in three to five years. For example, it does make sense to plant trees even though they take more than 10 years to pay off because the risk there is very low.

Healthy, Active Research

An industry will need a large, healthy, active research community. A body of researchers must be in industry and academia to push the technology, people who will devote their lives to the field and delve deeply into the problems. Many graduate students (who are the lowest-paid miners in the mine) must be chipping away at the boundaries of knowledge. There are several ways to judge how strong the research community is in a given field. Is it possible for students to get degrees? Once students have degrees, is it possible for them to find work? This is an issue, for example, in theoretical physics right now. Other clues come from looking at journals and conferences. Does the world conference in this field draw 200 or 20,000 people?

Commercial Infrastructure

There must be a commercial infrastructure. The industry must be part of a complete economic system that provides real goods to customers. There must be delivery channels. It is easy to overlook this requirement and to think that, for example, biotechnology is made up entirely of a few people in white coats in university laboratories.

Three companies were there that did nothing but architectural and engineering consulting for building biotechnology laboratories. Five companies were there that did nothing but plumbing, and three of those made nothing but stain less steel valves. At least a dozen journals were represented there, several glassware makers, half a dozen universities trying to find someone to license their patents, many suppliers of materials, makers of instruments, and so on. Much of this was really quite mundane material, but it must there for the advanced material to happen. Some of the industries that we will be looking at do not have this infrastructure.

Systems Mentality

An assembler will be a very complicated device. It will be at least as complicated as a modem computer system, possibly as complicated as a modem factory. If this will be a designed object, then someone will have to design it. The designers must be able to describe, design, and analyze large systems with many components that interact with each other.

Most of the fields of technology we have today that can deal with such complexity can only handle it by breaking things down into different levels of description. At each level, general principles are abstracted from the lower levels, and the details of the lower levels are ignored so that larger issues can be addressed at the higher level.

For example, one level of complexity in the electronics industry is the design of *transistors*. Well below that level is device physics. Above that level is the design of gates. Above the level of gate design is the consideration of registers, blocks of functions, and finite-state machines. Still above that level is the level of asynchronous interfaces, memory blocks, and so forth.

Thus, an entire hierarchy exists where, at each level, it is not necessary to think about too many things. If designers must think about too many things at once, they get lost, and the result is bad designs.

Continuous Path to Nanotechnology

Finally, there ought to be, or at least appear to be, a clear path from present capabilities to something that looks like nanotechnology. It is not enough merely to be making things smaller. It must be the case that if this field is carried to its conclusion, it will lead to strong control over molecular structure, which will allow something like an assembler to be built.

Further, the path that is discerned from current technology to molecular nanotechnology- should be more or less continuous. It should be possible to develop nanotechnology by incremental improvements of current technology. No major technological discontinuities or radical paradigm shifts should be required.

Existing Fields and Industries

Many fields could be analyzed in terms of their potential for the development of nanotechnology. However, for this presentation, we will consider only four. These fields are themselves fairly broad fields, each composed of many activities that have been lumped together.

Micromechanics

The cover of the November 1992 issue of *Scientific American* spotlights an article on micromechanics. The general idea of this article is that a field called "*micromechanics*" deals with building really small mechanical objects which are divided in two classes.

The first class of micromechanical objects is small gear-shaped objects. A *micro-motor* with a 130-micron diameter has been made with an *electrostatic rotor*. This rotor can be made to spin at about 15,000 rpm. Unfortunately, it is not currently coupled to anything that does any useful work, but people are trying to make that happen.

Another branch of micromechanics is making objects that simply bend, such as *piezoelectric objects*. It is possible, for example, to make artificial cilia that will lie flat on a surface, but will curl up to form almost a 90° angle when an electric voltage is applied. Thus, you could think of making something with little hairs on the bottom that could walk along some surface merely by flexing these hairs, without having any rotating parts.

Looking at micromechanical technology from the perspective of the six points, reveal some problems for believing that microtechnology will lead to nanotechnology.

First of all, *micromechanics* is based on bulk technology. The manufacturing techniques that are used are mostly the same techniques that are used in microelectronics—cooking, etching, and so forth. Almost no one working in this field is familiar with molecular structure or considers molecular technology to be part of micromechanics.

Also, the infrastructure in this industry is very limited. Work in micromechanics is primarily riding on the coattails of the microelectronics industry. No one is building fabrication equipment for micromechanics. Instead, the micromechanics industry uses the same equipment that is built for microelectronics, and is thus dependent on progress made in *microelectronics*.

Further, not very many profitable applications exist yet in this industry. One of the few that does exist is in the area of tiny sensors. It is possible to make very small instruments for measurement of molecular properties. Thus, some hope does exist that this field will become an industry.

There is moderate research activity, but few jobs. Most of the research is funded by the government. The Defense Advanced Research Projects Agency (DARPA) is investing a few tens of millions of dollars in micromechanics this year. However, some signs indicate that DARPA is, as someone put it, "getting tired of the snappy gizmo of the month," and instead wants to see working systems.

Finally, the path from *micromechanics to nanotechnology* is discontinuous. There is no obvious way to extend these methods to the fabrication of components to atomic precision.

The conclusion drawn from these considerations is that micro technology is not likely to contribute to the development of nanotechnology, except perhaps by providing an interface between the macroscopic and the molecular scales. Intermediate stages must be used for communication between people and molecular-scale devices. Microtechnology might provide such an intermediate stage, for example, to control the positioning of the tip of a scanning probe microscope.

Microelectronics and Computers

Most people are familiar with computers in one way or another. In considering how it might lead to the development of nanotechnology, we encounter some of the same issues that were encountered with micromechanics.

First of all, this industry, too, is based upon bulk technology. It will be possible for this industry to continue to advance for a while by further scaling down with the same bulk technology. However, the limits of physical law will eventually impose a discontinuity in the advance of this technology to molecular-scale devices. Although I do not want to be yet another person to have predicted hard scaling limit for electronics, and then subsequently been proven wrong, is does seems clear that some such scaling limit does exist, and that we cannot be that far away from it—probably not more than 20 years away from it.

The good thing about this industry is that it is extremely profitable. Diverse varied, and almost ubiquitous applications do exist. In conjunction with this profitability, a broad, solid infrastructure exists, and there is a great deal of research is the area. Therefore, much money will be invested in solving the problems encountered as the technology nears the limits imposed by scaling laws.

When the limits imposed by scaling are reached, the large amount of money invested in research in this area will create a tremendous amount of pressure to find solutions. This pressure may lead to new advances. For example, there ha been talk that diamond semiconductors will provide another order of magnitude increase in device performance. There has even been talk that it will be possible to build vacuum tubes out of diamond, making possible 10 million triodes on a chip inside of a piece of glass.

Another major advantage of this industry is that it is routine for the most competent teams to design systems that have more than 10 million components. This is a very difficult art, and it is being done in several different domains—both hardware and software.

Biotechnology

The major strength of *biotechnology* is that this industry is clearly comfortable with molecules, even very large ones. The ability to control molecular structure is vital to success, and the industry is moving ahead at full speed. However, the molecular expertise in biotechnology is largely directed toward the analysis of natural molecular systems, rather than the design of artificial systems of equal complexity.

Other strong points include a well-developed infrastructure and plenty of research activity. This is a healthy, profitable industry, although the advance of the industry has been slowed in some ways by the governmental regulatory process.

Fundamentally, however, biotechnology is already building nanometer-scale products in the real world, and it is solving real-world problems in doing so. Analogies to those problems do exist in biotechnology—such as purity, and whether your protein forms inclusion-bodies in bacterial expression. Much crucial empirical knowledge already has been accumulated about what works and does not work, what gives good results and what does not.

Another positive trend is that considerable effort is being invested in the necessary computational support for biotechnology. This effort has produced extensive chemical and genetic databases. The effort is particularly intense in the computer-aided design of small molecules, which is akin to entering schematics for a computer. Not much in the way of design automation exists yet, but research in universities is trying to attack this problem.

One example is the Invention system, done by Todd Wipke and his students at University of California at Santa Cruz. Invention allows a designer to specify only the desired position of key atoms at the active site of a molecule. The system will then try to find a backbone that will hold those atoms in place. Already, the software has proposed several new morpheme analogs.

In terms of manufacturing ability, biotechnology has processes that are equivalent to what are called "*Numerically Controlled*" (NC) tools used in the manufacture of items such as automobiles. Such NC tools in conventional manufacturing include tools in which a computer guides the cutting tip of a lathe. Biotechnology has similar capabilities in the form of PCR (*polymerase chain reaction*) and ribosomes, in which digital information in the form of nucleic acid sequences can be used to manufacture genes and to manufacture proteins. This is so easily done these days that you can order custom DNA, RNA, or proteins over the phone and pay for them with your credit card. The problem is that no one knows what to design.

The conclusion about the role of biotechnology as a developmental path toward nanotechnology seems to be that a conceptual shift from natural to artificial molecular systems will be required, but perhaps not much change in technology will be required.

Chemistry

The major contribution that chemistry offers is that chemists have the best understanding of small molecules. Furthermore, many profitable industries are based upon chemistry, such as the pharmaceutical industry. The *pharmaceutical industry* may be shifting from what we think of as classical chemistry toward biotechnology, but there is still a lot of money to be made from small molecules. The applications of low-molecular-weight drugs have not yet been fully explored.

The real problem with this field is one of vision. Chemists think of chemistry as a branch of pure science, not as an engineering discipline. Generally, as a group, they seem to be unwilling to deal with designing large systems or with developing principles for the design of large molecules to perform specified functions. In short, the path toward nanotechnology is not being pursued.

Chemistry provides the substrate for the development of nanotechnology, just as solid-state physics provided the substrate for the development of the semiconductor industry. Theoretical chemistry will provide the models of molecular structure and behaviour, based on quantum mechanics, which will underlie nanotechnology's computer-aided design tools. Chemistry also will provide the building blocks for molecular devices. The major advantage here is that synthesizing building blocks a mole (6.02×10^{23} items) at a time is much more efficient than synthesizing them one at a time (for example, using a scanning probe microscope), at least until nano technology becomes quite advanced.

4

Prospect of Nanomedicine

The history of disease is vastly older than that of humankind itself. Indeed, disease and parasitism have been inseparable companions to life since the dawn of life on Earth. Fossilized bacteria similar to those responsible for many infections that afflict people today have been found in geological formations that are 500 million years old. Fossil shells dating from an era almost equally remote show clear evidence of disturbance by injury and parasites. Examination of the skeletons of long-extinct dinosaurs and other great reptiles show that these creatures suffered from fractures, bone tumours, arthritis, osteomyelitis, dental caries and other diseases that still plague us in the 20th century. While available fossil evidence is largely limited to changes observable in bones and teeth, it is probably safe to assume that disease processes were equally prevalent in the soft organs and tissues that have not been geologically preserved, and that the general pattern of disease has not changed in its essentials during the hundreds of millions of years that animal life has existed on this planet.

Since its first appearance on the prehistoric stage millions of years ago, the human body has also been constantly subject to assault and injury, invasion by parasites, extremes of heat and cold, and infections. Early man probably suffered from a number of diseases due to nutritional factors and body chemistry disorders. For example, the poor condition of the teeth of an 18-year old Australopithecus who lived 1.75 million years ago, found at Olduvai Gorge by Louis Leakey in 1959, suggests that the hominid had a disease that lasted for many months—most likely gastro-enteritis due to malnutrition—with three major attacks of the disease at the ages of two, four, and four and a half. Diseases that may date back more than 25 million years to the ape ancestors of modern apes and man are amoebic dysentery, malaria, pinworm infections, syphilis, yaws, and yellow fever. Diseases which may have appeared and evolved with man include leprosy and typhoid; certain modern diseases such as cholera, measles, mumps, smallpox, whooping cough, and the common cold require large concentrated populations to support them, thus probably could not have existed in the prehistoric era.

As for cancer, Java man, first discovered by Dutch anatomist Eugene Dubois in 1891 and considered to be half a million years old, had a morbid growth on his femur. It is likely that bone cancer and other forms of cancer have existed from the earliest times. The remains of *Neanderthal*, a competing species to *Homo sapiens* that roamed through Europe, Africa, and the Near East during the last glacial period ~75,000 years ago, show clear evidence of arthritis, tooth loss, and suppurative bone disease. (The high rate of broken bones and early death suggests that *Neanderthals* engaged in more close-quarter combat with large animals than did modern humans, who had figured out safer strategies.) Human bones unearthed from the New Stone Age period reveal that Neolithic man suffered from arthritis, congenital dislocations and fractures, sinusitis, tuberculosis of the spine, and tumours.

In considering the question of how our earliest ancestors dealt with these conditions, we are on somewhat uncertain ground, since little direct evidence has been preserved. Anthropologists point out that in pre-*Neanderthal* hunter-gatherer tribes, a sick or lame person is a serious handicap to a group on the move. In the event of major illness or mortal wounds, sufferers may either leave the group, be abandoned, or, as with lepers in medieval Europe, may be ritually expelled, becoming "culturally dead" before they are biologically dead. Early hunter-gatherer hominid bands were more likely to abandon their seriously sick than to succor them, although *Cro-Magnons* and *Neanderthals* evidently were the first to care for their wounded and disabled, and to bury their dead.

But moderate wounds, bruises, fractures, or foreign bodies such as arrowheads or thorns are tangible things that demand attention. Thus the art of surgery must first have originated as a response to immediate crises. The first and most obvious course of action of a wounded man would be to protect the site of injury from the influence of external forces or agents. For this there was, and remains to this day, only one means—the application of a dressing. Many observations were made and many substances tried, and in time a body of experience was accumulated and passed on orally to others for use in similar emergencies, eventually creating a considerable sum of inherited empirical knowledge. It is believed that the art of dressing wounds long constituted the whole of medicine—the use of internal remedies or herbs, and use of the knife or fire, came much later.

Present-day primitive and folk medicines provide additional clues to early medical practice. For example, many early peoples developed effective methods to control bleeding—the use of cobwebs is an ancient folk remedy, as is the application of tourniquets, packing with absorbent materials, the laying on of snow or, at the other temperature extreme, the application of cautery by hot knife or spear. The *Masai* and *Akamba* tribes treated sword wounds by slapping on a poultice of cow dung and dust. The suturing of wounds is practiced by some primitive peoples and may have been known to prehistoric man. Bone needles furnished with an eye have been found in paleolithic deposits in France and England, and some Indian tribes suture with threads of sinew or bone needles (the needles are left in and the thread is twisted around them). One of the strangest suturing techniques, observed among primitive tribal cultures in such widely separated places as India, East Africa and Brazil, is the sealing of wounds, especially abdominal wounds, using termites or ants. The edges of the wound are drawn closely together and the insect is allowed to bite through them both, firmly securing the flesh on two sides. Once attached, the insect's body is severed, allowing only the jaws to remain in place, holding the wound shut.

While able to deal with wounds, early man did not admit the existence of disease from "natural causes". Internal diseases were generally ascribed to malevolent influences exercised by a supernatural entity or a human enemy. As centuries passed, many internal diseases were eventually recognized and simple treatments empirically established—for example, Celsus (ca. 30 AD), a Roman medical writer, described ligature (tying off blood vessels), suturing the large intestine, eye operations such as couching for cataract, tonsillectomy, and bladder surgery to remove stones. But the rational basis for internal medicine awaited basic physiological knowledge such as the circulation of the blood, finally proven by William Harvey (1578-1657) in 1628, and the discovery and acceptance of the theory of infection by microorganisms in the 1800s. Throughout most of human history, life has been short indeed. Even by the 17th century, more than half of all children never lived past the age of ten, often succumbing to diseases such as cholera, diphtheria, scarlet fever, or whooping cough, or being scarred for life by smallpox. It was a common and widely accepted fact of life that people would die at all ages, although wealthy families could leave town every year during the cholera season.

One of the constant missions of human civilization has been the avoidance and elimination of animals that prey on humans. Cave-dwelling carnivorous saber-toothed tigers, having occupied the upper echelons of the food chain for 30-35 million years, finally became extinct not more than about ten thousand years ago, most likely at the hands of newly-arrived human hunters crossing the Siberian land bridge into post-Pleistocene North America. As late as medieval times, wolves ranged freely over Europe, remaining abundant in France through 1500 AD—in winter, audacious wolf packs would enter Paris and eat children, dogs, and even adults who were alone on the streets. With technological advances, extant tiger and wolf species are now largely confined to artificial habitats or isolated nature preserves and no longer pose any serious threat to human health. People in most places are not eaten by wolves; indeed today, a few venturesome individuals actually keep and breed wolves with dogs, as pets.

Bacteria are among the last remaining "wild animals" on Earth that threaten man. As our instrumentalities continue to progress from the macroscale to the microscale, and finally to the molecular or nanoscale, all of the remaining natural "wild things" that endanger human life and health—whether viruses, bacteria, protozoa, metazoan parasites, or even our own pathological native cells—will be confined and tamed, reconstructed or eliminated. Using smaller tools, we hunt smaller prey. As with sabertooths in the post-Neolithic era, and with wolves in post-medieval times, people of the 21st century will no longer fear or need suffer predation by wild microbes or tumour cells run amok.

Humanity is poised at the brink of completion of one of its greatest and most noble enterprises. Early in the 21st century, our growing abilities to swiftly repair most traumatic physical injuries, eliminate pathogens, and alleviate suffering using molecular tools will begin to coalesce in a new medical paradigm called *nanomedicine*. Nanomedicine may be broadly defined as the comprehensive monitoring, control, construction, repair, defense, and improvement of all human biological systems, working from the molecular level, using engineered nanodevices and nanostructures.

Current Medical Practice

In order to fully appreciate the changes that nanomedicine will inevitably bring, it is useful first to review the history and development of current medical practice. As Winston Churchill once remarked: "The further backward you look, the further forward you can see." The author unapologetically favours what medical anthropologists would regard as the "Western" healing tradition. The great medical theorist Otto E. Guttentag agreed:"Contemporary Western medicine involves a type of healing that potentially and actually exceeds all other approaches in maintaining that optimal status of selfhood we call being healthy and in eliminating that reduced status of selfhood we call being sick."

Evolution of Scientific Medicine

A study of the history of "scientific" or Western medical practice suggests a continuity in certain aspects of the medical paradigm even since ancient times—as for instance the basic precepts of observation and diagnosis, followed by treatment—but reveals changes in attitudes, techniques, and instrumentalities, and a gradual evolution in the epistemology and ontology of medical logic. Note that 2010 is marked as a possible date for the first applications of *nanomedicine*. These will most likely be *ex vivo* applications only *in vivo* nanomedical treatments may come much later.

Another important reason to study the history of medicine is to gain a deeper appreciation of the long, hard struggle to improve human health, a struggle that is expected finally culminate in victory in the 21st century. If the ~10 billion people that have ever lived survived an average of 40 years and spent 5% of their lives in sickness or physical misery from disease, then ~200 trillion man-hours of suffering have been paid to achieve this remarkable result, a not inconsiderable price. There is no pretense to completeness here. For example, Chinese, Indian and Islamic contributions are omitted, not because they are unimportant, but because they did not significantly alter the evolutionary pathway of the Western medical paradigm. A great deal of physiology, pathology, neurology, systematics, and many other important medical-related disciplines are also neglected in the interest of brevity.

Prehistoric medicine

The elementary nature of prehistorical medical practice has already been mentioned. There is direct evidence that *Stone Age* man used a natural fungus as a treatment for intestinal parasites. Besides basic wound-tending, the practice of circumcision is an age-old form of simple surgery with a rational, if sometimes controversial, hygienic basis.

An even more dramatic surgical operation, for which there is considerable prehistoric fossil evidence, is trephining the skull to remove a round piece of bone from it. This can be done using flint instruments, either by gradually scratching through the skin and bones of the skull, getting gradually deeper and deeper, or by drilling a series of small holes in a circle in the skull, then cutting the small bridges between to remove a disk of bone. Skulls have been found with multiple holes. The additional fact that some hole edges show callus formation (evidence of healing) indicates survival of the patient. Trephined skulls have been found in Western Europe, including England, North Africa, Asia, the East Indies, New Zealand, and the Americas from Alaska in the north down through the continent to Peru in the south. The practice was probably regarded as therapeutic, either to remove a depressed fracture, to try to cure mental illness, or to relieve severe headache or epilepsy, presumably by letting out the demon possessing the patient.

Ancient mesopotamian medicine

The first written records, which came from ancient Babylon and Egypt, contain the earliest references to medical care,obviously codifying earlier practices no record of which has survived. The medicine of the Sumerians of Mesopotamia from ca. 3000 BC was primarily religious. The Mesopotamian peoples saw the hands of the gods in everything. Disease was caused by spirit invasion, sorcery, malice, or the breaking of taboos, and sickness was both judgement and punishment. An Assyrian text circa 650 BC describes epileptic symptoms within a demonological framework: "If at the time of the possession, his mind is awake, the demon can be driven out; if at the time of his possession his mind is not so aware, the demon cannot be driven out." Headaches, neck pain, intestinal ailments and impotence were read as omens. The appropriate remedy was to identify the demons responsible and expel them by spells or incantations.

But medicine also had an empirical component, with some sicknesses being ascribed to cold, dust and dryness, putrefaction, malnutrition, venereal infection and other natural causes. The Babylonians drew on an extensive pragmatic materia medica—some 120 mineral drugs and twice that number of vegetable items are listed in surviving tablets. Alongside various fats, oils, honey, wax, and milk, were many active ingredients that included mustard, oleander and hellebore (a plant in the buttercup family that is a violent gastro-intestinal poison, hence acts as a powerful purgative, though it is lethal in high doses). Colocynth, senna and castor oil were used as laxatives, while wound dressings were compounded with dried wine dregs, salt, oil, beer, juniper, mud or fat, blended with alkali and herbs. With the discovery of distillation, the Mesopotamians made essence of cedar and other volatile oils. Turpentine, asafetida, henbane, myrrh, mint, poppy, fig, and mandrake are also mentioned. Dog dung and other fecal ingredients were used to drive off demons.

Surgical conditions such as wounds, fractures and abscesses were also treated by Mesopotamian surgeons. Practitioners were priests, and after 2000 BC, they were ruled by the strict laws included in the Code of Hammurabi. The Code laid down rewards for success and severe punishment for failure, and contained laws relating to medical practice which show that medicine and surgery were highly organized professions. Fees were regulated on a sliding scale of rewards based on the patient's rank, and severe penalties were laid down for failure:

"Concerning the wounds resulting from operations it is written: if a physician shall produce on anyone a severe wound with a bronze operating knife and cure him, or if he shall open an abscess with the operating knife and preserve the eye of the patient, he usually shall receive 10 shekels of silver [more than a craftsman's annual pay]; if it is a slave, his master shall usually pay 2 shekels of silver to the physician."

"If a physician shall make a severe wound with an operating knife and kill him, or shall open an abscess with an operating knife and destroy the eye, his hands shall be cut off."

"If a physician shall make a severe wound with a bronze operating knife on the slave of a free man and kill him, he shall replace the slave with another slave. If he shall open an abscess with a bronze operating knife and destroy the eye, he shall pay the half of the value of the slave."

The Hammurabic Code also mentions the Gallabu, or barber-surgeons, whose province was minor surgery, including dentistry and the branding of slaves. If Herodotus (ca. 485-425 BC) may be believed, Babylonian medicine must have declined in the 5th century BC. Herodotus states that there were no physicians, but that the people brought their sick into the marketplace in order that passers-by might make suggestions or offer cures.

Ancient Egyptian medicine

The civilization of ancient Egypt dates from around 3000 BC. As in Mesopotamia, ancient Egyptian medicine was religious-empirical. An examination of the preserved bodies of members of the Royal family has provided much information on the diseases of ancient Egypt, including congenital deformities such as clubfoot, dental caries, gallstones, bladder and kidney stones, rheumatoidarthritis, mastoiditis, numerous eye diseases, and bone fractures, some of which show treatment by quite sophisticated splinting. Other evidence suggests that the average ancient Egyptian was extremely diseased. For example, a weaver who died in the 11th century BC, aged 14-18 years, evidently suffered from *schistosomiasis*, tapeworm likely associated with malnutrition, *anthracosis* of the lungs presumably due to environmental pollution from cooking and heating, *pulmonary silicosis*, and possibly malaria and fleas.

The Ebers Papyrus (ca. 1550 BC), deriving from Thebes, is the principal medical document and may be the oldest surviving medical book. Over 20 meters long, it deals with scores of diseases and proposes remedies involving spells and incantations, but also includes many rational treatments. The Ebers Papyrus covers 15 diseases of the abdomen, 29 of the eyes, and 18 of the skin, and, perhaps unsurprisingly to the modern consumer, lists no fewer than 21 cough treatments. About 700 drugs and 800 formulations are mentioned, mainly involving herbs but also including mineral and animal remedies. For example, to cure night blindness the patient should eat fried ox liver—possibly a tried-and-tested procedure, since liver is rich in Vitamin A, lack of which causes the illness. Eye disorders were also common, for example:

"To drive away inflammation of the eyes, grind the stems of the juniper of Byblos, steep them in water, apply to the eyes of the sick person and he will be quickly cured. To cure granulations of the eye, prepare a remedy of cyllyrium, verdigris, onions, blue vitriol, powdered wood, and mix and apply to the eyes."

For stomach ailments, a decoction of cumin, goose-fat and milk was recommended, but other remedies sound more exotic, including a drink prepared from black ass testicles. A mixture of vulva and penis extracts and black lizard was supposed to cure baldness. Also good for hair growth was a compound of hippopotamus, lion, crocodile, goose, snake and ibex fat—merely assembling these ingredients might promise a hair-raising experience! Egyptian medicine credited many vegetables and fruits with healing properties, and also tree products such as sycamore bark and resins such as myrrh, frankincense and manna. As in Mesopotamia, plant extracts—notably senna, colocynth and castor oil—were employed as purgatives, and oil of camomile to improve digestion. Recipes included ox spleen, pig's brain, stag's horn, honey-sweetened tortoise gall, and the blood and fats of various animals.

Antimony, copper salts, alum, and other minerals were recommended as astringents or disinfectants. Containing ingredients from leeks to lapis lazuli—including garlic, onion, tamarisk, cereals, spices, condiments, resins, gums, dates, hellebore, opium and cannabis—compound drugs were administered in the form of pills, ointments, poultices, fumigations, inhalations, gargles and suppositories; they might even be blown into the urethra through a tube. Among the most interesting of the healers whose names have been recorded for posterity are Peseshet, head female physician or overseer, proof of the existence, as in Mesopotamia, of female healers; and Iri, Keeper of the Royal Rectum, presumably the pharaoh's enema expert.

The Edwin Smith Papyrus (ca. 1600 BC), discovered by its American namesake at Luxor in 1862, may be the world's earliest surviving surgical text, and contains material probably derived from even more ancient times. The "*book of wounds*" comprises 48 case reports, which commence with the top of the head and proceed systematically downward —nose, face, ears, neck and chest, mysteriously stopping in mid-sentence at the spine, presumably when the scribe was interrupted at his work. The only surgical conditions treated were wounds, fractures, abscesses, and circumcisions. In the Papyrus, the method of presentation is first to set out the title or the chief symptom, followed by the further symptoms, and then the examination, diagnosis, prognosis, and treatment. The advice given is entirely rational, as will be seen by the following directions for the treatment of a fractured humerus:

"Instructions concerning a Break in his Upper Arm. If thou examinest a man having a break in his upper arm, and thou findest his upper arm hanging down, separated from its fellow, thou shouldst say concerning him: One having a break in his upper arm. An ailment which I will treat."

"Thou shouldst place him prostrate on his back, with something folded between his two shoulder blades; thou shouldst spread out with his two shoulders in order to stretch apart his upper arm until that break falls into its place. Thou shouldst make for him two splints of linen, and thou shouldst apply for him one of them both on the inside of his arm, and the other of them both on the underside of his arm. Thou shouldst bind it with ymrw (an unidentified mineral substance), and treat it afterward with honey every day until he recovers."

An interesting point in the case histories is that after the diagnosis the writer gives a decision about his further course of action. The verdict may take one of three forms: (i) an ailment which I will treat;

(ii) an ailment with which I will contend; and (iii) an ailment not to be treated. Like modern military triage, the "hopeless" patient is left to his inevitable fate. This guarded attitude on the part of the medical man was widespread in antiquity. While present-day doctors generally do everything possible to alleviate symptoms to the very end, even when the patient has no chance of recovery, the view in ancient times was that hopeless cases were not to be touched. This attitude was entirely practical. A doctor in attendance at the courts of ancient Egypt might expect rich rewards if his patient recovered, but if a patient died under his care, the unfortunate physician ran a grave risk of impalement.

Ancient Greek medicine

By 1000 BC the communities later collectively known as the Greeks were emerging around the Aegean Sea. How much medical knowledge they took from Egypt remains controversial, but the contrasts between the two are striking. Little is known of Greek medicine before the appearance of written texts in the 5th century BC. Archaic Greece had its folk healers, including priest healers employing divination and herbs. From early times (the first Olympic games were recorded in 776 BC), the love of athletics produced instructors in exercise, bathing, massage, gymnastics and diet. The Homeric epics (ca. 600 BC) offer glimpses of early Greek medicine. Scholars count 147 cases of battle wounds in the Iliad, including 106 spear thrusts, 17 sword slashes, 12 arrow shots, and 12 sling shots. Among the arrow wound survivors was King Menelaus of Sparta, whose physician extracted the arrow, sucked out the blood and applied a salve. As with other medical interventions in Homer, this shows no Egyptian influence, supporting the idea that, even if Greek practice owed much to Egypt, it rapidly went its own way.

Various Greek gods and heroes were identified with health and disease, the chief being *Asclepius*, who even had the power to raise the dead. A heroic warrior and blameless physician, Asclepius was the son of Apollo, sired upon a mortal mother, who was taught herbal remedies by Chiron and then generously used them to heal humans. Incensed at being cheated of death, Hades, the ruler of the underworld, appealed to the supreme god, Zeus, who obligingly dispatched Asclepius with a thunderbolt, though he was later elevated to godhood. A different version appears in Homer, who portrays Asclepius as a tribal chief and a skilled wound healer whose sons became physicians and were called Asclepiads, and from whom all Asclepian practitioners descended. As the tutelary god of medicine, Asclepius is usually portrayed with a beard, staff and snake—the origin of the caduceus symbol of the modern physician, with its two snakes intertwined, double-helix like, on a winged staff. The god was often shown accompanied by his daughters, Hygeia (health or hygiene) and Panacea (cure-all).

For all that, *Hippocratic medicine*, the foundation of Greek written medicine, explicitly grounds the art upon a quite different basis—a healing system independent of the supernatural and built upon natural philosophy. The beginnings of true medical science in the West were established when the reliance on superstition that underpinned tribal medicine was replaced by civilized and rational curiosity about the cause of illness. The growth of civilized thought allowed for argument on medical cause and cure, with great doctrinal multiplicity. The separation of medicine from religion reveals another distinctive feature of Greek healing: its openness, a quality characteristic of Greek intellectual activity in general, owing to political diversity. There was no imperial Hammurabic Code and, unlike *Egypt*, no state medical bureaucracy, nor were there examinations or professional qualifications. Those calling themselves doctors (iatroi) had to compete with bone-setters, exorcists, root-cutters, incantatory priests, gymnasts, and showmen, exposed to the quips of playwrights and the criticism of philosophers. Medicine was open to all.

Empedocles (fl. 450 BC) may have been the first to advance some of the key physiological doctrines in Greek medicine, including innate heat as the source of living processes such as digestion, the cooling function of breathing, and the notion that the liver makes the blood that nourishes the tissues. His contemporary, *Alcmaeon of Croton* (fl. 470 BC), believed that the brain, not the heart, was the chief organ of sensation. Alcmaeon's examination of the eyeball led him to discern the optic nerve leading into the skull, a genuine observational basis. He gave similar explanations for the sensations of hearing and smelling, because the ear and nostrils suggested passages leading to the brain. Most of such knowledge depended heavily on wound observation and animal dissection, for in the classical period the dignity of the human body forbade dissection.

All we know for sure of *Hippocrates* (ca. 460-377 BC), who "taught all that were prepared to pay," is that he was born on the island of Cos and lived a long and virtuous life. The sixty or so works comprising the *Hippocratic Corpus* (ca. 440-340 BC) derive from a variety of hands, and, as with the books of the Bible, they became jumbled up, fragmented, then pasted together again in antiquity. What is now called the *Corpus* was gathered around 250 BC in the Library at Alexandria, with further texts added later still. Some volumes are philosophical, others are teaching texts or case notes. What unites them all is the conviction that health and disease are capable of explanation by reasoning about nature, independently of supernatural interference. Man is governed by the same physical laws as the cosmos, hence medicine must be an understanding, empirical and rational, of the workings of the body in its natural environment. Anticipating modern medicine, appeal to reason, rather than to rules or to supernatural forces, gives Hippocratic medicine its distinctiveness. It was also patient- rather than disease-centered; the Hippocratics specialized in medicine by the bedside, prizing trust-based clinical relations:

"Make frequent visits; be especially careful in your examinations, counteracting the things wherein you have been deceived at the changes. Thus you will know the case more easily, and at the same time you will also be more at your ease. For instability is characteristic of the humours, and so they may also be easily altered by nature and by chance."

"...Keep a watch also on the faults of the patients, which often make them lie about the taking of things prescribed. For through not taking disagreeable drinks, purgative or other, they sometimes die. What they have done never results in a confession, but the blame is thrown upon the physician."

The Hippocratics promulgated the idea of "*vis medicatrix naturae*," or the power of nature to cure itself, and thus the belief that there was a natural tendency for things to get better on their own. This tendency could be aided by providing a beneficial environment for the patient and by improving physical function with a regimen of suitable diet, lifestyle, and exercise—the diatetica. In extreme cases, further aids to recovery could be sought. Stubbornly offending "*humors*" could be removed with the help of venesection (phlebotomy or bloodletting) and purgatives, sudorifics applied to induce sweating, and diuretics to increase urination. But the Hippocratic physician was extremely reluctant to administer drugs of any kind, because the physician's goal was to aid nature in healing the body.

Hippocratics scorned heroic interventions and left risky procedures to others. Their Oath explicitly forbade cutting, even for stones, and other texts reserved surgery for those used to handling war wounds. Surgery was regarded as an inferior trade, the work of the hand rather than the head, a fact reflected in its name: "surgery" derives from the Latin "*chirurgia*," which comes from the Greek "cheiros" (hand) and "ergon" (work); surgery was handiwork. Hippocratic surgical texts were thus conservative in outlook, encouraging a tradition in which doctors sought to treat complaints first through management, occasionally through drugs, and finally, if need be, by surgical intervention.

The art of diagnosis involved creating a profile of the patient's lifestyle, work and dietary habits, partly by asking questions and partly by the use of trained senses:

"When you examine the patient, inquire into all particulars; first how the head is...then examine if the hypochondrium [abdomen beneath lower ribs] and sides be free of pain, for...if there be pain in the side, and along with the pain either cough, tormina [painful intestinal colic] or bellyache, the bowels should be opened with clysters [enema].... The Physician should ascertain whether the patient be apt to faint when he is raised up, and whether his breathing is free..."

Hippocratics prided themselves on their clinical acuity, being quick to pick up telltale symptoms, as with the facies hippocratica, the facial look of those dying from long-continued illness or cholera: "a protrusive nose, hollow eyes, sunken temples, cold ears that are drawn in with the lobes turned outward, the forehead's skin rough and tense like parchment, and the whole face greenish or black or blue-grey or leaden." Experience was condensed into aphorisms, as for instance: "When sleep puts an end to delirium, it is a good sign."

The technique most prized among Hippocratics was the art of prognosis—a secular version of the priestly and oracular prognostications of earlier medicine, and bearing some analogy to the 20th century weatherman, who can give a bright or gloomy forecast but is powerless to change it. Noted one Hippocratic text:

"It appears to me a most excellent thing for the physician to cultivate Prognosis; for by foreseeing and foretelling, in the presence of the sick, the present, the past, and the future, and explaining the omissions which patients have been guilty of, he will be the more readily believed to be acquainted with the circumstances of the sick; so that men will have confidence to entrust themselves to such a physician....Thus a man will be the more esteemed to be a good physician...from having long anticipated everything; and by seeing and announcing beforehand those who will live and those who will die."

The ultimate significance of Hippocratic medicine was twofold. First, it carved out a lofty role for the selfless physician which would serve as a lasting model for professional identity and conduct. Second, it taught that an understanding of sickness required an understanding of nature.

Ancient Alexandrian medicine

Soon after the death of *Aristotle* (d. 322 BC) and his most famous pupil, *Alexander* the Great (d. 323 BC), a great medical school was founded in Egypt at the court of King Ptolemy, at his capital, *Alexandria*, at the mouth of the Nile. The King's main cultural creations were the Alexandrian Library and the Museum (Sanctuary of the Muses), which installed Greek learning in a new Egyptian environment—Archimedes, Euclid, and the astronomer Ptolemy were soon to teach there. The Library became a wonder of the scholarly world, eventually containing, it was said, 700,000 manuscripts, and other facilities including an observatory, zoological gardens, lecture halls and rooms for research.

The two earliest teachers at the Alexandrian medical school were also its greatest—*Herophilus* of Chalcedon (ca. 330-260 BC) and his contemporary, *Erasistratus* of Chios (ca. 330-255 BC). Their writings having been lost, we know about them only through later physicians.

Herophilus was the first to dissect cadavers in public. He was a student of *Praxagoras* of Cos (fl. 340 BC), who had improved *Aristotelian* anatomy by distinguishing arteries from veins, but who saw the arteries as air tubes, similar to the trachea and bronchi, a common error because arteries are devoid of blood in corpses. Herophilus observed that the coats of arteries were much thicker than those of the veins, thus he speculated that the arteries were filled not with air but with blood. *Herophilus* wrote at least eleven treatises, discovering and naming the prostate and the duodenum (from the Greek for twelve fingers, the length of gut he found). He also wrote on the pulse as a diagnostic guide and on therapeutics, ophthalmology, dietetics, and midwifery. He recognized the brain as the central organ of the nervous system and the seat of intelligence, extending the knowledge of the parts of the brain, certain of which still bear titles translated from those given by him. He was the first to grasp the nature of the nerves, which he distinguished as motor and sensory, though he did not separate them clearly from tendons.

Erasistratus surmised that every organ is formed of a threefold system of "vessels" —veins, arteries, and nerves, dividing indefinitely. These, plaited together, were postulated to make up the tissues. In the brain, *Erasistratus* observed convolutions, noting that they were more elaborate in man than in animals and associating this with higher intelligence. He distinguished between cerebrum and cerebellum, and is often regarded as an early mechanist because of his model of bodily processes—for instance, digestion involved the stomach grinding food. He was opposed to intrusive remedies such as venesection, and his favourite therapeutic measures were regulated exercise, diet, and the vapour bath—very much in the Hippocratic tradition.

Ancient Roman medicine

Roman tradition held that one was better off without doctors. According to Cato (234-149 BC), citizens had no need of professional physicians because Romans were hale and hearty, unlike the effete Greeks. Apparently Romans enjoyed bad-mouthing Greek physicians. Thus the Romans despised medicine as a profession but this did not prevent them from making use of Greek physicians or even of the services of their own slaves. *Galen of Pergamum* (130-200 AD) tells us that in his time, large cities such as *Rome* and *Alexandria* swarmed with specialists who also travelled about from place to place. *Martial* (40-104 AD) mentions some of them in an epigram: "Cascellius extracts and repairs bad teeth; you, Hyginus, cauterize ingrowing eyelashes; Fannius cures a relaxed uvula without cutting; Eros removes brand marks from slaves; Hermes is a very Podalirius for ruptures." Under the Empire, military medicine was highly organized—every cohort had its surgeon, and surgeons of a higher grade were attached to the legions as consultants. Army surgeons ranked as noncombatants and enjoyed many privileges.

In the Roman empire, the earliest scientific teacher was a Greek, Asclepiades of Bithynia (124-40 BC). *Asclepiades* ridiculed the Hippocratic expectant attitude as a mere "meditation on death," and urged active measures that the cure might be "seemly, swift and sure." Though outside the mainstream, his medical practice is interesting as a modification of the atomic or corpuscular theory, according to which disease results from an irregular or inharmonious motion of the corpuscles of the body. His pupils were numerous, constituting the Methodical school, but his available therapeutic tools were few—he trusted mainly to changes of diet, friction, bathing, exercise, and occasionally emetics, bleeding, and wine. He was also the first to use music in the treatment of the insane.

Back in the mainstream, *Galen of Pergamum* (130-200 AD) provided the final medical synthesis of antiquity and the effective medical standard for the next 13 centuries. His first medical appointment was as surgeon to the Roman gladiators. He later traveled to Rome and wrote extensively on anatomy, physiology and practical medicine. His fame is due in part to his prolific pen some 350 authentic titles ranging in topic from the soul to bloodletting polemics survive, about as much as all Greek medical writings together. Galen was a flamboyant character. One of his party tricks, revealing his genius for self-advertisement as well as experiment, was to sever the nerves in the neck of a pig. As these were severed, one by one, the pig continued to squeal; but when Galen cut one of the laryngeal nerves the squealing stopped, impressing the crowd.

Galen justified venesection in terms of his elaborate pulse lore. Written in the early 170s, his sixteen books on the pulse were divided into four treatises, each four books long. In one of these treatises, he explains how to take the pulse and to interpret it, raising key questions, for example: How was it possible to tell whether a pulse was full, rapid, or rhythmical? Such questions he resolved partly from experience and partly by reference to earlier authorities. Galen developed a characteristic physiological scheme that remained in vogue until the 17th century. It supposes three types of so-called spirits associated with three types of the activity of living things. These were the natural spirits formed in the liver and distributed by the veins, the vital spirits formed in the heart and distributed by the arteries, and the animal spirits formed in the brain and distributed by the nerves. Galen's system was an admirable if factually flawed working hypothesis, based on much experimental evidence, and he presented his work as "perfecting" the legacy of Hippocrates.

As in Greece, medicine remained personal in Rome. No medical degrees were conferred or qualifications required. In the absence of colleges and universities, the private face-to-face nature of medical instruction encouraged fluidity and diversity, Students attached themselves to an individual teacher, sitting at his feet and accompanying him on his rounds. Many different sorts of medical care were available, and self-help was universal. Celsus' *On Medicine* was written for a non-professional readership as willing to wield the scalpel as the plough. Disease explanations changed little. Public authorities still ascribed famines and pestilences to the gods, and Galen was silent on contagion. The essential trilogy of classical medicine remained dietetics, exercise, and drugs, accompanied by light surgery.

Medicine in the Middle ages

The passage from the glorious days of Rome to the Middle Ages was often violent, especially in the West, with wave after wave of barbarian onslaughts from the East. These culminated in the sack of the Eternal City by Alaric's Goths in 410 AD, which effectively ended the western empire and frayed the thread of learned medicine. Thus Galen had no effective successor. Indeed, medieval medicine may be summed up as a corrupted version of Galenism. The true scientific tradition did not reappear in the West until the 16th century, after a lengthy incubation in the Islamic world. For example, in England the Venerable Bede (ca. 672-735) and his monks possessed many medical writings, and knowledge of plant remedies was extensive, but the English healer used chants and charms, predicated on the belief that certain diseases and bad luck were caused by darts shot by elves, while other ailments involved a "great worm," a term applied to snakes, insects, and dragons.

By contrast with the naturalist focus of *Hippocratic* and *Galenic medicine*, healing became more authoritarian and intertwined with religion, for the rising Church taught that there was a supernatural plan and purpose to everything, including sickness and death. *Christian* and *Jewish* healing traditions became more prominent. Disease could be cured by prayers or by invoking the names of saints, by exorcism, by amulets or number magic, or by transferring the sickness to animals, plants, or to the soil. Certain maladies such as leprosy were associated with the Almighty's punishments for sin, according to the Book of Leviticus:

"When a man shall have in the skin of his flesh a rising [a swelling], a scab, or bright spot, and it be in the skin of his flesh like the plague [the spots] of leprosy...and the plague in sight be deeper than the skin of his flesh it is a plague of leprosy: and the priest shall look on him, and pronounce him unclean."

Such polluting diseases were curable by the Lord alone, encouraging some Jewish people to reject human medicine in favour of the divine, citing the fate of *King Asa* (ca. 914-874 BC), who "sought not the Lord, but his physicians," and whose foot sores consequently worsened, and he died. On the other hand, while *Jewish* dietary rituals (e.g. kosher food) are principally expressions of religious precepts about pollution and purification, a useful practical effect is to limit exposure to foodborne diseases such as trichinosis.

Many of the distinguished surgeons of the Middle Ages were clerics, but the practice of medicine and surgery by members of the Church was not favoured by the hierarchy. Many laws were passed against the practice of medicine for worldly profit. In 1215 the Fourth

Lateran Council forbade all sub-deacons, deacons, or priests to practice that part of surgery that had to do with burning and cutting. Finally, Pope *Honorius III* (d. 1216) prohibited all persons in holy orders from practicing medicine in any form, which meant that the educated classes were prohibited from performing any type of surgery.

In Europe in the Middle Ages, practitioners unsuccessfully sought to cure wounds by treating only the weapons that caused them, using an ointment on the offending knife or sword known as a weapon salve. Physicians practiced urine-gazing (uroscopy), but proper chemical analysis of urine did not develop until the 18th century. Medieval medicine is the source of many humorous "remedies" in contemporary diatribes against modern medicine—such as being strapped into a halter and walked around a pigsty three times for mumps, drinking water from the skull of a bishop to treat rheumatism, or nuzzling a mouse to cure the common cold. But we must recognize that such treatments were not put forth as reasonable science. Rather, they serve as a reminder of the results of injecting mystical or religious elements into medicine as a substitute for good science.

The Middle Ages are also remarkable for their plagues, which became more numerous as people crowded closer and closer together in urban centers of increasing population density. *Plague* records from China during this period often show 50%-70% mortality rates, and China probably served as the original source of the greatest of the bubonic plagues that swept through Europe in the 14th century, known as the Black Death. The disease broke out in 1346 among the armies of a Mongol prince who laid siege to the trading city of Caffa in the Crimea. This compelled his withdrawal, but not before the disease had entered Caffa, whence it rapidly spread by ship throughout the Mediterranean.

The initial shock in 1346-1350 was severe. Die-offs varied widely, with some small communities experiencing total extinction and others, such as Milan, being spared entirely. The lethal effect of the plague may have been enhanced by the fact that it was propagated not solely by the bites of fleas carried on infected rodents, but also was transmitted person to person as a result of inhaling bacillus-filled droplets that had been coughed or sneezed into the air by an infected individual. Lung infections of this kind were observed to be 100% lethal in Manchuria in 1921, the only time modern medicine has directly observed airborne plague communication, so it is tempting to assume a similar mortality for pneumonic plague in 14th century Europe.

Mortality rates for sufferers from bubonic infection transmitted by flea bite varies from 30% to 90%. All told, the best estimate of European plague-induced mortality is that about one-third of the population died during the initial five-year period.

The plague returned in the 1360s, the 1370s, and thereafter, and European population declined irregularly as a result, reaching a low point in England sometime between 1440-1480. People learned to minimize infection risk via quarantine, a practice which stemmed from Biblical passages prescribing the ostracism of lepers. Plague sufferers were treated as though they were temporary lepers, with a standard 40-day quarantine for people and incoming ships at all major ports. However, since the role of fleas and rats in disease propagation remained unknown until the end of the 19th century, quarantine measures were often ineffectual. Through the 17th century, occasional plague outbreaks that carried off up to a third or a half of a city's population in a single year were considered normal. For example, Venetian statistics show that in 1575-77 and again in 1630-31, a third or more of the city's population died of plague.

Pre-modern medicine was powerless against bubonic plague. Before antibiotics reduced the disease to triviality in 1943, the average mortality rate was 60%-70% of those affected, despite all that the best hospital care could accomplish. The last recorded outbreak of plague that ran its course without benefit of penicillin and related antibiotics (which destroy the infection rapidly) occurred in Burma in 1947, with 78% lethality.

Renaissance and Pre-modern medicine

While internal medicine languished through the Middle Ages, the surgeons of the Renaissance gained wide experience during the many religious wars of the period. They had many new problems to face, including the treatment of wounds caused by firearms. The French had employed gunpowder at the siege of Puiguillaume in 1338, and cannon were used by the English at the Battle of Crecy in 1346. Gunshot wounds at that time were caused by large missiles of low velocity that caused ragged wounds and carried pieces of clothing into the tissues. These wounds were severe and very liable to become septic. The universal belief among contemporary surgeons was that gunpowder itself was venomous. To neutralize the effect of this venom, the general practice was to cauterize the wound by injecting boiling oil.

The first man to break away from this old doctrine was *Ambroise Pare* (1510-90). Pare came to Paris in 1532 as an apprentice to a barber-surgeon and then moved to the great Hotel Dieu as resident surgeon. In that immense medieval hospital, the only one in Paris at the time, he gained great experience, and in 1536 he began his career as a military surgeon. As described in his book *The Apologie and Treatise*, he recounts how, during his first campaign as a greenhorn military surgeon in Turin in 1537, he had run out of boiling oil, the established treatment for gunpowder wounds, just after French troops had captured the castle of Villaine. So in place of boiling oil, he applied: "...a digestive of yolk of eggs, oil of roses, and turpentine. In the night I could not sleep in quiet, fearing some default in not cauterizing, that I should find those to whom I had not used the burning oil dead impoisoned; which made me rise very early to visit them, where beyond my expectation I found those to whom I had applied my digestive medicine, to feel little pain, and their wounds without inflammation or tumour, having rested reasonable well in the night; the other to whom was used the said burning oil, I found them feverish with great pain and tumour about the edges of their wounds. And then I resolved with myself never so cruelly to burn poor men wounded with gunshot."

Pare also went on to show that bleeding after amputations should be arrested, not by the terrible method of the indiscriminate use of the red-hot cautery, but by simple tying of the blood vessels. His most famous phrase, so reminiscent of the old Hippocratic school of thought, was: "I dressed the wound, and God healed him."

In 1633 appeared the earliest book on first-aid for the injured, by one *Stephen Bradwell*, although the proffered advice sounds impractical and a bit odd to modern ears. For example, the treatment for the "Biting of a Madde Dogge" is to throw the patient into water. "In doing this, if he cannot swim, after he hath swallowed a good quantity of water, take him out again. But if he be skilful in swimming, hold him under the water a little while till he have taken in some pretty quantity." This procedure may not be wholly irrational—standard 20th century first aid for dog bites includes a thorough cleansing of the wound with water.

Venesection remained a popular 17th century universal remedy. As described by the surgeon *Richard Wiseman* (1622-1676), "a gentleman of about thirty years of age coming out of *Hertfordshire* through Tottenham and riding upon the causeway near an inn, one emptying a chamber pot out of the window as he was passing by, his horse started and rushed violently between a signpost and a tree which supported part of the sign. The poor gentleman was beaten off his horse and lay stunned upon the ground." A barber-surgeon was hastily summoned but nothing much was done for the injured man until Wiseman arrived, whereupon:

"I found the gentleman lying upon the ground, the people and chirurgeon gazing upon him. I felt his pulse much oppressed, the right brow bruised and inquired whether they had bled him blood. The chirurgeon replied that he had opened a vein in his arm but it would not bleed. I replied, we must make him bleed through it by splitting his veins. Turning his head on one side, I saw the jugular vein on the bruised side turgid and opened it. He bled freely. After I had taken about twelve ounces, the blood ran down from his arm which had been opened before and would not bleed. We bled him till he came to life, and then he raved and struggled with us."

The patient's injuries were dressed and he was subjected to further bleedings, but evidently made a good recovery.

Even by the 18th century, the traditional surgeon's day-to-day business eschewed high-risk operations like amputations; rather, it was a round of minor procedures such as venesection, lancing boils, dressing skin abrasions, pulling teeth, managing whitlows, trussing ruptures, and treating skin ulcers. The fatality rates of these procedures were low, for surgeons understood their limits, and the repertoire of operations they attempted was small, because of the well-known risks of trauma, blood loss, and sepsis. Internal disorders were treated not by the knife but by medicines and management, since major internal surgery was unthinkable before anesthetics and antiseptic procedures. Improvements did occur in certain operations such as lithotomy. *William Cheselden* (1688-1752), a great British surgeon of the 18th century, perfected a technique which enabled him to remove a stone in the bladder in one minute (his record time was 54 seconds), thus reducing mortality from about 50% to under 10%. Cheselden's results were not bettered until almost the end of the 19th century.

In the 17th century, internal medical treatment was frequently overdone on those affluent enough to afford it. Critics often denounced physicians as meddlesome, capriciously practicing an often dangerous polypharmacy—a blunderbuss approach. The deathbed of *Charles II* (1630-1685) of England was a conspicuous case of such medical overkill; after the king had suffered a stroke, his doctors moved in, and *Sir Raymond Crawfurd* (1865-1938) recreated the scene:

"Sixteen ounces of blood were removed from a vein in his right arm with immediate good effect. As was the approved practice at this time, the King was allowed to remain in the chair in which the convulsions seized him. His teeth were held forcibly open to prevent him biting his tongue. The regimen was, as Roger North pithily describes it, first to get him to wake, and then to keep him from sleeping. Urgent messages had been dispatched to the King's numerous personal physicians, who quickly came flocking to his assistance; they were summoned regardless of distinctions of creed and politics, and they came. They ordered cupping-glasses to be applied to his shoulders forthwith, and deep scarification to be carried out, by which they succeeded in removing another eight ounces of blood. A strong antimonial emetic was administered, but as the King could be got to swallow only a small portion of it, they determined to render assistance doubly sure by a full dose of Sulphate of Zinc. Strong purgatives were given, and supplemented by a succession of clysters. The hair was shorn close, and pungent blistering agents were applied all over his head. And as though this were not enough, the red-hot cautery was requisitioned as well."

One of the dozen attending physicians noted with pride that "nothing was left untried"; the King graciously apologized for being "an unconscionable time a-dying."

Meanwhile, the common citizen experienced poor health exacerbated by the many new dangers attending the Industrial Revolution. In 1775, *Percivall Pott* (1714-1788) pointed out that boy chimneysweeps developed scrotal cancer, due to soot irritation. In his *Condition of the Working Classes in England* (1844), *Friedrich Engels* (1820-1895), a Manchester factory owner as well as *Karl Marx's* collaborator, described workers who were "pale, lank, narrow-chested, hollow-eyed ghosts," cooped up in houses that were mere "kennels to sleep and die in." In 1832, the Leeds physician *Charles Turner Thackrah* (1795-1833) published *The Effects of Arts, Trades, and Professions on Health and Longevity*, documenting the diseases and disabilities of various occupations. Apart from factory workers, among those most exposed to harmful substances were cornmillers, maltsters, coffee-roasters, snuff-makers, rag-pickers, papermakers and feather-dressers. Tailors were so subject to anal fistulas that they set up their own "*fistula clubs*." Thackrah's overall verdict was bleak: "Not 10% of the inhabitants of large towns enjoy full health."

The single worst malady cultivated in populous cities was tuberculosis (TB), a disease characterized by fever, night sweats, and hemoptysis (coughing up blood), called "consumption" because victims were almost literally consumed. By 1800 TB was proclaimed the most common disease, and in 1815 *Thomas Young* (1773-1829) surmised that tuberculosis brought a premature death to one in four in the general population. Autopsies conducted in the chief Paris hospitals recorded TB as the cause of death in some 40% of cases. In the continental U.S. as late as 1890, the corresponding percentage was about 13%, and tuberculosis was still the leading cause of death, though the data are partially suspect because cases of lung cancer were sometimes reported as "consumption".

One important 18th century improvement in internal medicine which decisively saved many lives was the introduction of inoculation and vaccination against smallpox. Smallpox, "the speckled monster," had become virulent throughout Europe and in bad years accounted for about 10% of all deaths; *Queen Mary of England* (1662-1694), *Louis XV of France* (1710-1774), and Queen Anne's son and sole surviving direct heir (d. 1700) died of it. Doctors had long been aware of the immunizing properties of an attack, and smallpox inoculation seems to have been known and practiced for centuries at a folk level throughout Arabia, North Africa, Persia, and India. Reports of a more elaborate Chinese method, involving the insertion of a suitable infected swab of cotton inside the patient's nostril, reached London in 1700. But it was a report from *Mary Wortley Montagu* (1689-1762), wife of the British consul in Constantinople, that Turkish women held smallpox parties at which they routinely performed inoculations with the aim to induce a mild dose so as to confer lifelong protection without pockmarking, that hastened acceptance in the rural medical community. The usual method was to transfer the infection by introducing matter from a smallpox pustule into a slight wound made in the patient's skin. Occasionally the patient developed a severe case of smallpox from such treatment, and some died. But usually the symptoms were slight—a few score of pox only—and immunity proved equivalent to that resulting from contracting the disease naturally.

Edward Jenner (1749-1823), an English country doctor who performed such inoculations, noticed that cowpox, a cattle disease occasionally contracted by humans, particularly dairy maids, also conferred immunity against smallpox. Suspecting that it might be possible to produce this immunity by arm-to-arm inoculation from the cowpox pustule, and surmising it would be safer than inoculation from smallpox pustules directly, since in humans cowpox was benign, Jenner tried the experiment, and it worked. In 1798 he published his discovery in *An Inquiry into the Causes and Effects of the Variolae Vaccinae*.

By 1799 over 5000 individuals had been vaccinated in England and abroad the practice was taken up remarkably swiftly, being made compulsory in Sweden and supported by *Napoleon*, who had his army vaccinated. For the first time in history, organized medicine began to contribute to human population growth in a statistically significant fashion.

The Napoleonic Wars spurred new attempts to treat battle-wounded soldiers in a more timely manner. Traditionally, the wounded were left on the field unattended until the end of battle, but Napoleon's chief surgeon *Dominique Jean Larrey* (1766-1842) introduced the use of rudimentary carts called ambulances volantes (little more than horsedrawn rickshaws) as the first "ambulances" to evacuate and transport wounded soldiers from the field to nearby aid stations, even while the battle raged on. In 1792, Larrey organized the first air evacuation, by hot air balloon.

The use of ambulances didn't catch on until the late 1800s; until then, anyone injured in the streets of Paris, London, New York or Boston depended on the kindness of strangers or a nearby business shop for a place to rest until a doctor could be summoned. The first "modern" ambulance appeared in the city of Cincinnati in 1865, but the first true city ambulance system was developed in association with Bellevue Hospital in New York City in 1866, receiving 1500 requests for transport in its first three years of service. These horse-drawn ambulances, usually provided by local mortuaries, carried a driver and a surgeon, who was on board mainly to pronounce a patient's death at the scene or upon arrival at the hospital, since little could be done for the seriously injured. The surgeon kept meticulous notes on the ride, recording the time of the call, transport and arrival times, and any other details that "a coroner's jury might possibly require".

The period also saw the development of many simple diagnostic tools that are taken for granted today. For example, a French physician, *Rene Theophile Hyacinthe Laennec* (1781-1826), in his *Treatise On Mediate Auscultation* (1819), described pathological lesions found in the chest at autopsy and showed how they correlated with disease detected in living patients, establishing for the first time the modern concept of clinicopathological correlation, the cornerstone of modern diagnosis. Laennec also developed an instrument that he named a "stethoscope" to assist him in his examination of patients, especially female patients, against whose chests the direct placing of a male ear was socially taboo. The original device was a straight wooden tube; by mid-century rubber tubing was introduced to create a flexible monaural stethoscope, and in 1852 an American physician, *George P. Cammann* (1804-1863), devised our familiar two-ear instrument.

The clinical thermometer is of like vintage. *Galileo* (1564-1642) invented the first thermometer in the late 16th century, but it was not applied to medicine. Early medical thermometers in the 18th century were a foot long and difficult to use at the bedside, and were reportedly carried under the arm "as one might carry a gun." The short clinical thermometer was devised by *Sir Clifford Allbutt* (1836-1925) in the 1860s, and was widely used during the American Civil War (1861-1865). The classical work in temperature diagnostics was *Carl Wunderlich's* (1815-1877) *The Temperature in Diseases*, published in 1868, which presented data on nearly 25,000 patients and analyzed temperature variations in 32 diseases, showing that temperature readings could differentiate fevers. Other devices emerged later to measure pulse and blood pressure. In 1854, *Karl Vierordt* (1818-1884) created the sphygmograph, a pulse recorder usable for routine monitoring on humans. Blood pressure was measured using the familiar inflatable band wrapped around the upper arm, called the sphygmomanometer, whose basic design was established in 1896 by *Scipione Riva-Rocci* (1863-1937). The hypodermic syringe was invented in 1853.

Biochemistry also began to play an increasing diagnostic function. In the 18th century, *Matthew Dobson* (1784) developed tests for diabetes. In 1827, *Richard Bright* (1789-1858) showed show the kidney complaint subsequently called Bright's disease could be diagnosed by a single, simple chemical test. Chemical analysis was crucial to Alfred Becquerel's (1814-1862) urinalysis studies in 1841, establishing the average amounts of water, urea, uric acid, lactic acid, albumin, and inorganic salts secreted over 24 hours, and correlating these with various disease conditions. In 1859, *Alfred Garrod* (1819-1907) devised a simple chemical test pathognomonic for gout.

Fully invasive surgery

The fully invasive surgery of the 19th and 20th centuries rests upon the triple foundation of anatomy, anesthesia, and asepsis. Each has an interesting story, described below, presaging the emergence of the rational-scientific approach to medical treatment.

Anatomy

In surgical practice, lack of accurate anatomical knowledge had long been a great obstacle. Dissection of the human body was first practiced systematically at the great medical school of Alexandria, which flourished from about 300 BC until the death of the last ruler of Ptolemaic Egypt, Cleopatra, in 30 BC. After the decline of Alexandria, dissection was carried on at a few other centers in the Middle East, but in the first two centuries of the Christian era human bodies were replaced on the dissection table by those of apes and other animals. The anatomical knowledge gained by Galen and others from the dissection of animals was an adequate guide for the simple operative procedures carried out at this time because the abdomen, the chest, and the head were rarely opened by the surgeon's knife.

Anatomical demonstrations of a kind were introduced into some Italian medical schools early in the 14th century but their main purpose was to serve as an aid in memorizing what Galen had written a thousand years before. During the late Middle Ages, and for a long time after, the procedure was for the professor to read from some second- or third-hand manuscript version of Galen while a demonstrator pointed to the part under discussion with a wand. As the text of Galen was often based on the dissection of an ape or pig, there were naturally many occasions when the anatomical structure under examination did not correspond with Galen's description.

Some of the great artists took up the scientific study of anatomy, but the true founder of modern anatomy was *Andreas Vesalius* (1514-1564), a native of Brussels who studied medicine in Paris and later taught surgery and anatomy at Padua and Bologna. Vesalius was filled with a passionate desire for anatomical study, and many stories are told of the great risks which he took in obtaining material, including, it is reported, graverobbing. On one occasion he stole the skeleton of a criminal which was hanging on a gallows outside the city wall of Louvain and the trophy proved of great value in his studies. His public dissections during his seven years in Padua drew enormous crowds of students. In 1543 he published his 355-page great work, *De Humani Corporis Fabrica Libri Septem (On the Fabric of the Human Body)*, the outstanding and precise illustrations in which were copied and plagiarized by scholars for more than 100 years, and could be used for teaching even today. For the first time in the history of medicine, doctors had at their disposal a detailed and accurate anatomical text with illustrations from the hand of a great artist. This book is the foundation stone of anatomy—indeed of all modern medicine—because without a sound knowledge of the structure of the body there can be no real understanding of the body's functions in health and disease.

A tremendous shortage of dead human bodies for dissection, to teach anatomy, remained a problem for centuries. In Edinburgh in

1827, an old man died in *William Hare's* (1792-1870) boardinghouse; assisted by his lodger William Burke (1792-1829), the two men bypassed the grave and sold the body directly to anatomists. Spurred by success, they turned to murder, luring victims and suffocating them to avoid signs of violence. Sixteen were done to death and their bodies sold, fetching 7 pounds apiece, before Burke and Hare were brought to justice in 1829. Hare turned King's evidence and Burke was hanged. The last cadaver was found in the dissecting room of a respected anatomist, *Robert Knox* (1791-1862). Despite his cries of innocence, an incensed crowd burned down his house and he fled to London, his career in ruins, and eventually *Knox* died in obscurity.

To help pass the English Anatomy Act of 1832 (which awarded to doctors the "unclaimed bodies" of paupers) and to dispel public concern about dissection, the body of the great English philosopher and jurist, *Jeremy Bentham* (1748-1832), in accordance with his directions, was dissected in the presence of his friends. The skeleton was then reconstructed, supplied with a wax head to replace the original (which had been mummified), dressed in Bentham's own clothes and set upright in a glass-fronted case. Both this effigy and the head are preserved at University College, London, to this day.

Anesthesia

Pain did not prevent surgery but made it almost unbearable, and the accompanying trauma often proved dangerous. Before the anesthetic era, which commenced in the late 1840s, surgical operations were agonizing. Of course, if the patient had a broken leg or amajor wound, there was no choice but submit to a surgeon's knife. But non-emergency elective operations would only be undergone if the condition itself was so painful or life-threatening that the victim could even consider allowing surgery. In this event, patients would choose a surgeon with the best reputation for quickness—a limb might be removed or a bladder stone evacuated in a couple of minutes. Progress in technique was often rapid. For example, in 1824, *Astley Cooper* (1768-1841) took 20 minutes to amputate a leg through the hip joint; ten years later, *James Syme* (1799-1870) was doing it in just 90 seconds.

In pre-anesthetic days, operations were rushed through at lightning speed and under conditions of appalling difficulty. The most hardened surgeons had to steel themselves to perform operations which they knew would cause agony to their patients and nerve-wracking distress to themselves. It is hard for any 20th century inhabitant of an industrialized nation to imagine what a major surgical operation must have meant to the patient in the days before anesthesia. The following is a personal account by a male patient who suffered the removal of a stone from the bladder by *Henry Cline* (1750-1827), surgeon to St. Thomas's Hospital and one of the leading operators of the day, on 30 December 1811, just three decades before the widespread adoption of anesthesia:

"My habit and constitution being good it required little preparation of body, and my mind was made up. When all parties had arrived I retired to my room for a minute, bent my knee in silent adoration and submission, and returning to the surgeons conducted them to the apartment in which the preparations had been made. The bandages &c. having been adjusted I was prepared to receive a shock of pain of extreme violence and so much had I overrated it, that the first incision did not even make me wince although I had declared that it was not my intention to restrain such impulse, convinced that such effort of restraint could only lead to additional exhaustion. At subsequent moments, therefore I did cry out under the pain, but was allowed to have gone through theoperation with great firmness."

"The forcing up of the staff prior to the introduction of the gorget gave me the first real pain, but this instantly subsided after the incision of the bladder was made, the rush of urine appeared to relieve it and soothe the wound."

"When the forceps was introduced the pain was again very considerable and every movement of the instrument in endeavoring to find the stone increased. Still, however, my mind was firm and confident, and, although anxious, I was yet alive to what was going on. After several ineffectual attempts to grasp the stone I heard the operator say in the lowest whisper, "It is a little awkward, it lies under my hand. Give me the curved forceps," upon which he withdrew the others. Here, I think, I asked if there was anything wrong—or something to that purport — and was reanimated by the reply conveyed in the kindest manner, "Be patient, Sir, it will soon be over." When the other forceps was introduced I had again to undergo the searching for the stone and heard Mr. Cline say, "I have got it." I had probably by this time conceived that the worst was over; but when the necessary force was applied to withdraw the stone the sensation was such as I cannot find words to describe. In addition to the positive pain there was something peculiar in the feel. The bladder embraced the stone as firmly as the stone was itself grasped by the forceps; it seemed as if the whole organ was about to be torn out. The duration, however, of this really trying part of the operation was short and when the words "Now, Sir, it is all over" struck my ear, the ejaculation of "Thank God! Thank God!" was uttered with a fervency and fulness of heart which can only be conceived....I never heard what was the precise duration of the operation but conceive it to have been between twelve and fifteen minutes."

And now, a woman's point of view. In 1810, Napoleon's famed military doctor Dominique *Jean Larrey* performed a radical mastectomy without anesthetic on the popular female novelist *Fanny Burney* (1752-1840). Burney later wrote a long account of the operation which, despite the excruciating agony, she believed had nevertheless saved her life:

"M. Dubois placed me upon the Mattress, & spread a cambric handkerchief upon my face. It was transparent, however, & I saw through it that the Bed stead was instantly surrounded by the 7 men and my nurse. I refused to be held; but when, bright through the cambric, I saw the glitter of polished steel — I closed my eyes..."

"Yet — when the dreadful steel was plunged into the breast — cutting through veins—arteries—flesh—nerves—I needed no injunctions not to restrain my cries. I began a scream that lasted unintermittingly during the whole time of the incision—& I almost marvel that it rings not in my Ears still! so excruciating was the agony."

"When the wound was made, & the instrument was withdrawn, the pain seemed undiminished, for the air that suddenly rushed into those delicate parts felt like a mass of minute but sharp & forked poniards [small pointed daggers], that were tearing at the edges of the wound. But when again I felt the instrument, describing a curve, cutting against the grain, if I may so say, while the flesh resisted in a manner so forcible as to oppose & tire the hand of the operator, who was forced to change from the right to the left —then, indeed, I thought I must have expired, I attempted no more to open my eyes....The instrument the second time withdrawn, I concluded the operation over—Oh no! presently the terrible cutting was renewed—& worse than ever, to separate the bottom, the foundation of the dreadful gland from the parts to which it adhered...yet again all was not over...."

When did the practice of anesthesia begin? The Herbal of Dioscorides (ca. 40-90 AD) contains specific directions for giving a decoction (boiled extraction) of mandragora "to such as shall be cut or cauterized," one of the earliest references to surgical anesthesia. Bernard de Gordon (ca. 1260-1308) tells us that the Salernitans rubbed up poppy seed and henbane and used them as a plaster to deaden the sensibility of a part to be cauterized. *Arnold of Villanova* (1235-1311) gives the following recipe:

"To produce sleep so profound that the patient may be cut and will feel nothing, as though he were dead, take of opium, mandragora bark, and henbane root equal parts, pound them together and mix with water. When you want to sew or cut a man, dip a rag in this and put it to his forehead and nostrils. He will soon sleep so deeply that you may do what you will. To wake him up, dip the rag in strong vinegar."

Some of the surgical textbooks of the Middle Ages contain references to anesthetic sponges which were prepared by soaking them in various herbs reputed to have soporific properties. The favourite herb for this purpose was the mandrake. Another simple method of producing analgesia used intermittently from early times was compression. Writing in 1564 about the various uses of the tourniquet, the French surgeon *Ambrose Pare* noted that "it much dulls the sense of the part by stupefying it." Amusingly, reliable witnesses claim that as late as the 19th century, a method of anesthesia practiced at the Imperial Court of China was to "knock the patient out by a sudden blow on the jaw."

The almost complete absence of any mention of pain-relieving drugs in medical literature of the post-medieval period is not easy to explain. It is, however, probable that the action of crude concoctions employed in early times was very uncertain, and that drugged sleep often ended in death. The active ingredients of the many herbs used in medicine had not been isolated and it would have been very difficult to regulate dosages reliably.

The story of inhalation anesthesia begins in 1799 when *Sir Humphry Davy* (1778-1829) recorded the effects produced by the inhalation of nitrous oxide. He breathed various concentrations of the gas and noted that a headache and the pain associated with the cutting of a wisdom tooth were relieved. Demonstrations of the effects of nitrous oxide were frequently given, bladders filled with "laughing gas" being passed around at lectures. Gas inhalation became a popular party game. A little book of 1839 contains adescription of the "irresistibly ridiculous" sight of a large room filled with persons each of whom was sucking from a bladder. As the gas began to take effect, "some jumped over the tables and chairs; some were bent on making speeches; some were very much inclined to fight; and one young gentleman persisted in attempting to kiss the ladies." At about the same time, "ether frolics" became equally popular.

In January 1842, *William E. Clarke* (b. 1818), a young American physician of Rochester, New York, who had acquired some knowledge of ether by attendance at ether frolics, administered the chemical on a towel to a Miss Hobbie who then had one of her teeth extracted painlessly. So far as is known this was the first use of ether for a dental or surgical operation. In March 1842, *Crawford W. Long* (1815-1878) of Danielville, Georgia, who had also witnessed ether frolics "enjoying sweet kisses from the girls," successfully removed a small tumour from the neck of a patient under the influence of ether. *Horace Wells* (1815-1848), a dentist of Hartford,Connecticut, attended a public demonstration of the effects of nitrous oxide in December 1844, and the day after he administered the gas to himself and had one of his own teeth pulled out by a colleague. Afterwards, Wells wrote: "I didn't feel it so much as the prick of a pin." His former partner, *William Thomas Green Morton* (1819-1868), also introduced ether into his dental practice in 1846. By February 1847, the Lancet and other medical journals were reporting anesthetic operations from all parts of Great Britain, and ether had been used in most European countries. In June 1847 the news reached South Africa and a leg was amputated painlessly by *W.G. Atherstone* of Grahamstown.

Sir James Young Simpson (1811-1870), professor of surgery at Edinburgh, introduced chloroform in 1847. Tradition has it that Simpson had been testing chemicals with his assistants when somebody upset a bottle of chloroform; upon bringing in dinner, Simpson's wife found them all asleep. Unlike ether, chloroform did not irritate the lungs or cause vomiting, and was powerful and easy to administer. In April 1853, *Queen Victoria* (1819-1901) took chloroform for the birth of Prince Leopold; *John Snow* (1813-1858) administered the anesthetic. Protests followed—some objections were religious (e.g., the Bible taught that women were supposed to bring forth in "travail and pain") but most were medical, putatively on grounds of safety but ringing with naturophilia: "In no case could it be justifiable to administer chloroform in perfectly ordinary labour," complained the Lancet. Incredibly, some early 19th century surgeons believed that using anesthesia for an operation would weaken a patient's character.

Of course, general anesthesia could indeed prove dangerous, and deep unconsciousness was unnecessary for less invasive procedures, so the search was on for substances that would numb a particular area for local surgery. Cocaine was isolated in 1859 and was first used in ophthalmologic procedures by Carl Koller (1857-1944). Cocaine became the first widely-used local anesthetic, synthesized in 1885 by the Merck drug company.

Germ theory and antisepsis

By the middle of the 19th century, pain had been banished from surgical operations, but one grave danger still faced every patient submitting himself to the surgeon's knife. This was the ever-present risk of sepsis (infection). Hospital diseases such as erysipelas, pyemia, septicemia and gangrene, were rife. In the 1850s the death rate after amputations varied from 25%-60% in different countries and in military practice it reached the appalling figure of 75%-90%. The first ovariotomies, which were the first abdominal operations performed on a fairly large scale, had a mortality rate of more than 30% even in the most expert hands. That all these diseases were due to some form of "contagion" had long been suspected, but the general view was that whatever agent was responsible was generated spontaneously in wounds. Alternatively, it was theorized that air itself was responsible for suppuration and many attempts were made to exclude the air from wounds by means of elaborate dressings.

Some medical men had postulated the existence of minute particles in the air which carried contagion, the so-called "germ theory." In 1546, *Girolamo Fracastoro* of Verona (1478-1553) proposed "seminaria, the seeds of disease which multiply rapidly and propagate their life," minute bodies passing unseen from the infector to the infected by contact, by clothing or utensils, and by infection at a distance through the air.

What made the germ theory of contagion so difficult to accept was that no one could see the supposed microbes. Magnifying lenses were used in ancient times, and by the beginning of the 17th century they had been combined in a tube to make the compound microscope. The first man to employ the microscope in investigating the causes of diseases was probably *Athanasius Kircher* (1601-1680), a learned Jesuit priest. In 1658 Kircher described experiments upon the nature of putrefaction, showing how maggots and other living creatures developed in decaying matter. He also claimed to have found in the blood of plague-stricken patients "countless masses of small worms, invisible to the naked eye." It is impossible that he could have seen the plague bacillus with the very low power microscopes at his disposal, but he may have seen some of the larger microorganisms and his statements about the doctrine of contagion are even more explicit than those of Fracastoro.

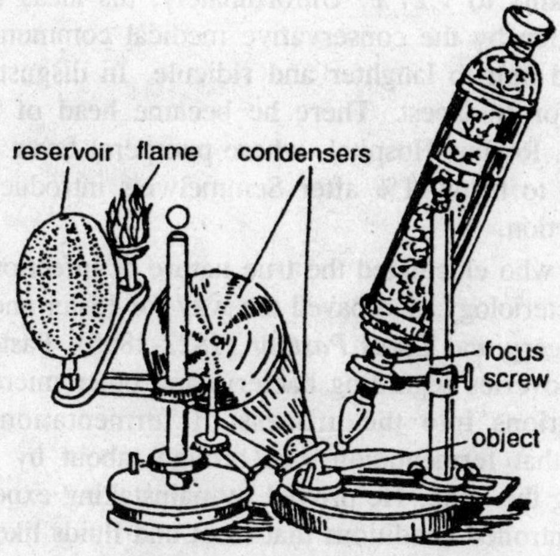

Fig. 4.1. Anthoni van Leeuwenhoek microscope

The great pioneer of modern microscopy was *Anthoni van Leeuwenhoek* (1632-1723), a Dutch linen draper, who ground his own lenses and made hundreds of microscopes. Few of his instruments provided a magnification of more than 160X but he is generally credited as making the first observations of germs, reported in his communications to the Royal Society in London. Leeuwenhoek was the first to describe spermatozoa and he gave the first complete account of the red blood corpuscles in 1674; he also found that the film from his own teeth contained "little animals, more numerous than all the people in the Netherlands."

In 1847, *Ignaz Philipp Semmelweis* (1818-1865), an assistant at the Vienna General Hospital in the maternity clinic (the world's largest at the time), was investigating the 29% postpartum mortality rate among women in Ward One, where births were handled by medical students, vs. a 3% rate in Ward Two where births were handled by midwifery pupils. Semmelweis noticed that the appearances of these deaths from puerperal fever looked the same as those observed in the body of an older colleague, forensic medicine Professor *Jakob Kolletschka* (1803-1847), who had died from adissection wound suffered in the clinic. He correctly surmised that the postpartum deaths were caused by infection from "putrid particles" carried on the hands of medical students who often shuttled back and forth between the labour wards, the obstetrical clinic, and the autopsy room where dissections were performed. Semmelweis instituted a simple routine of hand-washing with water solutions of chloride of lime, which promptly reduced mortality to 1.27%. Unfortunately, his ideas were met with fierce opposition by the conservative medical community in Vienna, who subjected him to laughter and ridicule. In disgust, Semmelweis left Austria for Budapest. There he became head of the obstetrical division of St. Rochus Hospital, where puerperal fever mortality rates were reduced to below 1% after Semmelweis introduced chlorinated water disinfection.

The man who elucidated the true nature of infection, founded the science of bacteriology, and paved the way for Lister and the antiseptic system in surgery was *Louis Pasteur* (1822-1895). Pasteur was led to his great discoveries regarding bacteria and other microorganisms by his investigations into the process of fermentation. He showed conclusively that fermentation was brought about by some external agent entering the wine. He proved by painstaking experiments under rigorously controlled conditions that meat and fluids like blood did not putrefy if they were kept in such a way that all air was excluded from them. By taking samples of air at different levels Pasteur showed that contamination became less with increasing altitude. Then he proved that the contaminating agents were living organisms (bacteria) which were everywhere—in every room, in the air, on every article of clothing, on furniture, on the ground, and on the skin. He showed that putrefaction was caused by the presence of bacteria and that this applied to putrefaction in foods (milk, wine, and meat), urine, and in wounds.

The application of Pasteur's discoveries to surgical practice was the work of *Joseph Lister* (1827-1912), a young English surgeon who had concluded that it must also be bacteria that caused the suppuration, pus and gangrene which plagued the surgical wards of those days. He determined to prevent the access of organisms by killing them in or on the surface of the wound. Pasteur had shown that heat could kill microbes, but it was impossible to apply heat to a wound without burning the patient, so some chemical substance had to be found. After trying various chemical agents he finally selected carbolic acid, and he insisted that everything which touched the wound, the dressings, the instruments and the fingers, should be treated with this *antiseptic*. He even produced an antiseptic atmosphere by means of a carbolic spray. The clinical results of Lister's first antiseptic system in 1865 included 11 compound fracture cases with only one death, a 9% mortality rate, marking a watershed between the primitive and modern eras of surgery.

Throughout the 19th century, the arguments continued as to whether microorganisms seen in a sick patient were merely coincidental with the illness, or resulted from the changes brought about by the illness itself. In 1882 the German microbiologist *Robert Koch* (1843-1910) formulated three famous postulates to guide scientists searching for disease-causing microbes. Koch argued that to prove an organism causes a disease, microbiologists must show that the organism occurs in every case of the disease; that it is never found as a harmless parasite associated with another disease; and that once the organism is isolated from the body and grown in laboratory culture, it can be introduced into a new host and produce the disease again. (An oft-stated fourth postulate, that the microbe must be isolated again from the second host, was not part of Koch's original formulation.) Koch and his pupils discovered specific bacillary causes for various diseases, including anthrax, cholera, tuberculosis, gonorrhea, diphtheria, leprosy, typhoid, trypanosomiasis, and malaria.

The theory that specific germs could cause specific disease remained contentious until the beginning of the 20th century. Many scientists of great repute rejected Koch's conclusions, with one scientist confidently asserting that "no microbe found in the living blood of any animal was pathogenic." In one celebrated case, *Max von Pettenkofer* of Bavaria (1818-1901), a distinguished 19th century experimental hygienist, induced Koch to send him a sample of his cholera vibrios culture and then wrote a letter back to Koch in 1892, as follows:

"Herr Doctor Pettenkofer presents his compliments to Herr Doctor Professor Koch and thanks him for the flask containing the so-called cholera vibrios, which he was kind enough to send. Herr Doctor Pettenkofer has now drunk the entire contents and is happy to be able to inform Herr Doctor Professor Koch that he remains in his usual good health."

Apparently Pettenkofer, aged 74 at the time he wrote the letter, survived this cholera exposure quite well, perhaps possessing the high stomach acidity which sometimes neutralizes the bacillus, though he shot himself to death in Munich 9 years later.

Skeptics notwithstanding, microbes were key. Enormous new vistas now lay open, as surgeons could confidently make an incision through intact skin without incurring an extreme risk of wound infection. The next step was to progress beyond killing wound bacteria with chemical antiseptics, to the prevention of bacterial contamination by eliminating bacteria in the operating theater—aseptic surgery. The use of steam sterilization of instruments, dressings and gowns, the wearing of masks, caps and gloves, air filtration and the other rituals of the operating theater of today were introduced over the decades following Lister's efforts.

Anesthesia and antisepsis enabled surgeons to carry out procedures that had formerly been quite beyond them, including long operations inside the head, the abdomen and the pelvis. *Theodor Billroth* of Vienna (1829-1894) resected the esophagus in 1872, parts of the intestines in 1878, and the pyloric end of the stomach in 1881. Billroth also made the first complete excision of the larynx. The first successful repair of a gunshot wound on a major artery was performed in Chicago by *John B. Murphy* (1857-1916) in 1897; *Ludwig Rehn* (1849-1930) of Frankfurt am Main performed the first successful repair of a cardiac injury in Germany in the same year.

Writing in 1874, *Sir John Eric Erichsen* (1818-1896), Professor of Surgery at University College, London, had predicted that "the abdomen, the chest, and the brain would be forever shut from the intrusions of the wise and humane surgeon." By the time of Erichsen's death 22 years later, surgeons had successfully removed from patients the stomach and large parts of the intestines, a whole lung had been excised, and a brain tumour had been extirpated. These operations were not mere feats of surgical showmanship; they saved the lives and restored the health of thousands of human beings.

Cells and tissues

The doctrine of the essential cellular nature of living things was established by 1840. Modern cell theory began in botany. The Jena botanist *Matthias Schleiden* (1804-1881) observed that plants were aggregates of cells, existing as self-reproducing living units. Exploring analogies between animals and plants in structure and growth, Theodor *Schwann* (1810-1882) took up the idea, maintaining that all these phenomena could also be demonstrated in animal structures. Thus living cells were basic to living things, and cells incorporated a nucleus and an outer membrane.

Jacob Henle (1809-85) applied cell biology to man. His three-volume *Handbuch der systematischem Anatomie des Menschen (Handbook of Systematic Human Anatomy)* (1866-1871) addressed the body from an architectural standpoint, describing its macro- and microscopic structure. Henle discovered kidney tubules and was the first to describe the muscular coat of the arteries, the minute anatomy of the eye and various skin structures, earning him a reputation as the Vesalius of histology (the study of tissues). Physiologists stressed organ and tissue function—*Claude Bernard* (1813-1878) emphasized the experimental

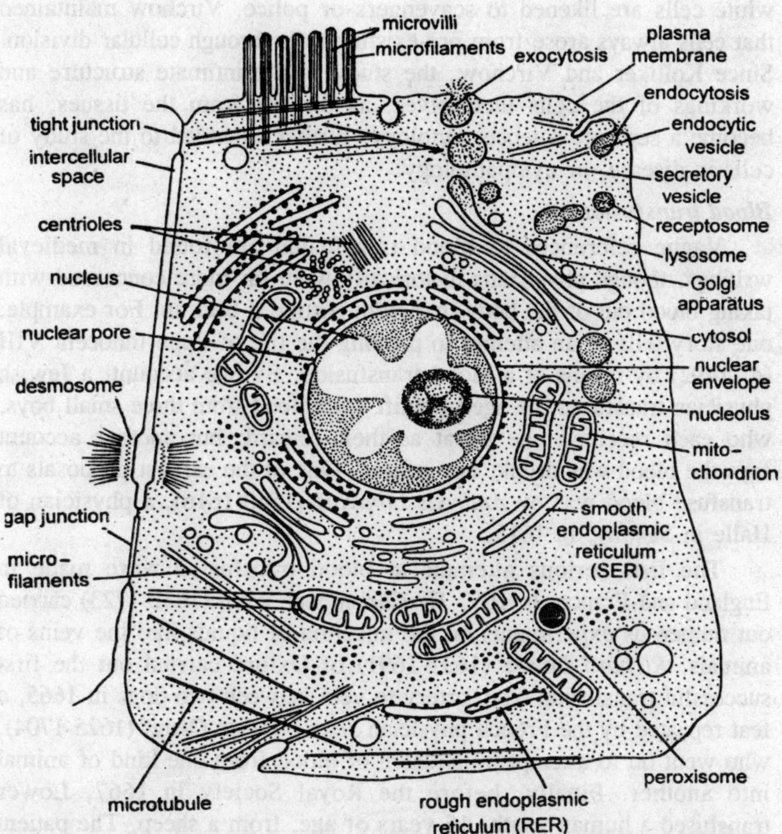

Fig. 4.2. Ultrastructure of a typical animal cell as seen in the electron microscope.

method in establishing biological knowledge and urged that medical practice should be grounded in such knowledge.

Histology was raised to the status of an independent science by the Swiss microanatomist *Albert von Kolliker* (1817-1905), who wrote the first textbook on the subject, *Handbuch der Gewebelehre des Menschen (Handbook of the Tissues of Man)* (1852). The medical implications of cell theory were taken up by *Rudolph Virchow* (1821-1902), who dominated German biomedical research for half a century. Virchow extended the cell concept to diseased tissues; his *Die Cellularpathologie (Cellular Pathology)* (1858) analyzed such tissues from the point of view of cell formation and cell structure. Virchow initiated the idea that the body may be regarded as a "cell state in which every cell is a citizen." Disease is often but civil war, and white cells are likened to scavengers or police. Virchow maintained that cells always arose from pre-existing cells through cellular division. Since Kolliker and Virchow, the study of the intimate structure and workings of the cells themselves, as distinct from the tissues, has become a separate science, cytology, further extended to the study of cells in disease, or cytopathology.

Blood transfusions

Vague references to blood transfusion are found in medieval writings, though physicians historically were far more concerned with taking blood out of the body than with putting it back in. For example, one story tells of an attempt to prolong the life of Pope Innocent VIII (d. 1492) by means of a blood transfusion. By one account, a Jewish physician transfused the aged Pontiff with blood from three small boys, who each received one ducat as their reward; but another account says the blood was drunk, not infused. One of the earliest proposals to transfuse blood was by *Andreas Libavius* (1540-1616), a physician of Halle in Saxony, in 1615.

The first serious attempts at blood transfusion were made in England and France. In 1657 *Sir Christopher Wren* (1632-1723) carried out numerous experiments on the injection of liquids into the veins of animals. *Richard Lower* (1631-1691) of Oxford carried out the first successful transfusion from artery to vein between two dogs in 1665, a feat repeated by the French physician *Jean-Baptiste Denys* (1625-1704), who went on to attempt transfusion of blood from one kind of animal into another. Finally, before the Royal Society in 1667, Lower transfused a human youth, 15 years of age, from a sheep. The patient was greatly improved and the only ill effect was a feeling of great heat along his arm; subsequent patients were not so lucky. Blood transfusions from lambs and calves to humans continued to be tried in the mid 17th century, but were not very successful, and in 1670 were finally forbidden by law in England.

The first known transfusion of human blood into an already moribund person was attempted by *James Blundell* in 1818, and on several other occasions during the 1820s, with poor results. Transfusion was carried out on a small scale during the American Civil War, but technical difficulties connected with premature clotting and the occurrence of accidents arising from the use of incompatible blood prevented the rapid acceptance of the process. The cause of many of the untoward effects of blood transfusion was finally explained in 1901 when the presence of agglutinins and iso-agglutinins in the blood was

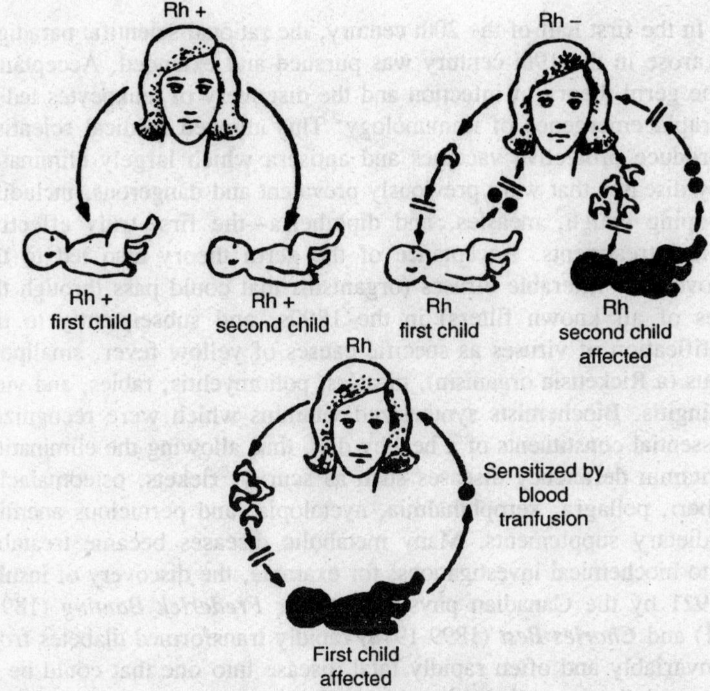

Fig. 4.3. Blood rhesus factor.

demonstrated by *Karl Landsteiner* (1868-1943) in Vienna, winning him the Nobel Prize in 1930; in 1907, the four main blood groups were determined by *Jan Jansky* (1873-1921) of Prague. *Rhesus factor* was later identified by two American scientists, *Philip Levine* (1900-1987) and *Rufus Stetson* (1886-1967). These advances were of fundamental importance and it became possible in the 20th century to eliminate most of the fatalities due to incompatibility, allowing blood transfusions to be practiced reliably and with uniformly good results.

20th century medicine

The 20th century saw more discoveries and advances in medical science than all previous centuries combined. In this period, medicine became more powerful than ever before as scientists gained knowledge of matters and processes of illness that, at the beginning of the century, were still unknown or mysterious. Unlike the mere palliatives of earlier eras, 20th century physicians could actually cure some diseases, reverse some physical traumas, and save many lives that could not be saved before.

In the first half of the 20th century, the rational scientific paradigm that arose in the 19th century was pursued and extended. Acceptance of the germ theory of infection and the discovery of leukocytes led to the rapid emergence of immunology. This allowed medical scientists to produce protective vaccines and antisera which largely eliminated many diseases that were previously prevalent and dangerous, including whooping cough, measles, and diphtheria—the first truly effective medical treatments. Acceptance of the germ theory also led to the discovery of filterable viruses (organisms that could pass through the pores of all known filters) in the 1890s, and subsequently to the identification of viruses as specific causes of yellow fever, smallpox, typhus (a Rickettsia organism), measles, poliomyelitis, rabies, and viral meningitis. Biochemists synthesized vitamins which were recognized as essential constituents of a healthy diet, thus allowing the elimination of vitamin deficiency diseases such as scurvy, rickets, osteomalacia, beriberi, pellagra, xerophthalmia, nyctalopia, and pernicious anemia, via dietary supplements. Many metabolic diseases became treatable due to biochemical investigations; for example, the discovery of insulin in 1921 by the Canadian physiologists Sir *Frederick Banting* (1891-1941) and *Charles Best* (1899-1978) rapidly transformed diabetes from an invariably and often rapidly fatal disease into one that could be at least partially controlled, allowing sufferers many years of good life.

Blood group specification made transfusions convenient, facilitating dramatic advances in many branches of medicine,especially surgery. The first report of a successful autotransplant of a kidney into the neck of a dog was performed by *Emerich Ullmann* (1861-1937) in 1902. The first successful kidney graft between identical twins was performed in 1954 by *Joseph E. Murray* (b. 1919), who received a Nobel Prize for his work. The first heart transplantation in man was achieved by *Christiaan Barnard* (b. 1922) in 1967. Results were poor at first, but the use of new anti-rejection techniques such as the drug cyclosporin in the late 1970s, aided by advances in immunology, greatly improved the success of this and other organ transplants. By 1987, a total of ~7000 human hearts had been transplanted. At the close of the 20th century, heart, lung, heart-lung, and liver transplants were standard procedures, while grafts of small intestine and pancreas (the latter for refractory diabetes) were under active clinical investigation.

Two pivotal events transformed scientific medicine from a merely rational basis to a molecular basis, thus laying the groundwork for 21st century nanomedicine. The first pivotal event was the drug revolution, among which the most useful and spectacular were the antibiotics introduced between 1935-1945 and widely used ever since. *Antibiotics* are significant because they actively interfere with microbial metabolism and growth at the molecular level. The first antibiotic drugs were the *sulphonamides*, in 1935. Then penicillin became available in the 1940s, initially in very small quantities, then mass-produced by Pfizer during and after World War II, as a result of the research work of *Alexander Fleming* (1888-1955), *Howard Florey* (1898-1968) and *Ernst Chain* (1906-1979). In 1943, *Selman A. Waksman* (1888-1973) discovered streptomycin, the first effective anti-tuberculosis drug, for which he received the 1952 Nobel Prize. For the first time, physicians had true cures for many diseases,especially the most common bacterial diseases. Antifungal, anti-parasitic, and antiviral drugs of more limited effectiveness soon followed.

The 20th century also produced drugs that altered mood and levels of consciousness. *Barbiturates* were first introduced in 1903 (e.g. barbitone or Veronal), followed by phenobarbitone (Luminal) in 1912 and *Evipan*, the barbiturate anesthetic, in 1932. By mid-century these highly addictive drugs began to be replaced by the somewhat less-addictive *benzodiazepines*, including *Valium* and *Librium*. Tranquilizers, largely the phenothiazines such as *chlorpromazine* (*Thorazine*) and antimanics such as *lithium carbonate*, came to be widely used in psychiatry as effective medications for major mental illnesses including schizophrenia and manic depression.

The second pivotal event was the genetics revolution, starting with the discovery in 1953 of the information-carrying double-helix structure of DNA by *Francis Crick* (b. 1916) and *John B. Watson* (b. 1928), followed in the 1980s by the ability to chemically read the genetic code, isolate specific genes and clone them for further study. In the mid-1980s, the Human Genome Project was launched, with the objective of fully sequencing every gene in the human genome. The first phase of this project neared completion as the 20th century drew to a close.

Molecular biology became the premier scientific discipline of the latter 20th century. By 1998, the compositions of organs, tissues, cells, organelles, and membranes had been defined, and the biosynthesis and catabolism of hundreds of compounds had been elucidated. The regulation of body processes was described at progressively finer levels in biochemical language. Many pharmacologic agents were finally understood in terms of specific molecular loci and mechanisms of action. Advances were particularly rapid inimmunology, virology, cellular biology, peptide research, and structural biology. A beginning was made in explaining human behaviour in mechanistic terms, as more and more chemical mediators and pharmacologic modifiers were discovered. In biology, these disciplines developed porous boundaries with related disciplines such as physiology, pharmacology, neurosciences, biochemistry and biophysics. All entered a phase of confluence, employing the common language of chemistry.

Thus the late 20th century is best regarded as the molecular age of basic biological science. The molecular influence pervades all the traditional disciplines underlying clinical medicine. As of February 1999, one source listed exactly 1446 genetic disorders, most of which could be linked to a specific human chromosome; another source stated that ~4000 genetic disorders were known in 1998. There were more than 575 known abnormal human hemoglobins, and for each of these the precise structural defect in the DNA of the mutant gene could be defined. Knowledge of membrane, cytoplasmic, and nuclear receptors for hormones and drugs was exploding, with old as well as new diseases being defined in terms of receptor abnormalities—for example, type II hyper-cholesterolemia and nephrogenic diabetes insipidus. Recognition of opiate receptors led to the discovery of endogenous peptides (endorphins) with analgesic activity. Their localization promised further understanding of the limbic system, affective states, and addictions. Defects in a subcellular organelle, the peroxisome, were known to be responsible for a growing list of important genetic afflictions such as Refsum disease, Zellweger syndrome, and X-linked adrenoleukodystrophy. The genetic defect responsible for Huntington's disease was discovered, and in the 1990s the genetic defects responsible for many other important neurological disorders were becoming known, including forms of Alzheimer's disease, Charcot-Marie-Tooth disease, and common blinding retinal degenerative disorders such as retinitis pigmentosa, Leber's hereditary optic neuropathy, Norrie's disease, and choroideremia.

DNA sequencing techniques and restriction endonucleases permitted precise identification of the exact structural alteration of the gene in an increasing number of hereditary diseases. For example, Burkitt's lymphoma is characterized by a translocation of the distal end of the long arm of chromosome 8 to loci on chromosomes 14, 22, or 2. Gene therapy—both pharmacologic modification of specific gene action and physical replacement of damaged genetic segments—became possible in experimental systems. Complete maps of the entire genomes of 18 microbial species had been compiled and published by the end of 1998, with more than 60 others in progress. In 1992, wrote the conservative *Cecil Textbook of Medicine*: "The expansion of the knowledge bank of the past quarter century justifies great optimism for the eventual control and cure of major diseases and the possible elimination of premature death from illness."

Global changes in progressive aging dysfunction were shown to be strongly related to declining secretion of growth hormone by Rudman in 1990. Shortly thereafter, anti-aging medicine was recognized as a distinct discipline and was promoted by the American Academy of Anti-Aging Medicine (A^4M), a group that claimed >6000 physician and scientist members worldwide by 1998 and had held numerous conferences. In a book on the subject, *Ronald Klatz*, A^4M's president, first comprehensively documented the implications of Rudman's work.

While biological science was vaulting forward into the molecular realm at a blistering pace, biomedical engineering lagged considerably behind, though many remarkable successes had been achieved since mid-century. Radiology expanded with sophisticated radiotherapy, ultrasound, scanning and imaging techniques (e.g., CAT, PET, NMR), with submillimeter resolution in living tissues. Surgical instruments became less damaging and less invasive; minimally invasive techniques used OK (orifice and keyhole) surgery and surgery performed under the "eye" of a scan. Fetuses could be screened for abnormality and surgery then performed upon them inside the womb, or they could be removed temporarily from the womb and then returned to continue gestating, after surgery. Prospective parents with fertility difficulties could make use of a wide range of therapies including *in vitro* fertilization. Implantable pacemakers, defibrillators, and ventricular assist devices were commonplace, and in 1998 artificial wearable/implantable and full/partial replacements for lungs, heart, kidney, liver, and pancreas were either available, in clinical trials, or under development. Electron microscopes with 200,000X magnifications allowed many details of internal cellular and viral structures to be resolved as early as 1946; by 1998, atomic-force (AFM) and scanning-tunneling microscopy (STM) permitted direct tactile examination of individual biomolecules in fixed cells.

As the 20th century drew to a close, a few preliminary efforts had been made to apply the molecular approach to medical diagnostic and clinical tools. Biotechnology was one avenue being pursued, with the rational design of artificial enzymes and specified-ligand binding sites having already been achieved in certain limited cases, and the beginnings of gene therapy as noted earlier. Carbon fullerenes had been used to create a water-soluble inhibitor of HIV protease, and had other biological applications. AFM-based force-amplified biological sensors could detect defined biological species such as cells, proteins, toxins, and DNA at concentrations as low as 10^{-18} M ($\sim 1/mm^3$), and automated laboratory systems for sorting and handling individual cells and viruses were commonplace. Biotech companies such as Physiome Sciences of Princeton NJ had developed three-dimensional computer models of the heart and other organs. Physiome's heart model was based on detailed molecular, biochemical, cellular and anatomical information, including submodels of all the different cell types found in the heart embodying knowledge of the function of each cell type in healthy and diseased hearts, and information on gene function and the causes and effects of congestive heart failure, arrhythmias and heart attacks. Other cell biochemistry simulators included E-CELL and the Virtual Cell. There was also much progress and interest in nanostructure analysis and nanomaterials fabrication for medical and biological purposes, with research groups emerging at major universities such as the Cornell Nanofabrication Facility, theUniversity of Michigan Center for Biologic Nanotechnology, the Rice University Center for Nanoscale Science and Technology, the CalTech Materials and Process Simulation Center, the Washington University Nanotechnology Center (St. Louis, MO), the USC Laboratory for Molecular Robotics, the UCLA Exotic Materials Center, the Institute for Molecular Medicine of the University of Oxford, and at many biotechnology-oriented corporations such as Nanogen and Affymetrix.

However, the greatest medical revolution of all awaits the ability to engineer and fabricate whole devices and systems at the molecular scale. Along this course lies nanotechnology and molecular manufacturing, a deep well from which nanomedicine will inevitably spring.

21st century medicine

It is always somewhat presumptuous to attempt to predict the future, but in this case we are on solid ground because most of the prerequisite historical processes are already in motion and all of them appear to be clearly pointing in the same direction.

Medical historian *Roy Porter* notes that the 19th century saw the establishment of what we think of as scientific medicine. From about the middle of that century the textbooks and the attitudes they reveal are recognizable as not being very different from modern ones. Before that, medical books were clearly written to address a different mind-set.

But human health is fundamentally biological, and biology is fundamentally molecular. As a result, throughout the 20th century scientific medicine began its transformation from a merely rational basis to a fully molecular basis. First, antibiotics that interfered with pathogens at the molecular level were introduced. Next, the ongoing revolutions in genomics, proteomics and bioinformatics provided detailed and precise knowledge of the workings of the human body at the molecular level. Our understanding of life advanced from organs, to tissues, to cells, and finally to molecules, in the 20th century. By the early 21st century, the entire human genome will be mapped. This map will inferentially incorporate a complete catalog of all human proteins, lipids, carbohydrates, nucleoproteins and other molecules, including full sequence, structure, and much functional information. Only some systemic functional knowledge, particularly neurological, may still be lacking by that time.

This deep molecular familiarity with the human body, along with simultaneous nanotechnological engineering advances, will set the stage for a shift from today's molecular scientific medicine in which fundamental new discoveries are constantly being made, to a molecular technologic medicine in which the molecular basis of life, by then well-known, is manipulated to produce specific desired results. The comprehensive knowledge of human molecular structure so painstakingly acquired during the 20th and early 21st centuries will be used in the 21st century to design medically-active microscopic machines. These machines, rather than being tasked primarily with voyages of pure discovery, will instead most often be sent on missions of cellular inspection, repair, and reconstruction. In the coming century, the principal focus will shift from medical science to medical engineering. Nanomedicine will involve designing and building a vast proliferation of incredibly efficacious molecular devices, and then deploying these devices in patients to establish and maintain a continuous state of human healthiness.

The very earliest nanotechnology-based biomedical systems may be used to help resolve many difficult scientific questions that remain. They may also be employed to assist in the brute-force analysis of the most difficult three-dimensional structures among the 100,000-odd proteins of which the human body is comprised, or to help ascertain the precise function of each such protein. But much of this effort should be complete within the next 20-30 years because the reference human body has a finite parts list, and these parts are already being sequenced, geometered and archived at an ever-increasing pace. Once these parts are known, then the reference human being as a biological system is at least physically specified to completeness at the molecular level. Thereafter, nanotechnology-based discovery will consist principally of examining a particular sick or injured patient to determine how he or she deviates from molecular reference structures, with the physician then interpreting these deviations in light of their possible contribution to, or detraction from, the general health and the explicit preferences of the patient.

In brief, nanomedicine will employ molecular machine systems to address medical problems, and will use molecular knowledge to maintain human health at the molecular scale.

Volitional Normative Model of Disease

What, exactly, is "medicine"? Dictionaries give several definitions, ranging from the very restrictive to the most general, as follows: "a drug or remedy"; "any substance used for treating disease"; "any drug or other substance used in treating disease, healing, or relieving pain"; "in a restricted sense, that branch of the healing art dealing with internal diseases"; "treatment of disease by medical, as distinguished from surgical, treatment"; "the branch of this science and art that makes use of drugs, diet, etc., as distinguished especially from surgery and obstetrics"; "the study and treatment of general diseases or those affecting the internal parts of the body"; "the science of treating disease, the healing art"; "the art and science of preventing or curing disease"; "the act of maintenance of health, and prevention and treatment of disease and illness"; "the department of knowledge and practice dealing with disease and its treatment"; or, most generally, "the science and art of diagnosing, treating, curing, and preventing disease, relieving pain, and improving and preserving health". In this book, we shall adopt the latter, maximally-inclusive, definition of "medicine".

The contemporary physician might at first be inclined to relegate molecular approaches to some minor subfield, perhaps "nanoanalytics," "nanogenomics," or "nanotherapeutics." This would be a serious mistake, because the application of molecular approaches to health care will significantly impact virtually every category of laboratory and clinical practice across the board. Thus we are led to the broadest possible conception of nanomedicine as "the science and technology of diagnosing, treating, and preventing disease and traumatic injury, of relieving pain, and of preserving and improving human health, using molecular tools and molecular knowledge of the human body."

This brings us to the question of "disease," a complex term whose meaning is still hotly debated among medical academics. Illnesses due to microorganisms, or conditions in which the doctor's contribution to the diagnosis was important, were most likely to be called a disease, but if the cause was a known physical or chemical agent the condition was less likely to be regarded as disease; general practitioners also had the broadest definition of disease. No less than eight different types of disease concepts are held by at least some people currently engaging in clinical reasoning and practice, including:

Disease nominalism

A disease is whatever physicians say is a disease. This approach avoids understanding and forestalls inquiry, rather than furthering it.

Disease relativism

A disease is identified or labeled in accordance with explicit or implicit social norms and values at a particular time. In 19th century Japan, for example, armpit odor was considered a disease and its treatment constituted a medical specialty. Similarly, 19th-century Western culture regarded masturbation as a disease, and in the 18th century, some conveniently identified a disease called drapetomania, the"abnormally strong and irrational desire of a slave to be free." Various non-Western cultures having widespread parasitic infection may consider the lack of infection to be abnormal, thus not regarding those who are infected as suffering from disease.

Sociocultural disease

Societies may possess a concept ofdisease that differs from the concepts of other societies, but the concept may also differ from that held by medical practitioners within the society itself. For instance, hypercholesterolemia is regarded as a disease condition by doctors but not by the lay public; medical treatment may be justified, but persons with hypercholesterolemia may not seek treatment, even when told of the condition. Conversely, there may be sociocultural pressure to recognize a particular condition as a disease requiring treatment, such as alcoholism and gambling.

Statistical disease

A condition is a disease when it is abnormal, where abnormal is defined as a specific deviation from a statistically-defined norm. This approach has many flaws. For example, a statistical concept makes it impossible to regard an entire population as having a disease. Thus tooth decay, which is virtually universal in humans, is not abnormal; those lacking it are abnormal, thus are "diseased" by this definition. More reasonably, a future highly-aseptic society might regard bacterium-infested 20th century humans (who contain in their bodies more foreign microbes than native cells) as massively infected. Another flaw is that many statistical measurables such as body temperature and blood pressure are continuous variables with bell-shaped distributions, so cutoff thresholds between "normal" and "abnormal" seem highly arbitrary.

Infectious agency

Disease is caused by a microbial infectious agent. Besides excluding systemic failures of bodily systems, this view is unsatisfactory because the same agent can produce very different illnesses. For instance, infection with hemolytic *Streptococcus* can produce diseases as different as erysipelas and puerperal fever, and Epstein-Barr virus is implicated in diseases as varied as Burkitt's lymphoma, glandular fever, and naso-pharyngeal carcinoma.

Disease realism

Diseases have a real, substantial existence regardless of social norms and values, and exist independent of whether they are discovered, named, recognized, classified, or diagnosed. Diseases are not inventions and may be identified with the operations of biological systems, providing a reductionistic account of diseases in terms of system components and subprocesses, even down to the molecular level. One major problem with this view is that theories may change over time —almost every 19th century scientific theory was either rejected or highly modified in the 20th century. If the identification of disease is connected with theories, then a change in theories may alter what is viewed as a disease. For example, the 19th century obsession with constipation was reflected in the disease labelled "autointoxication," in which the contents of the large bowel were believed to poison the body. Consequently much unnecessary attention was paid to laxatives and purgatives and, when surgery of the abdomen became possible toward the end of the century, operations to remove the colon became fashionable in both England and America.

Disease idealism

Disease is the lack of health, where health is characterized as the optimum functioning of biological systems. Every real system inevitably falls short of the optimum in its actual functioning. But by comparing large numbers of systems, we can formulate standards that a particular system ought to satisfy, in order to be the best of its kind. Thus "health" becomes a kind of Platonic ideal that real organisms approximate, and everyone is a less than perfect physical specimen. Since we are all flawed to some extent, disease is a matter of degree, a more or less extreme variation from the normative ideal of perfect functioning. This could be combined with the statistical approach, thus characterizing disease as a statistical variation from the ideal. But this view, like the statistical, suffers from arbitrary thresholds that must be drawn to qualify a measurable function as representing a diseased condition.

Functional failure

Organisms and the cells that constitute them are complex organized systems that display phenomena (e.g. homeostasis) resulting from acting upon a program of information. Programs acquired and developed during evolution, encoded in DNA, control the processes of the system. Through biomedical research, we write out the program of a process as an explicit set (or network) of instructions. There are completely self-contained "closed" genetic programs, and there are "open" genetic programs that require an interaction between the programmed system and the environment, e.g. learning or conditioning. Normal functioning is thus the operation of biologically programmed processes, e.g. natural functioning, and disease may be characterized as the failure of normal functioning. One difficulty with this view is that it enshrines the natural as the benchmark of health, but it is difficult to regard as diseased a natural brunette who has dyed her hair blonde in contravention of the natural program, and it is quite reasonable to regard the mere possession of an appendix as a disease condition, even though the natural program operates so as to perpetuate this troublesome organ. A second weakness of this view is that disease is still defined against population norms of functionality, ignoring individual differences. As a perhaps overly simplistic example, 65% of all patients employ a cisterna chyli in their lower thoracic lymph duct, while 35% have no cisterna chyli—which group has a healthy natural program, and which group is "diseased"?

A ninth view of disease is also proposed, a new alternative which seems most suitable for the nanomedical paradigm, called the "volitional normative" model of disease. As in the "*disease idealism*" view, the volitional normative model accepts the premise that health is the optimal functioning of biological systems. Like the "functional failure" view, the volitional normative model assumes that optimal functioning involves the operation of biologically programmed processes.

However, two important distinctions from these previous views must be made. First, in the volitional normative model, normal functioning is defined as the optimal operation of biologically programmed processes as reflected in the patient's own individual genetic instructions, rather than of those processes which might be reflected in a generalized population average or "Platonic ideal" of such instructions; the relative function of other members of the human population is no longer determinative. Second, physical condition is regarded as a volitional state, in which the patient's desires are a crucial element in the definition of health. This is a continuation of the current trend in which patients frequently see themselves as active partners in their own care.

In the volitional normative model, disease is characterized not just as the failure of "optimal" functioning, but rather as the failure of either (a) "optimal" functioning or (b) "desired" functioning. Thus disease may result from:

1. A failure to correctly specify desired bodily function (specification error by the patient);
2. A flawed biological program design that doesn't meet the specifications (programming design error);
3. Flawed execution of the biological program (execution error);
4. External interference by disease agents with the design or execution of the biological program (exogenous error); or
5. Traumatic injury or accident (structural failure).

In the early years of nanomedicine, volitional physical states will customarily reflect "default" values which may differ only insignificantly from the patient's original or natural biological programming. With a more mature nanomedicine, the patient may gain the ability to substitute alternative natural programs for many of his original natural programs. For example, the genes responsible for appendix morphology or for sickle cell expression might be replaced with genes that encode other phenotypes, such as the phenotype of an appendix-free cecum or a phenotype for statistically typical human erythrocytes. Many persons will go further, electing an artificial genetic structure which, say, eliminates age-related diminution of the secretion of human growth hormone and other essential endocrines. (The graduated secretion of powerful proteolytic enzymes, perhaps targeted for gene-expression in appropriate organs, may reverse and control the accumulation of highly crosslinked collagenaceous debris; by 1998, many members of the mainstream medical community were already starting to regard aging as a treatable condition.) On the other hand, a congenitally blind patient might desire, for whatever personal reasons, to retain his blindness. Hence his genetic programs that result in the blindness phenotype would not, for him, constitute "disease" as long as he fully understands the options and outcomes that are available to him. (Retaining his blindness while lacking such understanding might constitute a specification error, and such a patient might then be considered "diseased.") Whether the broad pool of volitional human phenotypes will tend to converge or diverge is unknown, although the most likely outcome is probably a population distribution (of human biological programs) with a tall, narrow central peak (e.g., a smaller standard deviation) but with longer tails (e.g., exhibiting a small number of more extreme outliers).

One minor flaw in the volitional normative model of disease is that it relies upon the ability of patients to make fully informed decisions concerning their own physical state. The model crucially involves desires and beliefs, which can be irrational, especially during mental illness, and people normally vary in their ability to acquire and digest information. Patients also may be unconscious or too young, whereupon default standards might be substituted in some cases.

Nevertheless, the volitional normative view of disease appears most appropriate for nanomedicine because it recognizes that the era of molecular control of biology could bring considerable molecular diversity among the human population. Conditions representing a diseased state must of necessity become more idiosyncratic, and may progressively vary as personal preferences evolve over time. Some patients will be more venturesome than others— "to each his own." As an imperfect analogy, consider a group of individuals who each take their automobile to a mechanic. One driver insists on having the carburetion and timing adjusted for maximum performance (the "racer"); another driver prefers optimum gas mileage (the "cheapskate"); still another prefers minimizing tailpipe emissions (the "environmentalist"); and yet another requires only that the engine be painted blue (the "aesthete"). In like manner, different people will choose different personal specifications. One can only hope that the physician will never become a mere mechanic even in an era of near-perfect human structural and functional information; an automobile conveys a body, but the human body conveys the soul. Agrees theorist Guttentag: "The physician-patient relationship is ontologically different from that of a maintenance engineer to a machine or a veterinarian to an animal."

Treatment Methodology

The availability of advanced nanomedical instrumentalities should not significantly alter the classical medical treatment methodology, although the patient experiences and outcomes will be greatly improved. Treatment in the nanomedical era will become faster and more accurate, efficient, and effective. In clinical practice, patient treatment customarily includes up to six distinguishable phases: examination, diagnosis, prognosis, treatment, validation, and prophylaxis. Let us consider each of these, in turn.

Examination

The first step in any treatment process is the examination of the patient, including the individual's medical history, personal functional and structural baseline, and current complaints. In classical medicine, interview and observation have long been the cornerstone of examination. In ancient times this was limited to obvious manifestations and simple constellations of observables, such as the Hippocratic facies, the four signs of inflammation noted by Celsus, or pulse rate and fever. Clinicians recognize that the traditional taking and interpreting of oral medical histories from new patients is a subtle and complex art, although some aspects of this process might be automated using voice recognition and text preinterpretation software, somewhat easing the physician's burden.

Advancing technology has also brought a plethora of tests that contribute to accurate diagnosis, including auscultation, microscopy and clinical bacteriology in the 19th century, and radiological scanning, clinical biochemistry, genetic testing, and minimally invasive exploratory surgery in the 20th century.

In the 21st century, new tools for nanomedical testing and observation will include clinical *in vivo* cytography; real-time whole-body microbiotic surveys; immediate access to laboratory-quality data on the patient (e.g. blood tests such as blood counts, dissolved gases and solutes, vitamin and ion assays); physiological function and challenge tests; tissue composition including direct organelle counts in specified tissue populations; quantitative flowcharts of *in cyto* secondary messenger molecules, extracellular hormones and neuropeptides; per-compartment cytoglucose inventories; and so forth. Before a proper diagnosis can be made, the physician must also establish the patient's personal functional and structural baseline against which any deviations can be noted and corrected, in keeping with the volitional normative model of disease.

The capabilities of nanomedial testing are explored at length. By way of introduction, it is instructive to think about a trivial class of test procedures that might be used to diagnose a simple infectious disease at several different levels of technological competence. Let us consider a patient who presents with signs and symptoms that are nonspecific in nature but which suggest an infectious process— e.g. nasal congestion, mild fever, discomfort and cough. The initial signs are due in part to the body's inflammatory response and in part to the infectious agent itself. The diagnostic goal is to identify the infectious agent.

In the late 20th century, the usual procedure would be to culture a sample taken from the patient, in the microbiology laboratory, using various broths, petri plates, and biochemical tests. Some infectious agents are easy to demonstrate. Beta hemolytic streptococci from a throat swab will grow overnight on a blood agar plate, and colony counts for *E. coli* in a urine sample are available in 24 hours. A throat culture that the lab reports as a mixed culture causes no excitement, and a single isolate of *Staphylococcus epidermidis* in a blood culture is usually regarded as a skin contaminant.

Moving up to a higher level of technological competence, biotechnologists describe an ideal diagnostic scenario which takes a molecular approach to the diagnosis of infectious disease using recombinant DNA technology. This approach was not yet possible in 1996 when first suggested, but was regarded as a reasonable and likely future application of biotechnology in the early 21st century given rapid progress in single-molecule DNA assay techniques:

"A patient presents in the clinic with mild fever, nasal congestion, discomfort, and cough. A swab of his throat is taken. Instead of culture to identify abnormal microorganisms by their pattern of growth, the sample is analyzed by recombinant DNA techniques. The cotton throat swab is mixed with a cocktail of DNA probes. Enzymes that digest and release the DNA from both host cells and invading bacteria make the DNA in the sample immediately available for hybridization to the probes. The swab is swirled in the liquid mix of the prepackaged test kit for 1 minute. The liquid is then poured through a column that separates hybridized DNA molecules (bacterial target DNA sequences bound to probe DNA) from all other debris [taking several minutes]. A chemi-luminescence detection system for the probes shows two of several possible colours indicating mixed infection. The diagnostic result, available in 10 minutes, indicates a Rhinovirus of a strain known to be epidemic in the geographic area. A significant superinfection with a penicillin-resistant streptococcus is also identified. With a definitive diagnosis, the patient is started on the appropriate antibiotic."

How might nanomedicine handle this test? In the nanomedical era, taking and analyzing microbial samples will be much simpler for the practitioner. Such analysis will be as quick and convenient as the electronic measurement of body temperature using a tympanic thermometer in a late 20th-century clinical office or hospital. The physician faces the patient and pulls from his pocket a lightweight handheld device resembling a pocket calculator. He unsnaps a self-sterilizing cordless pencil-sized probe from the side of the device and inserts the business end of the probe into the patient's opened mouth in the manner of a tongue depressor. The ramifying probe tip contains billions of nanoscale molecular assay receptors mounted on hundreds of self-guiding retractile stalks. Each receptor is sensitive to one of thousands of specific bacterial membrane or viral capsid ligands. An acoustic echolocation transceiver provides gross spatial mapping. The patient says "Ahh," and a few seconds later a three-dimensional colour-coded map of the throat area appears on the display panel that is held in the doctor's hand. A bright spot marks the exact location where the first samples are being taken. Underneath the colour map scrolls a continuously updated microflora count, listing in the leftmost column the names of the ten most numerous microbial and viral species that have been detected, key biochemical marker codes in the middle column, and measured population counts in the right column. The number counts flip up and down a bit as the physician directs probe stalks to various locations in the pharynx to obtain a representative sampling, with special attention to sores or any signs of exudate. After a few more seconds, the data for two of the bacterial species suddenly highlight in red, indicating the distinctive molecular signatures of specific toxins or pathological variants. One of these two species is a known, and unwelcome, hostile pathogen. The diagnosis is completed, the infectious agent is promptly exterminated, and a resurvey with the probe several minutes afterwards reveals no evidence of the pathogen.

Diagnosis

Diagnosis is the determination of the cause and nature of a disease in order to provide a logical basis for treatment and prognosis. Traditionally the diagnostic process begins with a thorough history taken from the patient and a relevant physical examination. Often this sufficed to make a confident diagnosis, but the cause of some illnesses remained uncertain without recourse to additional information such as blood tests or radiological examinations. Nanotechnology-based diagnosis will consist principally of examining the patient to determine how he or she deviates from autogenous reference structures and functions, and then interpreting those deviations as healthy or unhealthy for that patient.

In the 20th century, diagnoses frequently involved a high degree of uncertainty, largely due to the general lack of comprehensive molecular diagnostic tools. Thus diagnosis would be guided by statistical analyses; one branch of decision analysis, called utility analysis, even allowed the patient to participate in the decisionmaking process. When the correct decision is unclear, urges one textbook, it is well to remember time-honored Hippocratic aphorisms such as "first, do no harm" and "common things occur commonly." The eminent Canadian physician Sir William Osler (1849-1919) lamented that "errors of judgement must occur in the practice of an art which consists largely in balancing probabilities." Most doctors would prefer to understand the root cause of medical problems rather than adopt mere statistical approaches.

Nanomedical tools will vastly reduce diagnostic uncertainty. Using nanomedical instrumentalities, doctors will gain access to unprecedented amounts of information about their patients including in-office comprehensive genotyping and real-time whole-body scans for particular bacterial coat markers, tumour cell antigens, mineral deposits, suspected toxins, hormone imbalances of genetic or lifestyle origin, and other specified molecules, producing three-dimensional maps of desired targets with submillimeter spatial resolution. Embedded *in vivo* nanomedical data archives can provide onboard storage of regularly updated self-diagnostic scans, reducing to a minimum the need for symptomatic interview data from patients who may be unconscious, inarticulate, or verbose, who may have limited powers of self-analysis or self-observation and who may have forgotten, suppressed, or amplified descriptions of symptoms. Physicians do not require an exhaustive survey of the entire body of each patient to molecular detail to make a valid diagnosis. In any particular case, it is the function of the trained medical mind to quickly ascertain where and where not to look in molecular detail. But in the nanomedical era, powerful tools will be available to allow the practitioner to examine almost any portion of a patient in as much detail as desired, right down to the molecular level, with results available in seconds or minutes, and at reasonable cost.

Prognosis and treatment

Prognosis is a judgement or forecast, based upon a correct diagnosis, of the future course of a disease or injury, and of the patient's prospects for partial or full recovery. Guttentag identifies prognosis as "the predicted course of the reduced state of the patient's psychosomatic freedom of action," and treatment as "the physician's ability to intervene."

But prognosis is a function of treatment as well as disease. From the post-Hippocratic era through the 18th century, treatments were almost purely empirical and often did more harm than good. During the 19th and early 20th centuries, treatments were scientific but largely homeostatic—the medical intervention was rational but served mainly to assist the body in healing itself. Throughout the remainder of the 20th century, truly curative treatments began to rescue some patients from conditions from which their unaided bodies would not have been able to recover. Although conventional biotechnology will enable some important tissue and cellular replacement treatments by the early 21st century, nanomedicine will enable major reconstructive and restorative procedures at the tissue, cellular and molecular levels and will employ active antibiotic devices. The prognosis will almost always be good, except in cases of severe neural damage and a few other specialized circumstances. Therapeutic treatments will be selected to reverse all pathological effects of disease or injury, with a minimum of pain, discomfort, side-effects, intrusiveness and time, and with a maximum of effectiveness, efficiency, and likelihood of success, though of course some tradeoffs will always exist. Nanomedicine also will excel in the correction of molecular defects of a kind which Nature has no predesigned tools—such as the breakdown and removal of intracellular lipofuscin (for which there appear to be no natural enzymes) and the removal of indigestible waste products which interfere with neuronal axon transport.

We may compare the therapeutic response to a simple infection at several different levels of technological competence. Consider a patient who has been diagnosed with eastern equine encephalitis, a mosquito- or tick-borne arbovirus. In the 20th century, there was no specific treatment for this disease. Care was generally supportive, with the doctor attempting to maintain the patient's heart and lung function while the infection ran its course. The prognosis was poor.

There was a 50%-75% mortality rate with frequent sequelae including seizures and paralysis, especially in children.

Biotechnologists proposed a molecular approach to therapeutics using recombinant DNA technology that was not yet possible in 1996 but was regarded as likely by the early 21st century:

"A patient enters the hospital with high fever and intense headaches. A spinal fluid tap is submitted to the molecular microbiology lab. After screening for several viruses, a species of equine encephalitis virus is identified that is endemic to a location recently visited by the patient. A call to the Centers for Disease Control results in the emergency delivery of a new antiviral agent. Antisense oligonucleotides are injected into the cerebral spinal fluid. These small DNA pieces bind directly to the virus and block its further proliferation. A temporary reservoir giving access to the cerebrospinal fluid is placed and infusion of this therapeutic molecular inhibitor of the virus continues for 5 days until signs of encephalitis have passed."

The nanomedical therapy? As before, a nanomedical cure for eastern equine encephalitis may be far simpler, less painful and a great deal quicker. A single therapeutic dose consisting of ~0.1 cm^3 of isotonic saline fluid containing ~10 billion active micron-size virucidal nanodevices, a 10% volumetric nanodevice suspension, is injected into the cerebrospinal fluid. Each therapeutic nanorobot has chemical sensors that can unambiguously recognize fluidborne or *in cyto* arbovirus particles and, once recognition has occurred, destroy them and also reverse the cellular damage. A nanorobot population of this size should be able to destroy all viral particles and effect needed repairs in at most an hour, after which the devices are programmed and equipped either to eliminate themselves from the body or to be manually exfused (e.g., nanapheresis).

Validation and prophylaxis

A proper therapeutic protocol will include a procedure for follow-up to ensure that the prescribed treatment was correctly executed with good results. This step is often neglected in order to save costs and may be considered unimportant by some practitioners because approximately 80%-90% of all illnesses which take patients to the doctor are self-curing or self-limiting. For example, the common cold, most infectious diseases and many minor injuries are problems that usually will resolve on their own even with no treatment. In these cases the purpose of treatment is not to provide a cure, but rather to speed the healing process, improve comfort, and avoid complications.

Many nanomedical treatments will require supervision and will run quickly to completion, thus follow-up may come back into vogue. Validation may also be viewed as a post-treatment re-diagnosis to ensure that no disease remains present in the patient.

Prophylaxis is the prevention of disease, typically including patient education, immunization programs, amelioration of occupational hazards, and other preventive and public health measures. In a treatment environment that is rich in effective antibacterial instrumentalities, those microbes which survive will evolve to produce only modest or negligible symptoms that are insufficiently annoying to motivate a patient to seek professional therapeutic relief. It is well-known that bacteria can modify their behaviour over time. For example, syphilis had a much more fulminating course in the Middle Ages than it has in the 20th century. Some future strain of the syphilitic microbe might produce negligible symptoms, but we should still insist on its eradication because of its potential to revert to its earlier virulence if allowed to spread unchecked in a more benign form. Preventative procedures may also be needed to discover, diagnose, and treat apparently symptomless diseases, and a variety of molecular-based physiological malfunctions and structural micropathologies may require nanoscale tools in order to detect them.

With medical conditions that require ongoing supervision and adjustment, such as maintaining optimum hormone balance and minimal accumulation of molecular debris (e.g. anti-aging medicine), nanoscale monitoring stations may act as onboard cellular guidance systems, stimulating or suppressing endocrine secretion as necessary to preserve an ideal state of equilibrium. In some cases, direct manufacture of compounds not easily produced by ribosomes or other biological organelles may be required.

Evolution of Bedside Practice

The relationship between physician and patient has been evolving in response to the rapidly changing medical environment. Two of the most important trends are the decline of traditional holism (with a concomitant increase in specialization) and the rise of therapeutic customization in medical practice.

Specialization and holistic medicine

Holism is a philosophy which holds that individuals function as complete units that cannot be reduced merely to the sum of their parts. It is unarguable that a simplistic reductionist view of the patient which ignores the complex interactions among the many cells, tissues, organs, and systems constituting the human body is deeply flawed. For example, an understanding of the molecular basis of the contractile proteins of heart muscle will not alone tell us how heart muscle cells will look or act; knowing only about the parts of something is not sufficient to predict the behaviour of the whole. However, traditional concepts of holistic medicine go well beyond such basic systems philosophy—incorporating requirements for a consideration of all physical, emotional, social, spiritual, environmental and economic needs of the patient.

N. Jewson and others have decried the modern shift away from holistic medicine, which is asserted to have taken place in three historical phases in the West. The initial phase is identified as the practice of "bedside medicine," where wealthy fee-paying clients in the 17th and 18th centuries helped shape their own diagnosis and treatment by medical practitioners in an holistic manner. Aspects of the patient's emotional and spiritual life were seen as central by the practitioner in making a diagnosis, since most physical treatments were only palliative and so the doctor had to focus on nonphysical supportive measures. This frame of reference was progressively replaced during the 19th century with the trend toward "hospital medicine," wherein physicians concentrated on generic classifications of diseases that were manifested in the patient, moving doctors away from the earlier focus on the individual as a whole person. The 20th century saw the development of "laboratory medicine," which moved diagnosis and therapy even further away from the whole patient, "who came to be medically conceived as little more than a depersonalized object, comprised of a complex of cells."

A more charitable view is that physicians have increasingly specialized in treating those physical diseases for which effective treatments may be readily specified, leaving nonphysical and nontreatable issues for psychiatrists, social workers, fitness coaches, priests, lawyers, or other professionals to deal with. In the 21st century this operational specialization may become complete, since nanomedicine phenomenologically regards the human body as an intricately structured machine with trillions of complex interacting parts, with each part (and each subsystem of parts) subject to individual scrutiny, repair, and possibly replacement by artificial technological means. In this new medical cosmology, the concept of the whole patient almost completely dissolves into a data-intensive whirlwind of molecular detail at the cellular, tissue, organ, and systemic levels.

And yet, as in a bygone era, patients once again will help to shape their own diagnosis and treatment at the hands of medical practitioners who begin to apply the volitional normative model of disease in their practices. This may breathe new life into the age-old medical school dictum to "treat the patient as a person" and to "focus on the sick person" rather than exclusively on the body.

The availability of extremely powerful and transforming molecular technologies also argues for a return to the romantic and perhaps quaint concept of a single doctor taking care of a single patient. The potential for interactions among highly potent nanorobotic instrumentalities argues for diagnostic and therapeutic "gatekeeping" by a single trusted practitioner in whom strategic treatment responsibility is vested—in partnership, of course, with the patient.

Customized diagnosis and therapeutics

In science the objective is to understand the individual occurrence by means of a general law; in medical practice, knowledge of what is generally the case does not tell the physician how to treat a particular patient. Thus the problem of how to proceed from the prototypic case to the individual instance remains to be solved in a systematic manner by the practitioner.

The practitioner of "bedside medicine" in the 17th and 18th centuries had few curative and almost no customized tools at his disposal—perhaps a few dozen basic surgical techniques, performed septically, hazardously, and without anesthesia, and a drug/herbal formulary consisting of a few hundred substances. For instance, the *Edinburgh Pharmacopoeia* of 1803 listed only 222 simples while the *London Pharmacopoeia* of 1809 listed fewer than 200 items, most of which had variable, uncertain, minimal, or nonspecific potency. A few more options became available to the 19th century medical practitioner, including complex surgical techniques for specific conditions that could be performed aseptically with anesthesia and a good chance for success, a somewhat broader and more efficacious pharmacopeia, vaccines targeted to several diseases, and an improving diagnostic ability. By the 20th century, the physician could prescribe from among tens of thousands of specific drugs to target specific bacterial, viral, fungal, or parasitic infections; select from among a very precise array of anesthetic agents, chosen to avoid allergic responses in particular patients; perform a wide variety of noninvasive tests and scans for diagnostic purposes to identify very specific conditions; and perform minimally invasive surgeries directed at many arbitrary tissue masses as small as 1 mm^3 in volume. The first glimmerings of personalized genetic therapies also began to appear.

With the arrival of nanomedicine in the 21st century, the treatment paradigm will complete its transition from coarse-grained, one-size-fits-all, slow-acting methods to molecularly-precise, completely customized, speedy and highly-efficacious procedures and instrumentalities. The irregular shotgun pattern of 18th century palliatives will evolve during the early 21st century into a penetrating and perfectly tailored hail of "magic bullets" each targeting an individual cell or group of cells unique to the individual patient. The 19th century herbalist John Ayrton Paris could have been describing the nanomedical future in his popular textbook *Pharmacologia*, 1840 edition, when he wrote:

"If [a physician] prescribes upon truly scientific principles, he will rarely in the course of his practice compose two formulae that shall, in every respect, be perfectly similar, for the plain reason that he will never meet with two cases exactly alike. Now let me ask what constitutes the essential difference between the true physician and his counterfeit between the philosopher and the empiric? Simply this that the latter exhibits the same medicine in every disease, however widely each may differ from the other in its symptoms and character; while the former examines, in the spirit of philosophic analysis, all the existing peculiarities of his patient, and of his discord...and then adapts with a sound discretion and with a correct judgement of his medicinal agents, such means as may best be calculated to control and correct the patient's morbid condition."

Physician-patient relationship

Many other aspects of the physician-patient relationship, especially as this relationship may evolve in the coming era of nanomedicine, are important and worthy of extensive discussion. One such issue is the obligation of both parties in the partnership to tell the truth. The patient as a fellow human being has every right to know the truth about his or her biological condition, but other considerations may enter into the fulfillment of this obligation. Patients have a quite natural anxiety about their own possible death, and it has been claimed that this anxiety implies that no one can be truly objective toward his or her own body. Guttentag observes that "telling an unwelcome truth to the unprepared is as ill-conceived as trying to hide the truth from the prepared."

In the nanomedical era, the sheer number of "truths" that may become available for disclosure will increase enormously even as the terminal prognosis becomes rare. For example, each human being is believed to possess at least 4-10 potentially serious genetic defects; up to 1% of human DNA is of exogenous viral origin, and as much as 10% of the genome consists of transposons, discrete sequences that are positionally mobile among the chromosomes. Should something be done about this, or not? What should the average patient make of the news that his physician has discovered exactly 57 submicron-scale lamellar defects scattered throughout the compact bone of the caudal epiphysis of the patient's right humerus? In the nanomedical era, people will gain the ability to specify their own physical structure to minute detail, but many patients will not be ready, willing, or able to assume responsibility for this knowledge. Thus there is no ideal substitute for the doctor's interpretative abilities and judgements on the patient's behalf as to the personal significance of specific diagnostic information. As the great clinician Thomas Addis observed in another context: "Honesty with patients requires thought and discipline and effort."

Perhaps the single most important aspect of the physician-patient relationship, in any century, is the humanistic quality of the good doctor. The patient seeks a physician who cares about him as a person and will diagnose and prescribe in a sensitive and compassionate manner, accepting some degree of obligation to the patient. Speaking to medical students, J.C. Bennett describes the implicit social contract between doctor and patient that will still apply in the nanomedical era, as it does today:

"To receive medical care, patients must trust their bodies and their very lives to physicians, and so to be in an honest position to give medical care, physicians must earn such radical trust. Mere technical treatment of disease does not suffice. Patients must be able reasonably to believe that their physicians care about them in an extraordinarily personal way. This exchange of care for trust, while not identical to friendship or love, is equally binding. From it develops an interdependence that is far from unwholesome; rather, it potentiates care and promotes healing. Our late twentieth century sophistication and technologic orientation have too often cost us warmth, humor, and humanity, leaving us in social isolation. We do far better as professionals to err on the side of being human with our patients, than to try to play deus ex machina, the god from the machine."

Changing View of the Human Body

How does a patient regard his or her own body, and how might this most intimate of all relationships change in the nanomedical era?

The so-called dualist theory of the human compound, as originally developed by Descartes and widely accepted today by the ordinary person, holds that the human being consists of two separate kinds of thing: the body and the mind or soul. The body acts as a host or receptacle for the mind. The mind, often called "the ghost in the machine," is manifested by the brain, which it uses (via the bodily senses) to acquire and store information about the world and to integrate this with its genetically-driven imperative to live, thus resolving internal

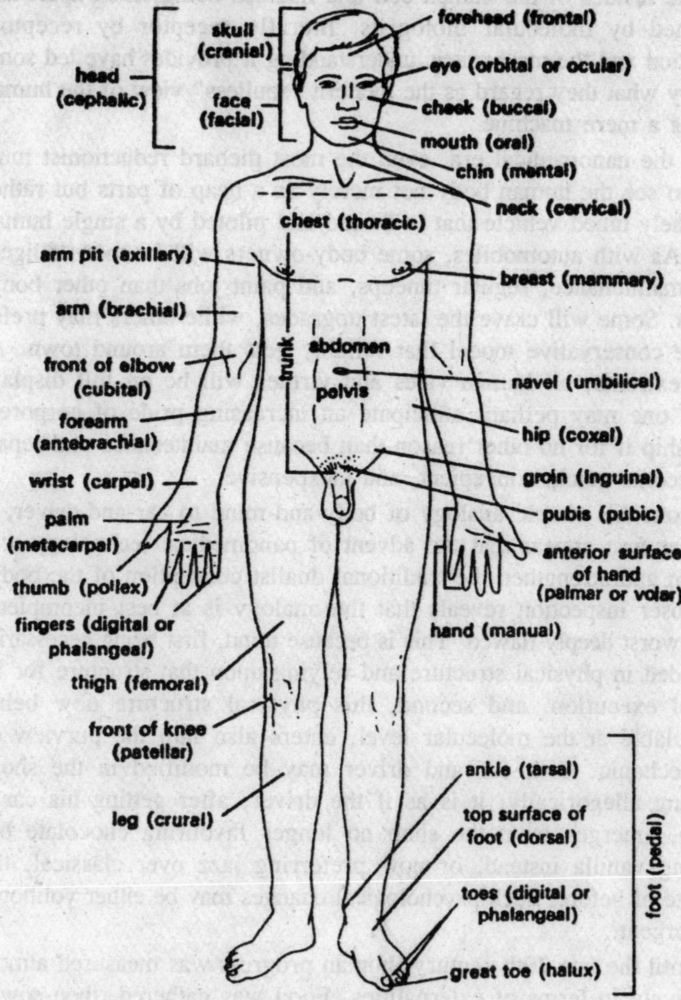

Fig. 4.4. Dorsal view showing the anatomical position.

conflicts among action-choices and expressing these in what we (in our consciousness) experience as decisive action.

Scientific medicine has concentrated primarily on the body. The ancient Roman physician Galen first dissected and vivisected a variety of animals to increase his knowledge of anatomy and physiology, and dissection became increasingly important in the training of physicians and surgeons, and in painting and sculpture, during the Renaissance. By the late 20th century, dissection had reached the molecular level, with the insides of the human cell and nucleus being taken apart and examined by molecular biologists, literally receptor by receptor. Dissection and the mechanistic understanding it provides have led some to decry what they regard as the modern "soulless" view of the human body as a mere machine.

In the nanomedical era, even the most diehard reductionist must come to see the human body not merely as a heap of parts but rather as a finely tuned vehicle that is owned and piloted by a single human mind. As with automobiles, some body-owners will be more diligent about maintenance, regular tuneups, and paint jobs than other body-owners. Some will crave the latest upgrades, while others may prefer a more conservative model that reliably gets them around town. At either extreme, all human vices and virtues will be on full display, though one may perhaps anticipate an increasing pride of corporeal ownership if for no other reason than because maintenance and repair will become quick, convenient, and inexpensive.

From this simple analogy of body-and-mind to car-and-driver, it might at first appear that the advent of nanomedical technology will confirm and strengthen the traditional dualist conception of the body. But closer inspection reveals that the analogy is at best incomplete, and at worst deeply flawed. This is because mind, first being necessarily embedded in physical structure and relying upon that structure for its faithful execution, and second, this physical structure now being manipulable at the molecular level, enters also into the purview of our mechanic. Both car and driver may be modified in the shop. Speaking allegorically, it is as if the driver, after getting his car a tuneup, emerges from the shop no longer favouring chocolate but enjoying vanilla instead, or now preferring jazz over classical, the opposite of before. Such psychological changes may be either volitional or emergent.

Until the late 20th century, human progress was measured almost exclusively in terms of externalities. Food was gathered, then sown, then manufactured. Shelters had no running water, then gained outhouses, then indoor plumbing. Natural lighting and campfires gave way to candles, then oil lamps, then electric illumination. Finger-counting yielded first to the abacus, then the mechanical adding machine, and finally to the digital computer. But throughout all of history, the human body itself has remained largely untouched by progress. We have always regarded our bodies, evolved by natural selection, as fundamentally inviolate and immutable—subject perhaps to various natural or traumatic degradations, but rarely to any significant intrinsic improvement on the timescale of human civilization.

Now we are set to embark upon an era in which our natural physiological equipment may for the first time in history become capable of being altered, improved, augmented, or rendered more comfortable or convenient, due to advances in medical technology. The physical human body may be one of the last bastions of "naturalness". It will also be one of the last elements in our common worldview to be modernized.

Our subjective experience of reality will shift by subtle degrees. For instance, all objective information about our physical surroundings has traditionally arrived in the conscious mind via the various natural senses such as hearing, sight, and smell. In the nanomedical era, machine-mediated sensory modalities may permit direct perception of physical phenomena well removed from our bodies in both time and space, or which are qualitatively or quantitatively inaccessible to our original natural senses. Perception will gradually expand to incorporate nonphysical phenomena including abstract models of mental software, purely artificial constructs of simulated or enhanced realities, and even the mental states of others. Such new perceptions will inevitably alter the way our minds process information.

But the winds of change will sweep deeper still, into our very souls. Like ants oblivious to the collective purpose of their colony, the billions of neurons in the human brain are all busily buzzing, wholly ignorant of the emergent plan. This is the physical, mechanical world of our electrochemical hardware. People also have thoughts, feelings, emotions, and volitions, a higher level in the data processing hierarchy which in turn is equally oblivious of the brain cells. We can happily think while being totally unaware of any help from our neurons. But nanomedicine will give us unprecedented systemic multilevel access to our internal physical and mental states, including real-time operating parameters of our own organs, tissues, and cells, and, if desired, the activities of small groups of (or even individual) neurons. Diverse parts of our selves previously closed to our attention may slowly conjoin and enter our conscious awareness.

Will this access promote an integrated identity or lead to hopeless confusion, or worse? Marvin Minsky, in his collection of essays *The Society of Mind*, persuasively argues that our selves or identities are in fact networks of semi-autonomous neurological "agencies" which sometimes cooperate and sometimes compete with one another. We think of ourselves as singular "persons," but we also experience "conflicting desires" and "differing viewpoints" within our minds that are, in Minsky's view, a direct experience of the multiplicity of our brain's neurostructures. Other models of the human mind suggest that our internal mental states, prospectively transparent via nanomedical augmentation, are diverse and intricate; Julian Jaynes is one of many writers who have drawn attention to profound dichotomies between the two cerebral hemispheres. The component-oriented personality models of Freud (e.g. ego/id/superego), Jung (e.g. archetypes), and Rank (e.g. will/counterwill), and the identification of 4541 distinct personality traits by Allport and Odbert warn us that full access to our brain's architecture could be perilous.

More seriously, most of us suppose that we are endowed with free will. But if choices by free will are simply the resolution of conflicts of neurological subsystems, and we become consciously aware of those subsystems and are able to intervene in their processes, do we run the risk of runaway instabilities at the deepest levels of what we presently call our "minds"? Will we find that these instabilities are profound counterparts to the maladies we currently designate as epilepsy, or psychosomatic illnesses? In any redesigns of our brains which would involve opening doors to, quite literally, the ultrastructure of our thoughts, we could become "naked to ourselves" in ways that we can only vaguely speculate about at present. Along with any other dangers we might encounter, this will raise entirely new issues of the proper role of psychotherapy and the sanctity of personal privacy.

Repairs to the brain may be carefully monitored to ensure quality control and to verify intended results, as already proposed in another context. Major modifications might be strictly regulated, both to prevent abuse by unscrupulous third parties and also to forestall accidental or volitional alterations that could render the patient a significant threat to society. Nanomedical alterations to the brain and other physical systems may give us vastly expanded freedom to be who we choose to be, along with increased responsibility to make wise and informed choices. The ethical and legal aspects of these questions, as well as the scientific and psychological ones, are extremely important and should be thoroughly debated in the years and decades that lie ahead.

Nanomedical Perspective

Nanomedicine and Molecular Nanotechnology

A mature nanomedicine will require the ability to build structures and devices to atomic precision, hence molecular nanotechnology and molecular manufacturing are key enabling technologies for nanomedicine. The prefix "nano-" (from the Greek root nanos, or dwarf) means one-billionth (10^{-9}) of something. The term "nanotechnology" refers most generally to technology on the scale of a billionth of a meter, or a nanometer (a nanometer is ~6 carbon atoms wide). Similarly, the words "nanomachine," "nanorobot," "nanomotor" and "nanocomputer" may refer to complex engineered objects fabricated by positioning matter with molecular control.

Molecular engineering was discussed as an extension of bulk technologies in the 1960s and 1970s by von Hippel, von Foester, and Zingsheim. The phrase "Nano-Technology" was first used in print in 1974 by N. Taniguchi to refer to the increasingly precise machining and finishing of materials, progressing from larger to smaller scales and ultimately to nanoscale tolerances, following in the path of Feynman's proposed "top-down" approach, a schemata which persisted in Taniguchi's thinking throughout the 1980s and 1990s. In 1981, K.E. Drexler described a new "bottom-up" approach involving molecular manipulation and molecular engineering in the context of building molecular machines and molecular devices with atomic precision, a fundamentally different mind-set. Drexler again described molecular technology in 1982 and molecular mechanical devices in 1983, first using the word "nanotechnology" in 1985 and 1986 as synonymous with molecular technology, finally settling upon "molecular nanotechnology" in 1991 and "molecular machine systems" in 1992 to clarify that his concept involved working devices constructed with atomic precision, as distinguished from nanostructured bulk materials, micromachinery, polymeric self-assembly, pure biotechnology, nanolithography, Langmuir-Blodgett thin films, and the like. Drexler's definition—molecular nanotechnology as the three-dimensional positional control of atomic and molecular structure to create materials and devices with molecular precision—is the usage. The first known use of the term "nanomedicine" was in 1991 by Drexler, Peterson, and Pergamit in their popular book *Unbounding the Future*.

Is molecular nanotechnology possible? This question is explicitly addressed, but the bottom line is that molecular nanotechnology violates no physical laws and there exist many possible technical paths leading to useful results. Even by 1985, for example, G. Yamamoto had reported on molecular gears, describing "compounds that exist in conformations which are regarded as static meshed gears with two-toothed and three-toothed wheels and some of them behave as dynamic gears." H. Iwamura prepared a system that formed a chain of beveled molecular gears with ~GHz rotation rates. In 1998, it was generally accepted that molecular nanotechnology would be developed, although there was still some disagreement about how long it would take.

It is often noted that molecular biological systems are themselves nanomachines, constituting an existence proof for molecular nanotechnology. Indeed, biotechnology is one possible implementation pathway for molecular nanotechnology that is being pursued. Such comparisons have a long history. For example, Marcello Malpighi (1628-1694), a professor of medicine in Pisa who discovered the fine structure of the lungs and the capillaries using the microscope, once observed that "our bodies are composed of strings, thread, beams, levers, cloth, flowing fluids, cisterns, ducts, filters, sieves, and other similar mechanisms."

Parallels to living systems as molecular machines were drawn by Changeau, McClaire, Laing, Drexler, and Mitchell, inspiring early thinking in molecular manufacturing. For example, in 1991 Drexler observed:

"Technology-as-we-know-it is a product of industry, of manufacturing and chemical engineering. Industry-as-we-know-it takes things from nature — ore from mountains, trees from forests — and coerces them into forms that someone considers useful. Trees become lumber, then houses. Mountains become rubble, then molten iron, then steel, then cars. Sand becomes a purified gas, then silicon, then chips. And so it goes. Each process is crude, based on cutting, stirring, baking, spraying, etching, grinding, and the like."

"Trees, though, are not crude. To make wood and leaves, they neither cut, stir, bake, spray, etch, nor grind. Instead, they gather solar energy using molecular electronic devices, the photosynthetic reaction centers of chloroplasts. They use that energy to drive molecular machines—active devices with moving parts of precise, molecular structure—which process carbon dioxide and water into oxygen and molecular building blocks. They use other molecular machines to join these molecular building blocks to form roots, trunks, branches, twigs, solar collectors, and more molecular machinery. Every tree makes leaves, and each leaf is more sophisticated than a spacecraft, more finely patterned than the latest chip from Silicon Valley. They do all this without noise, heat, toxic fumes, or human labour, and they consume pollutants as they go. Viewed this way, trees are high technology. Chips and rockets are not."

Contemplating applications of nanotechnology to medicine, Brian Wowk concluded that 20th century physicians were in a predicament similar to that which would be faced by 18th-century engineers trying to maintain a 20th-century automobile—repairs would be crude at best, and breakdowns inevitable:

"Like primitive engineers faced with advanced technology, medicine must 'catch up' with the technology level of the human body before it can become really effective. What is the technology level? Since the human body is basically an extremely complex system of interacting molecules (i.e., a molecular machine), the technology required to truly understand and repair the body is molecular machine technology—nanotechnology. A natural consequence of [our achieving] this level of technology will be the ability to analyze and repair the human body as completely and effectively as we can repair any conventional machine today."

In *Engines of Creation*, Drexler drew inspiration from the cell's eye view to recognize that nanotechnology could bring a fundamental breakthrough in medicine. Noting that 20th century physicians relied chiefly on surgery and drugs to treat illness, Drexler explained:

"Surgeons have advanced from stitching wounds and amputating limbs to repairing hearts and reattaching limbs. Using microscopes and fine tools, they join delicate blood vessels and nerves. Yet even the best microsurgeon cannot cut and stitch finer tissue structures. Modern scalpels and sutures are simply too coarse for repairing capillaries, cells, and molecules. Consider 'delicate' surgery from a cell's perspective. A huge blade sweeps down, chopping blindly past and through the molecular machinery of a crowd of cells, slaughtering thousands. Later, a great obelisk plunges through the divided crowd, dragging a cable as wide as a freight train behind it to rope the crowd together again. From a cell's perspective, even the most delicate surgery, performed with exquisite knives and great skill, is still a butcher job. Only the ability of cells to abandon their dead, regroup, and multiply makes healing possible."

"Drug therapy, unlike surgery, deals with the finest structures in cells. Drug molecules are simple molecular devices. Many affect specific molecules in cells. Morphine molecules, for example, bind to certain receptor molecules in brain cells, affecting the neural impulses that signal pain. Insulin, beta blockers, and other drugs fit other receptors. But drug molecules work without direction. Once dumped into the body, they tumble and bump around in solution haphazardly until they bump a target molecule, fit, and stick, affecting its function. Drug molecules affect tissues at the molecular level, but they are too simple to sense, plan, and act. Molecular machines directed by nanocomputers will offer physicians another choice. They will combine sensors, programs, and molecular tools to form systems able to examine and repair the ultimate components of individual cells. They will bring surgical control to the molecular domain."

By the end of the 20th century, mainstream military (DoD), NIH, NSF, and other international groups had begun to seriously consider the potential future applications of molecular nanotechnology in medicine. For example, in 1997 a panel of U.S. Department of Defense health science experts known as Military Health Service Systems (MHSS) 2020 concluded in its final report:

"If a breakthrough to a [molecular] assembler occurs within ten to fifteen years, an entirely new field of nanomedicine will emerge by 2020. Initial applications will be focused outside the body in areas such as diagnostics and pharmaceutical manufacturing. The most powerful uses would eventually be within the body. Possible applications include programmable immune machines that travel through the bloodstream, supplementing the natural immune system; cell herding machines to stimulate rapid healing and tissue reconstruction; and cell repair machines to perform genetic surgery."

The present book takes as its starting point the assumption that the mass production of nanomachines at modest cost is technically feasible, and then explores the medical implications of this assumption. Proposed systems presented in this trilogy are intended not as final engineering blueprints but merely as points of departure for further analysis and refinement. All designs and projected capabilities are, for the most part, conservatively drawn with generous safety margins, leaving a considerable volume of design space yet to be explored by more intrepid future investigators.

Nanomedicine: History of the Idea

Conclusive proof of the existence of atoms was not obtained until the close of the 19th century. This may explain why the idea of nanomedicine is an exclusively 20th century phenomenon. The first hint of it may be found in a famous 1929 essay written by J.D. Bernal:

"The discoveries of the twentieth century, particularly the micro-mechanics of the Quantum Theory which touch on the nature of matter itself, are far more fundamental and must in time produce far more important results. The first step will be the development of new materials and new processes in which physics, chemistry and mechanics will be inextricably fused. The stage should soon be reached when materials can be produced which are not merely modifications of what nature has given us in the way of stones, metals, woods and fibers, but are made to specifications of a molecular architecture. Already we know all the varieties of atoms; we are beginning to know the forces that bind them together; soon we shall be doing this in a way to suit our own purposes. The result—not so very distant—will probably be the passing of the age of metals and all that it implies—mines, furnaces, and engines of massive construction. Instead we should have a world of fabric materials, light and elastic, strong only for the purposes for which they are being used, a world which will imitate the balanced perfection of a living body."

Development of the concept of nanomedicine has followed two principal paths which Richard Smalley has termed "wet nanotechnology" in the biological tradition, and "dry nanotechnology" in the mechanical tradition. Both approaches were presaged in speculative fiction. The following abbreviated history focuses on nanomedicine largely to the exclusion of broader issues in molecular engineering, manufacturing and nanoscopy, and includes a number of inspirational, speculative, or fictional early references from non-refereed sources.

Biological tradition

The general idea of biological engineering stretches back at least to the mid-19th century, but the first science fiction story involving actual genetic engineering was "Proteus Island" (1936), written by the chemical engineer Stanley Weinbaum. Artificial engineered organisms subsequently appeared in minor roles in several stories, a notable example being the familiars employed by the fake witches in Fritz Leiber's "Gather, Darkness!" (1943), and A.E. van Vogt used "gene transformation" to create the superman in "Slan" (1940). The first artificially evolved creatures appeared in Theodore Sturgeon's

"Microcosmic God" (1941), wherein a biochemist established conditions allowing accelerated artificial evolution, creating the Neoterics, a submillimeter-sized race of intelligent hypermetabolic creatures which could accomplish tasks very rapidly. The first artificially designed microcreatures appeared in James Blish's "Surface Tension" (1952). In this story, crash-landed dying human astronauts create a completely new form of humanity—tiny men and women reduced to protozoan size—who are seeded in the pools and puddles at the surface of the new planet, and who go on to master the biotechnology necessary to travel from one water puddle to another.

The scientific tradition of biological nanomachines for medical purposes began in 1964 when Robert Ettinger, an early cryonics pioneer, suggested that cellular-level or even molecular-level repair might be developed for life extension. Ettinger speculated that "...surgeon machines, working 24 hours a day for decades or even centuries, will tenderly restore the frozen brains, cell by cell, or even molecule by molecule in critical areas."

In 1965, the synthesis of artificial life was publicly proposed as a national goal by the president of the American Chemical Society, Professor Charles Price, who pointed out that many new types of life might be made, not "mere imitations" of biology as we know it.

In 1967, Isaac Asimov suggested the future possibility of "factories...where the working machinery consists of submicroscopic nucleic acids" and that a "repertoire of hundreds or thousands of complex enzymes" could be used to "bring about chemical reactions more conveniently than any methods now used" and also for "helping to construct life."

In 1968, G.R. Taylor cited the possibilities for genetic engineering and genetic surgery: "The microsurgery of DNA may possibly be achieved by physical methods: fine beams of radiation (probably laser light or pulsed X-rays) may be used to slice through the DNA molecule at desired points." He also cited predictions that bacteria would soon be programmed.

In 1969, J. White suggested that a modified virus could be used as a cell repair machine: "It has been proposed that appropriate genetic information be introduced by means of artificially constructed virus particles into a congenitally defective cell for remedy; similar means may be used for the more general case of repair. The repair program must use means such as protein synthesis and metabolic pathways to diagnose and repair any damage... [Information] can be preserved by specifying that the repair program incorporate appropriate RNA tapes into itself..."

In 1970, Jeon et al carried out "the reassembly of *Amoeba proteus* from its major components: namely nucleus, cytoplasm, and cell membrane," taken from three different cells.

In 1972, Ettinger proposed using genetic engineering to make microscopic biorobots: "Genetic engineering's most sensational impact will concern the modification of humans, but it will have other uses as well. Some of the "robots" that will serve us will need to be nanominiaturized." Existing organisms could be modified to make biologically-based programmable biorobots for medical applications: "If we can design sufficiently complex behaviour patterns into microscopically small organisms, there are obvious and endless possibilities, some of the most important in the medical area. Perhaps we can carry guardian and scavenger organisms in the blood, superior to thc leukocytes and other agents of our human heritage, that will efficiently hunt down and clean out a wide variety of hostile or damaging invaders." Computerized cellular repair machines "must use means such as protein synthesis and metabolic pathways to diagnose and repair any damage...[Information] can be preserved by specifying that the repair program incorporate appropriate RNA tapes into itself...." Also in 1972, Danielli described various possibilities for generating new life forms via "life-synthesis" and genetic engineering, noting that "macromolecular engineering" might enable the development of very powerful and compact macromolecular computer systems.

In 1974, Halacy noted "some rather inglorious ways" to use "the miracle of artificial life," including potential capabilities for growing diverse items ranging from computers to airplanes. Morowitz suggested cooling cells to cryogenic temperatures in order to analyze and determine their structure. Artificial cells could also be assembled at such temperatures and then be set in motion by thawing. Morowitz further reported that microsurgery experiments on amoebas "have been most dramatic. Cell fractions from four different animals can be injected into the eviscerated ghost of a fifth amoeba, and a living functioning organism results."

In 1975, Richard Laing described the theoretical possibility of molecular machine self-replicators using molecular (data) tapes based on the idea of universal Turing machines, examining several ways that such "artificial organisms" might replicate themselves as "a vehicle for the exploration of broad biological possibilities."

In 1976, Donaldson presented the first detailed (and quite ambitious) list of biotechnological techniques that appeared necessary to achieve cell repair and might prove feasible, writing at a time that predated many current capabilities such as automated protein/DNA sequencing and synthesis, and most knowledge of restrictions on cellular developmental pathways and genetic programs and networks. For repair at the level of the cell, Donaldson's techniques would have included:

1. The ability to design enzymes to produce specific repair functions such as renaturing denatured proteins, joining broken lipoprotein complexes, annealing broken strands of DNA or RNA, reading proteins of existing or special types onto RNA and replicating them, and giving a cell the ability to metabolize new substrates, use novel cofactors, or construct essential amino acids;
2. Specially constructed bacteria or macrophages able to replicate themselves, spread throughout a specific target tissue, and carry out specific repairs according to the programs designed into their DNA/RNA; these could be designed to operate at unnatural temperatures or to utilize metabolic pathways not presently found in nature;
3. The abilities to re-introduce lost DNA or lost organelles such as mitochondria into a cell, to introduce entirely new forms of organelles perhaps to perform specific repair functions, and to introduce new metabolic capacities into a cell;
4. The ability to modify at will the developmental program of a cell, as for instance to induce postmitotic cells such as neurons to divide, according to a specific program, forming daughter cells with specific properties; and
5. Several different types of repair bacteria able to work together in an integrated fashion, and linked together by chemical means (e.g. hormones), so as to apply optimal repairs to every body cell in order to (A) diagnose the precise nature of the damage, call other repair bacteria to its location, and report to the attending doctor that new types of repair bacteria other than those already introduced are needed, and (B) identify structures which must be preserved (e.g. memory) and reconstruct them if necessary.

For repairs at the level of the whole organism, Donaldson offered the following rather aggressive biological nanotechnology techniques:

1. Understanding the physiology of aging combined with the ability to reverse it;

2. Control over growth and development, including the ability to program types of growth and development which do not naturally occur, such as growth of new eyes or other organs which have been lost or damaged, growth of an entire and well-formed body from a head alone, and regrowth of injured or lost brain tissue; and
3. Nonpermanent "substitute organs" which will take over from others which have been lost, including (A) the ability to keep a given tissue alive and healthy in vitro for an indefinite time and similar abilities for a body part, and (B) temporary replacements for any body organ which may have been lost. These would be used to support the body while new organs were growing, e.g. as "metabolic crutches." This capacity specifically includes the ability to make temporary replacements for diffuse systems such as the vascular or nervous systems. For instance, a specially created "plant" would grow an entire vascular system into the patient, down to replacement for the capillaries and venules from a single seed, always introducing its fibrils between cells and destroying none or very little of the original structure.

The last of these is a description of a (clearly speculative) "repair net," a concept which Donaldson may have been the first to propose in 1976. A whole-body repair net was later termed a "chrysalis", and twelve years later, in 1988, Donaldson provided an artist's conception and an additional description of these proposed biotechnological instrumentalities:

"Severe crushing or mangling injuries require us to provide a new vascular system. The repair device might resemble a fungus, growing mycelia into the injured tissue. A repair net would grow into a [crushed] limb, guided by recognition of the injured cells and a plan for how the limb should look after repair....[In the most severe injuries,] a chrysalis first envelops the patient, then enters in between all his cells. It disassembles the patient, surrounding each cell with its own repair machinery and vascular system. The geometry already preserves information about locations of the patient's cells. If necessary, morphogen chemical gradients could also retain this information. A patient would [locally] swell up to 10 times original diameter. After repair, the chrysalis withdraws the same way it entered."

In 1977, Darwin further developed the theme of tissue repair and cellular repair biomachines, independently proposing a modified white blood cell to perform repair functions.

In 1981, Asimov suggested that we "consider the bacteria. These are tiny living things made up of single cells far smaller than the cells in plants and animals....[We] can, by properly designing these tiniest slaves of ours, use them to reshape the world itself and build it close to our hearts' desire." More importantly, Donaldson elaborated on his earlier discussion of how cryonically suspended human beings might be repaired. He extended the earlier concepts of Ettinger and White, concluding that "with such hybrid technology as micro-miniature biological-mechanical machines the size of viruses" and related technology, "it seems unlikely that (to repair a single cell) there would be any difficulty at all in principle to carrying out any imaginable repair." He estimated that about 10 programmable cell-repair biomachines could be introduced into a cell that was under repair without causing too much mechanical disruption. In 1988, Donaldson described artificial macrophages that "can carry control machinery to recognize target cells, responding only to them, or responding differently depending upon cell type or cell conditions. They can still work even if the target cell isn't functioning (unlike viruses), rebuild target cell machinery other than the genes, and transfer many more genes up to an entire copy of the patient's genome." They can also "communicate with one another [and] release diffusible chemicals to guide one another's behaviour."

By the 1990s, bioengineered viruses of various types and certain other vectors were routinely being used in experimental genetic therapies as a means to target and penetrate certain cell populations, with the objective of inserting therapeutic DNA sequences into the nucleus of human target cells *in vivo*. Retrovirally-altered lymphocytes (T cells) began to be injected into humans for therapeutic purposes. Another example of an engineered cell in therapeutics was the use of genetically modified cerebral endothelial cell vectors to attack glioblastoma, which was being pursued by Neurotech (in Paris) in 1998. Engineered bacteria were being pursued by Vion Pharmaceuticals in collaboration with Yale University. In their "Tumour Amplified Protein Expression Therapy" program, antibiotic-sensitive *Salmonella typhimurium* (food poisoning) bacteria were attenuated by removing the genes that produce purines vital to bacterial growth. The tamed strain (cell line VNP20009) could not long survive in healthy tissue, but quickly multiplied ~1000-fold inside tumours, which are rich in purines. The next step would be to add genes to the bacterium to produce anticancer proteins that can shrink tumours, or to modify the bacteria to deliver various enzymes, genes, or prodrugs for tumour cell growth regulation. The engineered bacteria were available in multiple serotypes to avoid potential immune response in the host. Phase I human clinical trials were expected to begin in 1999 using clinical dosages produced in 50-liter fermenters, and other possible bacterial vector species were being examined.

Micro-biorobotics was still regarded by many in the biotechnology community as a highly speculative topic in 1998. Glen A. Evans described the possible construction of synthetic genomes and artificial organisms. His proposed strategy involves determining or designing the DNA sequence for the genome, synthesizing and assembling the genome, introducing the synthetic DNA into an enucleated pluripotent host cell, then introducing the host cell into an organism. Evans foresees the high throughput "read-out" of gene products discovered through genomic sequencing, rapid construction of designer genes and genomes, automated designer vector synthesis for gene replacement/insertional mutagenesis, and finally the construction of synthetic genomes and artificial organisms. Robert Bradbury has considered requirements and costs for genome synthesis and replacement. For example, Bradbury estimates genome synthesis costs by assuming a projected ~$0.03/base for DNA synthesis (compared to ~$0.80-$1.00/base-micromole in 1999), with ~20,000 expressed genes per biorobotic cell with an average gene size of ~3000 bases, giving a ~60 megabase expressed genome per biorobotic cell (1 chromosome) and a raw synthesis cost of ~$3 million per designer cell line (excluding design costs). There has also been discussion of "chemical reaction automata" as precursors for synthetic organisms, synthetic lifeforms and "nanobiology", "biorobotics", "cell rovers", microbial engineering, lymphocyte engineering, and artificial chromosomes, and cell engineering is rapidly gaining popularity.

Mechanical tradition

The late science fiction author Robert A. Heinlein nearly invented the concept of molecular nanotechnology in 1942 when he suggested a process for manipulating microscopic structures. Heinlein envisioned the extensive use of life-size teleoperator hands, called "waldoes," complete with sensory feedback for full, remote-controlled telepresence. His fictional hero, Waldo, used a collection of these mechanical teleoperated hands for building and operating a series of ever-smaller sets of such mechanical hands. The smallest mechanical hands, "hardly an eighth of an inch across," were equipped with micro-surgical instruments and stereo "scanners," and were used to "manipulate living nerve tissue, [to examine] its performance *in situ*," and to perform neurosurgery. Eric Frank Russell's 1947 story "Hobbyist" described a fabrication process with "atom fed to atom like brick after brick to build a house." In 1955, Russell's serial "Call Him Dead," also published in *Astounding Science Fiction*, had a virus-based alien intelligence that spread through contact with blood or saliva; the story features a "microforger," a man who makes "surgical and manipulatory instruments so tiny they can be used to operate on a bacillus." Also in the mechanical tradition, Isaac Asimov's "Fantastic Voyage" in 1966 took its miniaturized human crew in a miniaturized submarine through the bloodstream of a human patient on a mission of repair.

Heinlein, Russell, and Asimov overlooked the full implications of their ideas, and most scientists were unsympathetic—for example, in 1952 Erwin Schrodinger wrote that we would never experiment with just one electron, atom, or molecule. But by the early 1960s, several scientists had reinvented similar approaches to micromanipulation and miniaturization, this time extending their reach into the nanotechnology domain. The first and most famous of these scientists was the Nobel physicist Richard P. Feynman. In his remarkably prescient 1959 talk "There's Plenty of Room at the Bottom," Feynman proposed employing machine tools to make smaller machine tools, these to be used in turn to make still smaller machine tools, and so on all the way down to the atomic level. Feynman prophetically concluded that this is "a development which I think cannot be avoided." Such nanomachine tools, nanorobots and nanodevices could ultimately be used to develop a wide range of submicron instrumentation and manufacturing tools, i.e., nanotechnology. Feynman's suggested applications for these tools included producing vast quantities of ultrasmall computers and various micro- and nano-robots.

Feynman was clearly aware of the potential medical applications of the new technology he was proposing. After discussing his ideas with a colleague, Feynman offered the first known proposal for a nanomedical procedure to cure heart disease: "A friend of mine suggests a very interesting possibility for relatively small machines. He says that, although it is a very wild idea, it would be interesting in surgery if you could swallow the surgeon. You put the mechanical surgeon inside the blood vessel and it goes into the heart and looks around. (Of course the information has to be fed out.) It finds out which valve is the faulty one and takes a little knife and slices it out. Other small machines might be permanently incorporated in the body to assist some inadequately functioning organ." Later in his historic lecture, Feynman urged us to consider the possibility, in connection with biological cells, "that we can manufacture an object that maneuvers at that level!"

In 1961, K.R. Shoulders rejected the use of biological building blocks, even though biological "processes do work, and they can do so in a garbage can without supervision." He saw them as too limited environmentally and too difficult to control with available technology. Instead, he sought to directly produce much simpler, more powerful and rugged nanostructure arrays, operating at video frequency rates, which in turn could ultimately aid in their own replication. In 1965, Shoulders reported the actual operation of micromanipulators able to position tiny items with 10 nm accuracy while under direct observation by field ion microscopy.

In 1970, Volkenstein noted that "the creation of a nonmacromolecular system which would act as a model for living organisms is definitely possible" but could not arise by itself, and that the macromolecularity of present organisms is not essential, but due to their evolutionary origins. Taking some poetic license, he added: "Consequently, the cybernetic nonmacromolecular machine, which simulates life, could have been and can be created on earth only by man. Then it could perfect itself without limits." T. Nemes discussed artificial self-replicating machines and described how to construct "an automatic lathe able to reproduce itself," a concept apparently developed before von Neumann's work on machine replication.

In 1981, Drexler suggested the construction of mechanically deterministic nanodevices using biological parts; these devices could inspect cells at the molecular level and also repair cellular tissues that had been damaged during cryonic suspension. In 1982, Drexler described cell repair machines even more clearly in the mechanical tradition, in a popular publication.

By 1983, Drexler began privately circulating a draft technical paper entitled "Cell Repair Machines" which investigated for the first time, in some detail, whether an advanced mechanical-based nanotechnology would "permit construction of systems of molecular-scale sensors, computers, and manipulators able to enter and repair cells; the nature of the computational algorithms and magnitude of the computational resources needed to guide repairs; and the physical capabilities and constraints important to the repair process [in order to] sketch the conceptual design of a cell repair system based on a mature molecular technology."

Peterson notes that "medical applications were explored by Drexler during the early 1980s but the medical community was not ready for the concept." A technical paper by Drexler, invited by an editor at the *Journal of the American Medical Association*, was dismissed by a referee as "science fiction."

In 1985, G. Feinberg proposed using short-wavelength coherent laser energy to power and communicate with "nanosensors that could be implanted into the human body. They could...[monitor] various physiological functions from subcellular molecules up through tissues and organs...essential in determining some of the mechanisms involved in growth and aging."

In a 1985 book entitled *Robotics*, edited by Marvin Minsky (a well-known computer scientist and artificial intelligence pioneer), Minsky briefly described how fully-automatic cellular repair machines might work, following Drexler's vision:

"Suppose that we could design a repair machine so small that it could repair an artery from the inside! The first such machines might be the size of fleas. (There is room for a great deal of machinery in something the size of a flea—as the body of the flea itself testifies.) These micromachines could crawl into all but the smallest blood vessels, clear out debris, and reline the walls with suitable materials yet to be invented. Later generations of micromachines could be even smaller, perhaps no larger than the body cells they repair. These minuscule machines would be mass-produced by the billions, either by larger machines or by techniques of making them reproduce themselves. Perhaps these biological janitors would even be implanted in our bodies to remain there as permanent maintenance workers, just like many biological cells that already serve such purposes."

"Today the idea of such a technology may seem fantastic, yet many of the circuits in our computers are already smaller than many of our bodies' cells. Let's try, for a moment, to look ahead to: mass production of highly intelligent machines; gigantic advances in miniaturization; a technology so advanced that these machines reproduce themselves without our help. Fantastic? Not at all. Even the simplest algae and bacteria can do that. True, they're not intelligent, but each of them contains enough computerlike machinery and memory to do those things. So, to build bacteria-size computers should be perfectly feasible, once we have the necessary microtechnology. Some day we'll have the means to build artificial, cell-like machines with all those capabilities."

R.A. Freitas Jr. also contributed a chapter to Minsky's 1985 book, offering the following suggestion for remote-controlled incisionless nanosurgery: "One possibility is the concept of remote-controlled medical mites made feasible by modern micromachinery technology. Some medical mites would be like microminiature submarines, released inside the human body for internal sensing. Other mites could float, crawl, or swim through major arteries in the human body and perform on-site repairs from within, controlled by radio link under direction of a skilled telemicrosurgeon."

In 1986, Drexler published *Engines of Creation*, a popular text with two chapters devoted to discussions of cellular repair machines. Drexler further expounded upon the topic of cellular repair machines in articles published in 1985, 1986, 1987, and 1989, and his concepts were described by Brian Wowk in 1988 and in a Time-Life book in 1989.

In 1988, A.K. Dewdney reported an early nanomedical concept of an artery-cleaning nanorobot that he attributed to Drexler, accompanied by an artist's conception with the caption "a nanomachine swimming through a capillary attacks a fat deposit."

Nanomedicine in the 1990s

Despite the tremendous importance of molecular nanotechnology, a published 1993 literature review of robotics in health care included not a single reference to nanotechnology or to nanomedicine. At this writing in 1998, the list of original post-1989 technical and popular nonfiction works that deal with nanomedical topics in the mechanical tradition is short enough to permit virtually an exhaustive listing, including Beardsley, Bova, Coombs and Robinson, Crawford, Drexler, DuCharme, Emanuelson, Fahy, Fiedler and Reynolds, Freitas, Kaehler, Klatz and Kahn, Kurzweil, Lampton, Merkle, Merrill, Minsky, More, Ostman, Reifman, and Wowk and Darwin. The first nanomedical device design technical paper was published in 1998 by Freitas in the biotechnology journal *Artificial Cells*, and the present trilogy (*Nanomedicine*) is the first book-length technical treatment of the medical implications of molecular nanotechnology.

Biotechnology and Molecular Nanotechnology

Some pragmatic readers may be wondering what is so special about molecular nanotechnology, that existing or anticipated biotechnology could not accomplish just as well? After all, biotechnology is already an established medical capability. It has real applications and real products already on the market. Reflecting upon the future possibility of sophisticated mechanical medical nanorobots equipped with powerful nanocomputers, in 1989, one well-known cryobiologist mused that "what is not clear is just what need we would have for such devices." The simplest answer, is that each of the three contemporary branches of "nanotechnology" offers something of unique value to the practice of medicine.

Nanoscale materials technology has already found widespread use in medicine, including biocompatible materials and analytical techniques, surgical and dental practice, nerve cell research using intracellular electrodes, biostructures research and biomolecular research using near-field optical microscopy, scanning-probe microscopy and optical tweezers, and vaccine design, and also many 20th century bulk chemical and biochemical manufacturing techniques along with much of classical pharmacology.

As for "biotechnology," the original meaning of this word contemplated "the application of biological systems and organisms to technical and industrial processes". In recent times, the field has expanded to include genetic engineering and now takes as its ultimate goal no less than the engineering of all biological systems, even completely artificial organic living systems, using biological instrumentalities.

The third branch, molecular nanotechnology, takes as its purview the engineering of all complex mechanical systems constructed from the molecular level—potentially offering new tools for medical practice, the principal subject of this book. Observes G.M. Fahy, "the difference between nanotechnologists and biotechnologists is that the former do not restrict themselves to the biological limitations of the latter, and they are much more ambitious about the kinds of accomplishments that they want to achieve."

Doctors can utilize solutions to medical problems from all three approaches. As noted earlier, 80-90% of medical complaints resolve themselves via natural homeostatic processes, or without the necessary involvement of active biotechnological or molecular-nanotechnological agents. But by employing biotechnology, the range and efficacy of treatment options greatly increases. With molecular nanotechnology, the range, efficacy, comfort and speed of possible medical treatments again expands enormously. Molecular nanotechnology is essential when the damage to the human body is extremely subtle, highly selective, or time-critical (as in head traumas, burns, or fast-spreading diseases), or when the damage is very massive, overwhelming the body's natural defenses and repair mechanisms.

At every difficulty level, most classes of medical problems may be resolved with varying efficacy within the homeostatic/nanomaterials, biotechnological, or molecular-nanotechnological approaches. As the chosen technology becomes more precise, active, and controllable, the range of options broadens and the quality of the options improves. Thus the question is not whether molecular nanotechnology is required to accomplish a given medical objective. In many cases, it is not—though of course there are some things that only biotechnology and molecular nanotechnology can do, and some other things that only molecular nanotechnology can do. Rather, the important question is which approach offers a superior solution to a given medical problem, using any reasonable metric of treatment efficacy. For virtually every class of medical challenge, a mature molecular nanotechnology offers a wider and more effective range of treatment options than any other approach.

It is quite possible to imagine an advanced biotechnology that uses an engineered white cell, fibroblast, or macrophage chassis, energized by native oxygen and glucose and modified mitochondrial powerplants, driven by pseudopodia, cilia or flagella, communicating and navigating via biochemical signals, and even incorporating onboard digital biocomputers to make microscopic biorobots. Principal arguments favouring the biotechnology approach for medical purposes are: (1) that we are already somewhat familiar with such systems, after half a century of intensive molecular biology research; (2) that we have already "built" precursor systems, such as a whole living amoeba constructed from five distinct parts, bioengineered viruses and bacteria as DNA insertion devices, and natural replication stimulated in genetically engineered starter microbes; (3) that biocompatibility will not be a major issue since fibroblasts (which express no HLA Class II antigens, hence stimulate no rejection response) could be used as the starter material; (4) that both engineered viruses and bacteria are already in wide commercial and research use; and (5) the greater complexity of self-repair in mechanical systems, should it be needed. Many believe that the development pathway to early biorobots may be considerably shorter than for the mechanical nanorobots of the molecular nanotechnology approach, for which in 1998 not a single working prototype yet existed, even in research laboratories.

It is also possible to imagine a molecular nanotechnology that uses mechanical nanorobotic systems. Such systems will have many constitutional differences from biological-based systems. For instance, mechanical systems will transport parts, materials, energy and instructions via fixed channels, whereas most (but not all) biological systems operate by diffusion. Mechanical systems will have structures constrained by specific geometries, whereas biological systems have structures defined by patterns of containment and interconnection—the shape of a membrane compartment in a cell matters less than its continuity and the contents of the volume it defines. Mechanical systems will be deterministically manufactured by operations analogous to manual construction, whereas engineered ribosomes will self-assemble via diffusion and stochastic matching of complementary parts; in other words, biology uses recipes, while mechanical systems use blueprints. Cells grow, with their parts adapting to one another; mechanical nanorobots may be constructed from parts of fixed structure. Biology uses self-repair; mechanical systems generally do not largely because self-repair (by component exchange) will not be needed in molecular mechanical systems, whose designs may be made more simple by relying upon high component redundancy. These and other differences imply a number of important advantages that mechanical-based medical systems will enjoy over biological-based medical systems, which, taken together, strongly suggest that predominantly mechanical nanosystems may be the approach of choice for a mature medical nanotechnology. The advantages of molecular nanotechnology (e.g. the mechanical tradition) are many.

Speed of medical treatment

Doctors may be surprised by the incredible quickness of nanorobotic action when compared to the speeds available from fibroblasts or leukocytes. Normal homeostatic processes such as dermal wound repair via natural fibroblasts may require weeks to run to completion. Typical fibroblast movements occur at 0.1-1 microns/sec, but mechanical nanomanipulators can operate at 1-10 cm/sec speeds or faster, a speed advantage of 4-5 orders of magnitude. Even the strongest biological fibers (e.g. intermediate filaments) have a failure strength 3 orders of magnitude below the strongest mechanical fibers (e.g. fullerene nanotubes). Biological cilia beat at ~30 Hz while mechanical nanocilia may cycle up to ~20 MHz, though practical power restrictions and other considerations may limit them to the ~10 KHz range for most of the time. Flexible mechanical surfaces can complete a morphing motion in ~0.1 millisec, compared to the ~100 millisec snapback time for pinched red cell membrane, again a thousand-fold advantage in speed. Thus we expect that mechanical therapeutic systems can reach their targets up to ~10,000 times faster, all else equal, and treatments which require ~10^5 sec for a biological system may need only ~10^2 sec for a mechanical system; tachyiatria improves both patient and physician comfort. In either biological or mechanical systems, large numbers of devices of comparable physical size (e.g. ~microns) may be employed to do the work, so numbers alone cannot offset the mechanical speed advantage.

Power density and transduction

Biological cells typically employ power densities of 10^3-10^4 W/m^3, with maximum densities of ~10^6 W/m^3 in honeybee flight muscle cells and bacterial flagellar motors. By contrast, nanomechanical power systems can produce power densities of 10^9-10^{12} W/m^3, an advantage of 10^3-10^8 for mechanical over biological systems. By 1998, conducting polymer-based actuators generated 20-100 times the force for a given cross-sectional area as mammalian skeletal muscle. Additionally, amoebic locomotion in motile cells requires diffusion-limited cytoskeletal disassembly and reassembly to achieve movement; mechanical motility systems may employ simple cable-pulling, winches, or ratchets, which are faster and more direct. Some biological energy transducers are reversible, but muscle contraction is irreversible—not only cannot muscle actively re-expand, but stretching it doesn't make it produce much useful chemical energy. In contrast, electric motors can be run backwards to generate electricity, forcing pistons makes them pump, and loudspeakers can be used as microphones.

Superior building materials

Typical biological materials have tensile failure strengths in the 10^6-10^7 N/m^2 range, with the strongest biological materials such as wet compact bone having a failure strength of ~10^8 N/m^2, all of which compare poorly to ~10^9 N/m^2 for good steel, ~10^{10} N/m^2 for sapphire, and ~10^{11} N/m^2 for diamond and carbon fullerenes, again showing a 10^3-10^5 advantage for mechanical systems that use nonbiological materials. Nonbiological materials can be much stiffer, permitting the application of higher forces with greater precision of movement, and they also tend to remain stable over a larger range of temperature, pressure, salinity and pH. Proteins are heat sensitive in part because much of the functionality of their structure is due to noncovalent bonds involved in folding, which are broken more easily at higher temperatures; in diamond, sapphire, and many other rigid materials, structural shape is covalently fixed, hence is far more temperature-stable. Most proteins tend to become dysfunctional at cryogenic temperatures, unlike diamond-based mechanical structures. Biomaterials are not ruled out for all nanomechanical systems, but represent only a small subset of the materials that can be used in nanorobots. Mechanical systems can employ a wider variety of atoms and molecular structures in their design and construction, with novel functional forms that might be difficult to implement in a biological system such as steam engines or nuclear power. As another example, an application requiring the most effective bulk thermal conduction possible should use diamond, the best conductor available, not some biomaterial with inferior thermal performance.

Nondegradation of treatment agents

Diagnostic and therapeutic agents constructed of biomaterials generally are biodegradable in vivo, although there is a major branch of pharmacology devoted to designing drugs that are moderately non-biodegradable—anti-sense DNA analogues with unusual backbone linkages and peptide nucleic acids (PNAs) are difficult to break down. However, suitably designed nanorobotic agents constructed of nonbiological materials are not biodegradable. An engineered fibroblast may not stimulate an immune response when transplanted into a foreign host, but its biomolecules are subject to chemical attack *in vivo* by free radicals, acids, and enzymes. Even "mirror" biomolecules or "Doppelganger proteins" comprised exclusively of unnatural D-amino acids have a lifetime of only ~5 days inside the human body. Nonbiological materials such as diamond and sapphire are highly resistant to chemical breakdown or leukocytic degradation *in vivo*, and pathogenic biological entities cannot easily evolve useful attack strategies against these materials.

Control of nanomedical treatment

Present-day biotechnological entities are not programmable and cannot be switched on and off conditionally during task execution. A digital biocomputer, while possible in theory, represents a considerable conceptual departure from the usual biological paradigm. Even assuming that a digital biocomputer could be installed in a fibroblast, and that appropriate effector mechanisms could be attached, such a system would necessarily have slower clock cycles, less capacious memory per unit volume, and longer data access times, implying less diversity of action, poorer control, and less complex executable programs than would be available in nanoscale electromechanical computer systems. The mechanical approach emphasizes precise control of action, including control of physical placement, timing, strength, structure, and interactions with other (especially biological) entities. The biological approach emphasizes the use of poorly controlled natural structures, needlessly sacrificing huge blocks of the available functionality and design space.

Nanodevice versatility

Mechanical systems can readily incorporate biological elements if necessary, but artificial biological systems can incorporate nonbiological materials such as carbon nanotubes or diamond/sapphire structural elements only with difficulty, in part because biology has a more limited repertoire of "effector" mechanisms. Artificial biological systems cannot easily incorporate nonbiological materials where desired because natural biological assembly methods make no provisions for these materials either in the coded instructions in DNA or in the attachment chemistries. Rebuilding or reconstructing the human body with nonbiological components (e.g. fullerene encabled bone for bone damaged in an accident), or augmentation of human body function with unnatural abilities or features (e.g. autogenous paracrine control), will be very difficult or impossible to achieve using purely biotechnological means.

Avoiding overspecialization

M. Krummenacker notes that one of the most glaring shortcomings of bacteria and other naturally occurring molecular machinery—when viewed as systems subject to further engineering—is the rather limited range of molecular substrates they can utilize. For example, bacterial enzymes are highly specialized devices, with very narrow substrate specificities. Thousands of different enzymes are needed in each organism, and the substance classes capable of digestion are limited to some sugars, various amino acids including proteins that can be degraded by excreted proteases, lipids, and a few other smallish oxygen-functionalized carbon molecules such as glycerol and ethanol. Some bacteria can metabolize CO_2 and a handful of aromatic compounds, but there is a vast range of organic chemicals that most bacteria cannot degrade or manufacture. A much smaller set of substantially more general molecular tools can probably be designed using the mechanical approach; mechanosynthesis can fabricate and assemble, or disassemble, a far wider range of molecular structures than are available to the cellular machinery of life.

Faster and more precise diagnosis

The analytic function of medical diagnosis requires rapid communication between the injected devices and the attending physician. If limited to chemical messaging, biotechnology devices will require minutes or hours to complete each diagnostic loop. Nanomachines, with their more diverse set of input-output mechanisms, can outmessage the results of *in vivo* reconnaissance or testing literally in seconds. Such nanomachines can also run more tests of greater variety in less time. Mechanical nanoinstruments, including molecule-by-molecule disassemblers, will make comprehensive cell mapping and cell interaction analysis possible. Bacterial resistance can be assayed at the molecular level, allowing new treatment agents to more easily be composed, manufactured and immediately deployed.

More sensitive response threshold for high-speed action

Unlike natural systems, an entire population of nanobiotic devices can be triggered globally by just a single local detection of the target antigen or pathogen. The natural immune system takes $>10^5$ sec to become fully engaged after exposure to a systemic pathogen or other antigen-presenting intruder. A biotechnologically enhanced immune system that can employ the fastest natural unit replication time ($\sim 10^3$ sec for some bacteria) will require $\sim 10^4$ sec for full deployment post-exposure. By contrast, a nanobiotic immune system can probably be fully engaged (though not finished) in at most two blood circulation times, or $\sim 10^2$ sec.

More reliable operation

Engineered macrophages would probably individually operate less reliably than mechanical nanorobots. Many pathogens, such as *Listeria monocytogenes* and *Trypanosoma cruzi*, are known to be able to escape from phagocytic vacuoles into the cytoplasm; while biotech drugs or cell manufactured proteins could be developed to prevent this (e.g. cold therapy drugs are entry-point blockers), nanorobotic trapping mechanisms can be more secure. Proteins assembled by natural ribosomes typically incorporate one error per $\sim 10^4$ amino acids placed; current gene and protein synthesizing machines utilizing biotechnological processes have similar error rates. A molecular nanotechnology approach will improve error rates by at least a millionfold. Mechanical systems can also incorporate sensors to determine if and when a particular task needs to be done, or when a task has been completed. Finally, it is unlikely that natural organisms will be able to infiltrate mechanical nanorobots or to co-opt their functions. By contrast, a

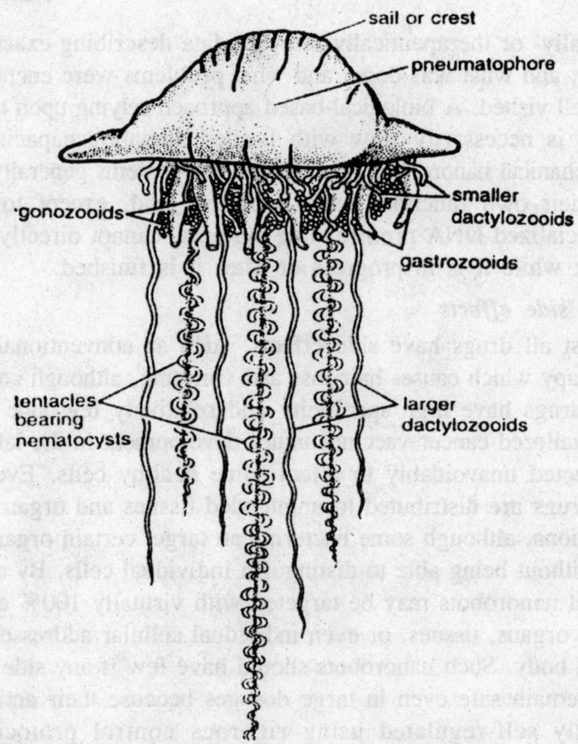

Fig. 4.5. Portuguese man-of-war.

biological-based robot could be diverted or defeated by microbes that can piggyback on its metabolism, interfere with its normal workings, or even incorporate the device wholesale into their own structures, causing the engineered biomachine to perform some new or different function than was originally intended. There are many examples of such co-option among natural biological systems, including the protozoan mixotrichs found in the termite gut that have assimilated bacteria into their bodies for use as motive engines, and the nudibranch mollusks (marine snails without shells) that steal nematocysts (stinging cells) away from coelenterates such as jellyfish (i.e. a Portuguese man-of-war) and incorporate the stingers as defensive armaments in their own skins, a process which S. Vogel has called "stealing loaded guns from the army."

Verification of progress and treatment

Using a variety of communication modalities, nanorobots can report back to the attending physician, with digital precision, a summary of diagnostically- or therapeutically-relevant data describing exactly what was found, and what was done, and what problems were encountered, in every cell visited. A biological-based approach relying upon chemical messaging is necessarily slow with limited signaling capacity. Also unlike mechanical nanorobots, biotechnological systems generally cannot monitor their own functions while working, and, except for a few highly specialized DNA proofreading systems, cannot directly inspect their work while it is in progress or after it is finished.

Minimum side effects

Almost all drugs have side effects, such as conventional cancer chemotherapy which causes hair loss and vomiting, although computer-designed drugs have high specificity and relatively few side effects. Carefully tailored cancer vaccines under development in the late 1990s were expected unavoidably to affect some healthy cells. Even well-targeted drugs are distributed to unintended tissues and organs in low concentrations, although some bacteria can target certain organs fairly reliably without being able to distinguish individual cells. By contrast, mechanical nanorobots may be targeted with virtually 100% accuracy to specific organs, tissues, or even individual cellular addresses within the human body. Such nanorobots should have few if any side effects, and will remain safe even in large dosages because their actions can be digitally self-regulated using rigorous control protocols that affirmatively prohibit device activation unless all necessary preconditions have been, and continuously remain, satisfied. G.M. Fahy has noted that these possibilities could transform "drugs" into "programmable machines with a range of sensory, decision-making, and effector capabilities [that] might avoid side effects and allergic reactions... attaining almost complete specificity of action....Designed smart pharmaceuticals might activate themselves only when, where, and if needed." Additionally, nanorobots may be programmed to excuse themselves from the site of action, or even from the body, after a treatment is completed; by contrast, spent biorobotic elements containing ingested foreign materials may have more limited post-treatment mobility, thus lingering at the worksite causing inflammation when naturally degraded or removed.

Reduced replicator danger

Drexler points out that living systems are evolved systems, while nanomechanical replicators would be designed: "The former are shaped to serve the goal of their own survival and replication in a natural environment, whereas the latter will be shaped (whether well or poorly) to serve human goals, perhaps in an artificial environment." Genetic engineering involves not design of replicators from scratch, but tinkering with the molecular machinery of existing bioreplicators. Since bioreplicators were not designed, they are not necessarily structured in ways that lend themselves to complete understanding, and processes based on diffusion and matching allow complex nonlocal interactions that can be hard to trace. Bioreplicators can be crippled, but having evolved in nature, they resemble systems that can survive in nature. Typically, they are able to exchange genetic information with wild organisms, raising the possibility of the introduction of new, unconstrained replicators in the natural environment. Having evolved to evolve, they have a capacity for further evolution—to serve their own survival, not human goals. For example, even when stripped of key pieces of DNA to interfere with its replication powers, a live attenuated AIDS vaccine can slowly recover its virulence and can attack immune cells. R. Bradbury suggests that artificial biorobots could incorporate multiple fail-safe mechanisms including required external essential nutrients or suicide suppressors, self-destruct triggers, countdown timers like telomeres, and engineering to reject foreign DNA (the basis for restriction enzymes), but it remains logically easier for a system with inchoate capacity to evolve to resume doing so, than for a system which has never had this capability to spontaneously develop it.

In contrast, nanomechanical replicators (e.g. assemblers) will be designed from scratch and thus will differ fundamentally from biological systems. The parts and structures of designed mechanical systems will be known, and the relationships among their parts will also be designed and fixed. More important, nanoreplicators will be fundamentally alien to the biosphere, unrelated to anything that has evolved to survive in nature. Certainly the capacity to fail can appear by accident, and emergent capabilities cannot be completely ruled out, but engineering experience shows that the ability to perform complex organized activities (such as replication in a natural environment) does not normally appear spontaneously. Also, and purely as a geometrical consideration, adding a new part inside a densely organized geometric structure (as would be found in a nanomachine) typically requires changes in the relative positions of many other parts, and hence corresponding adjustments in design. Adding a part inside a densely organized topological structure (as found in living systems) typically leaves topologies unchanged—room can be made by stretching and shifting other parts, with no change in their essential design, hence permitting easier modification to biological structures by exogenous agencies.

Note that mechanical medical nanodevices need not be capable of replication. There is no requirement for replication *in vivo*; such replication would be needlessly dangerous, and adding this capability would reduce effectiveness in carrying out the primary medical task. Analogously, viral vectors employed in genetic therapies are modified to be "incapable" of replication.

Assured patentability

Microscopic biorobots, unavoidably derived from natural biological material, may someday be deemed unpatentable under a general prohibition on "genetic colonialism" or other emerging legal doctrines. In contrast, mechanical nanorobots, being fully-artificial and designed machines, should always be patentable provided they satisfy the customary legal criteria.

Naturophilia

As already noted, living things in general and the human body in particular are awesome examples of a powerful and intricately woven natural molecular technology to which human engineers, in 1998, still aspire. But embracing Nature is not the same as finding her perfect. Today, the word "natural" has acquired a strong connotation of rightness, even of sanctity. For most of human history, notes biologist Steven Vogel, "the natural and human worlds stood opposed. Nature was something to be tamed and utilized; we had the ordinary attitude of organisms toward other species. Nowadays the natural world intrudes far less but gets venerated far more. And why not? When one's meat is bought in a store, when locusts don't threaten one's corn crop, when central heating and plumbing are the norm, the aesthetics of nature hold greater appeal." And so we embrace the natural rectitude or moral superiority of nature's ways, a kind of pantheism which may be called ethical naturalism, biophilia, or naturophilia.

Many great minds have fallen prey to naturophilia. In the 4th century BC, Aristotle wrote that "if one way be better than another, that you may be sure is Nature's way." In the 15th century, we have from Leonardo da Vinci: "Human ingenuity may make various inventions, but it will never devise any inventions more beautiful, nor more simple, nor more to the purpose than Nature does; because in her inventions nothing is wanting and nothing is superfluous."

However, as Virginia Postrel notes in *The Future and Its Enemies*: "If nature is itself a dynamic process rather than a static end, then there is no single form of 'the natural.' An evolving, open-ended nature may impose practical constraints, but it cannot dictate eternal standards. It cannot determine what is good. The distinction between the artificial and the natural must lie not in their source—human or not—but in their characteristics, in the way they relate to the world around them."

According to the dictionary, "artificial" usually means "made by man, rather than occurring in nature." More usefully, Herbert Simon defines the artificial as that which is designed, expresses goals, and possesses external purposes. The artificial is controlled and serves its creators' purposes, subject to the universal laws of physics. Kevin Kelly defines the natural as "out of control." Nature is evolved, not designed, and serves no goal or external purpose save its own survival. Nature, lacking intent, is amoral—it simply is. By building the artificial, observes Postrel, "we do not overthrow nature, but cooperate with it, using nature's own art to create new natural forms. Our artifice alters the path of nature, but it does not end it, for nature has no stopping point, no final shape. It is a process, not an end."

Some naturophilist writers have decried the increasing "medicalization of society" in which formerly natural functions have come to be regarded as medical conditions requiring intervention or treatment. However, history suggests that naturophilia is usually undermined by any new medical technology that offers clear, safe, and immediate benefits to patients. For example, prior to 1842, intense pain was viewed as the natural outcome of being cut with a scalpel during surgery. It had always been so—how could it ever be otherwise? The invention of anesthesia in 1842 suddenly altered this natural outcome and replaced it with a less painful artificial outcome, despite anguished cries from naturophiles within the medical community that eliminating pain might somehow diminish the human character.

Another example of a widely-accepted medicalization of normal function is childbirth, a quite natural activity that can nevertheless be very dangerous to the mother's health. Precise prehistoric death rates are unknown, although archeological evidence shows that Neandertal females tended to die before the age of 30 due to hazards of childbirth. In the worst 19th century maternity hospitals the natural death rate from childbirth was 9-10%, falling to a very artificial 0.4% rate in England by 1930 and to less than 0.01% in the U.S. during the 1990s. As a result, it now seems "natural" for a woman always to survive childbirth, even though the reverse may have been true for most of human history. Warns historian Roy Porter: "We should certainly not hanker after some mythic golden age when women gave birth naturally, painlessly, and safely; the most appalling Western maternal death rate today is among the Faith Assembly religious sect in Indiana, who reject orthodox medicine and practice home births; their perinatal mortality is 92 times greater than in Indiana as a whole."

A disease seems "natural" to those who suffer from it when no treatment exists. But once a treatment is discovered and is widely employed, the disease becomes rare and its absence now becomes "natural." To those in the past, writes K.E. Drexler, "the idea of cutting people open with knives painlessly would have seemed miraculous, but surgical anesthesia is now routine. Likewise with bacterial infections and antibiotics, with the eradication of smallpox, and the vaccine for polio: each tamed a deadly terror, and each is now half-forgotten history. What amazes one generation seems obvious and even boring to the next. The first baby born after each breakthrough grows up wondering what all the excitement was about." In the next century, says Charles Sheffield, "our descendants will look on angiograms, upper and lower GIs, and biopsies the way we regard the prospect of surgery without anesthetics."

Future generations who take for granted an all-pervasive nanomedicine in their lives may look aghast upon the 20th century, wondering among other things how we managed to retain productive focus given the constant annoyance of our numerous undiagnosed minor disease states. Most of these diseases are not yet recognized as such, and many are still regarded as "natural" and not worthy of treatment. In a few decades, this may change. Some examples:

Addictions

In 1998, many people laugh off seemingly harmless addictions to chocolate (*chocoholics*), fats or sugar (*sweet tooth*), food (*gluttony*), nicotine (*smokers*), caffeine (*coffee* and *cola drinkers*), work (*workaholics*), exercise (*runner's high*), telling falsehoods (*pathological liars*), gambling (*wagerphilia*), stealing (*kleptomania*), medical treatments (*hypochondria*), marriage (*polygamy*), power-seeking (*domination*), skydiving or bungee-jumping (*thrillseeking*), superstition (*astrology*), shopping (*spendoholics*), driving cars that kill 40,000 Americans per year (*mobilophilia*), unusual sexual preferences (bestiality), sexual activity (*nymphomania*, *satyriasis*), or pregnancy (*gravidophilia*). Without making any value judgements, it is highly likely that most or all of these addictions have genetic or physiological components which, once properly modified, can greatly reduce or eliminate the addiction if so desired. Many on the list are already suspected to have genetic components, much like *schizophrenia*, drug abuse, *bulimia*, and *alcoholism* (*dipsomania*).

Allergies and intolerances

A food allergy is an allergic reaction to a particular food, although true food allergies are much rarer than is generally believed. In the cases of milk, eggs, shellfish, nuts, wheat, soybeans, and chocolate, sufferers may lack an enzyme necessary for digesting the substance. In other cases, dust particles, plant pollens, pet danders, drugs, or foods may be allergens for natural IgE-mediated immunosensitivity. Intolerance, a much more common condition, is any undesirable effect of eating a particular food, including gastrointestinal distress, gas, nausea, diarrhea, or other problems. *Urticaria* (hives), *angioedema* and even mild *anaphylaxis* are common reactions to various drugs, insect stings or bites, allergy shots, or certain foods, particularly eggs, shellfish, nuts and fruits. Physical allergies to ordinary stimuli such as cold, sunlight, heat, pollen, pet dander, or minor injury can produce itching, skin blotches, pimples, and hives.

Minor physical annoyances

In a world where most major medical maladies are readily treated, numerous minor medical conditions which today escape our notice will rise up from obscurity and present themselves annoyingly to our conscious minds, demanding attention. These conditions may be of several kinds. First is cosmetics, including small moles, freckles and blemishes on the skin; broken fingernails or unevenly-growing cuticles; minor skin reddenings or pimples; old childhood scars, wrinkled skin, birthmarks or stretch marks; unwanted hair growth in unusual places, or differential hair colour or texture growing in patches; fingerprint patterns that are aesthetically unappealing; and mismatched leg lengths, hands with different left/right ring sizes, an asymmetrical face, or lopsided breasts. Second is minor aches and pains, which may include headaches; eyebrow hairs trapped in the eyeball conjunctiva; bent-hair pain (*folliculalgia*); dyspepsia; creaking limb joints and stomach growling; ingrown nails and hairs; earwax plugs and temporary tinnitis; chapped lips, canker sores and heat rashes; stuffy nose or gritty eyes upon rising in the morning; dermal chafing marks from elastic bands in clothing; minor flatulence; PMS (premenstrual syndrome); a leg or arm "falling asleep" in certain postures; nervous tics, itches, and twitches; uncracked knuckles, stiff neck, or backache; blocked middle ear following descent from high altitude; restless leg syndrome (*akathisia*); rotationally-induced dizziness, as on an amusement park ride; or rock-and-roll neck, wherein active musical performers or listeners bob their heads violently, rupturing small blood vessels in the neck. Third is minor physical or functional flaws, such as poor stream during male urination, female papillary leakage, colourblindness, snoring, unpleasant body odors, nosebleeds, declining visual or aural acuity, handedness (currently ~90% dextromanual, ~10% sinistromanual, mild strabismus (eyeball misalignment), bad moods (neurotransmitter imbalances), or post-intoxication hangover.

Undiscovered infectious agents

Peptic ulcers once were thought to result from a stressful life, a purely natural response to a lifestyle choice. Then it was found that the major cause of ulcers is the presence of *Helicobacter pylori* bacteria in the stomach. Bacteria have been implicated in some cases of atherosclerosis and *Alzheimer's disease*, and nanobacteria have been proposed as possible nucleation sites for kidney stones. Other seemingly natural but undesirable conditions may also be due to undiscovered microbial agents, especially since bacteria outnumber tissue cells in our highly infested 20th century bodies by more than 10:1.

Unwanted syndromes

Syndromes are groups of related symptoms and signs of disordered function that define a disease whose cause remains unknown, that is, idiopathic. A good example is irritable bowel syndrome (IBS), which affects up to 20% of the adult U.S. population and includes symptoms of abdominal distention and pain, with more frequent and looser stools. Many are unaware they are afflicted. In 1998 there was no known cause or simple complete treatment for this still "natural" disease. Even more mysterious than IBS is our general activity level—some people seem to have high-energy personalities, while others have more phlegmatic low-energy personalities. Either may be regarded as "natural," but nanomedicine can probably bring this ill-defined neurophysiological variable under human control. The need to sleep is another imperfectly understood syndrome. It is experienced by everyone and thus was universally regarded as "natural" in the 20th century. Physiological short sleepers were unusual, insomnia or asomnia was thought of as an abnormal state, and there were a few anecdotal but medically undocumented instances of total nonsomnia, such as the celebrated case of Al Herpin.

Psychological traits

Psychological traits which, if identified by a patient as undesirable, might be subject to genetic or physiological modification could include: sexual preference (6-10% of the adult population is homosexual); shyness or boldness; acquisitive or altruistic propensity; misanthropy or philanthropy; theistic or atheistic orientation; loquacity or dourness; childhood imprinting; criminal propensity (up to 1-5% of the population); various recognized personality disorders that affect ~10% of the population such as antisocial, paranoid, schizotypal, histrionic, narcissistic, avoidant, dependent, obsessive-compulsive, and passive-aggressive disorders; panic attacks (experienced at least once by ~33% of all adults each year); and phobic disorders such as social phobias (~13% of the population), specific phobias including fear of large animals (*zoophobia*), snakes (*ophidiophobia*), spiders (*arachnophobia*), needles (*belonephobia*, ~10%), the dark (*noctiphobia* or *scotophobia*), or strangers (*xenophobia*) (total ~5.7%), the fear of blood or *hemophobia* (~5%), *agoraphobia* (~2.8%), and other unusual phobias such as the fears of certain colours (*chromophobia*), daylight (*phengophobia*), girls (*parthenophobia*), men (*androphobia*), stars in the sky (*siderophobia*), the number thirteen (*triskaidekaphobia*), and even the fear of developing a phobia (*phobophobia*).

The above sampling of minor afflictions, almost all considered "natural" in 1998, may come to be regarded as commonplace correctable medical conditions in the nanomedical era. By the time such petty annoyances are deemed worthy of immediate treatment, biotechnology and nanomedicine already will have defeated the most fearsome illnesses of the late 20th century and will have moved on to other challenges. Naturophiles may dissent, but the emerging trend from medical biotechnology is to characterize health, not as a static standard, but rather as a condition defined by the lives that people want to lead. Affirming the volitional normative model of disease, Virginia Postrel concludes:

"Different goals will produce different choices about trade-offs and standards. What makes a condition unhealthy is not that it is unnatural but that it interferes with human purposes. Revering nature [would mean] sacrificing the purposes of individuals to preserve the world as given. It [would require] that we force people to live with biological conditions that trouble them, whether diseases such as cystic fibrosis or schizophrenia, disabilities such as myopia or crooked teeth, or simply less beauty, intelligence, happiness, or grace than could be achieved through artifice. In a world where it's no big deal to take hormone therapy, Viagra, or Prozac, to have a face lift, or to know a child's sex before birth, a world in which even such radical interventions as sex-change operations and heart transplants have failed to turn society upside down, it is extremely difficult to argue that medical innovations are dangerous simply because they fool Mother Nature."

5

MOLECULAR MANUFACTURING

Most contemporary industrial fabrication processes are based on "top-down" technologies, wherein small objects are sawn or machined from larger objects, or small features are imposed on larger objects, in either case by removing unwanted matter. The results of such processes may be small, such as micron-featured integrated circuits, or very large, such as jet aircraft, but in most cases the material is being processed in chunks far larger than molecular scale.

Molecular manufacturing, on the other hand, represents a "*bottom-up*" technology. Desired products will be built directly by "assembler" machines, molecule by molecule, making larger and larger objects with atomic precision. The results of such processes may also be very small or very large, much as biology builds both micron-sized bacteria and 100-meter tall sequoia trees. However, since assemblers add matter only where it is intended, little need be removed and hence there may be minimal waste during the process. By guiding with precision the assembly of molecules and supra-molecular structures, such a manufacturing system could construct an extraordinarily wide range of products of unprecedented quality and performance.

In 1959, the Nobel physicist Richard P. Feynman observed that:

"The principles of physics, as far as I can see, do not speak against the possibility of maneuvering things atom by atom. It is not an attempt to violate any laws; it is something, in principle, that can be done; but, in practice, it has not been done because we are too big ... Ultimately, we can do chemical synthesis."

"A chemist comes to us and says, 'Look, I want a molecule that has the atoms arranged thus and so; make me that molecule.' The chemist does a mysterious thing when he wants to make a molecule. He sees that it has got that ring, so he mixes this and that, and he shakes it, and he fiddles around. And, at the end of a difficult process, he usually does succeed in synthesizing what he wants."

"But it is interesting that it would be, in principle, possible (I think) for a physicist to synthesize any chemical substance that the chemist writes down. Give the orders and the physicist synthesizes it. How? Put the atoms down where the chemist says, and so you make the substance. The problems of chemistry and biology can be greatly helped if our ability to see what we are doing, and to do things on an atomic level, is ultimately developed a development which I think cannot be avoided."

Nearly 40 years after Feynman's famous "Plenty of Room at the Bottom" speech, and a decade after Drexler's original proposal for a bottom-up approach to machinebuilding using molecular assemblers, Nobel chemist Richard Smalley also largely agreed that this objective should prove feasible. Noted Smalley: "On a length scale of more than one nanometer, the mechanical robot assembler metaphor envisioned by Drexler almost certainly will work..."

Many skeptical questions arise when one first encounters the ideas of molecular nanotechnology and molecular assemblers. It is useful to keep in mind the proven feasibility of such systems in the biological tradition over billions of years of natural evolution. In one sense, molecular nanotechnology will be a refinement and expansion upon how nature works at the molecular scale. Nature's examples, such as human beings, certainly answer the most basic skeptical questions, such as: Can macroscopic objects be built from molecular scale processes? (Yes, thanks to cellular replication.) Are molecular objects stable? (Of course; the human population alone contains $>10^{29}$ reasonably stable and quite functional ribosomal "nanomachines"). Observes Nobel chemist Jean-Marie Lehn: "The chemist finds illustration, inspiration and stimulation in natural processes, as well as confidence and reassurance since they are proof that such highly complex systems can indeed be achieved on the basis of molecular components."

What about quantum effects? The uncertainty principle makes electron positions somewhat fuzzy, but the atom as a whole has a comparatively definite position set by the relatively great mass of the atomic nucleus. The quantum probability function of electrons in atoms tends to drop off exponentially with distance outside the atom, giving atoms a moderately sharp "edge". Mathematically, the positional uncertainty of a single carbon atom of mass $m_C = 2 \times 10^{26}$ kg bound in a single C-C bond of stiffness $k_C = 440$ N/m may be crudely estimated from the classical vibrational frequency $n_C = (k_C/m_C)^{1/2} = 1.5 \times 10^{14}$ Hz. This sets the zero-point vibrational bond energy $E_C = h\, n_C / 2 = 4.9 \times 10^{20}\ \mathrm{J} = k_C\, \Delta X_C^{\,2} / 2$ where $h = 6.63 \times 10^{34}$ J-sec (Planck's constant) and $\Delta X_C \sim 0.015$ nm is the maximum classical amplitude of the bound carbon atom (roughly the same as the 3 dB point for the gaussian wave-function, notes J. Soreff). Thus ΔX_C is just ~5% of the typical atomic electron cloud diameter of ~0.3 nm, imposing only a modest additional constraint on the fabrication and stability of nanomechanical structures. (Even in most liquids at their boiling points, each molecule is free to move only ~0.07 nm from its average position.)

How about the effects of Brownian bombardment on nano-machines? Describing a nanomechanical component as a harmonic oscillator embedded in a gas, Drexler notes: "At equilibrium, an impinging gas molecule is as likely to absorb energy as to deliver it, and so molecular bombardment has no net effect on the amplitude of vibration. How a system is coupled to a thermal bath can affect its detailed dynamics, but not the statistical distribution of dynamical quantities." Nano-machines must also obey the laws of thermodynamics. Nanorobots cannot be used as a "Maxwell's demon" —energy must be expended to do useful work; there can be no free lunch.

Will high-energy radiation damage nanomachines? Radiation can break chemical bonds and disrupt molecular machines. The annual failure rate for a properly designed ~1 micron^3 molecular machine (containing many billions of atoms) can be made as low as several percent. This estimate is derived from a consideration of all known thermal, photochemical, radiation, and other damage mechanisms, but the dominant error mechanism that appears difficult to substantially reduce is damage caused by background radiation. Failure rates are not zero, but are nonetheless remarkable by today's standards, especially given error detection and correction at the module level.

During nanodevice fabrication, the problems of crossbonding and reactive intermediates pose additional constraints that must be dealt with using appropriate assembler designs. Specifically, assemblers with rigid positioning components operating in vacuo can restrict reactive intermediates and molecular tool tips to desired locations with a precision of at least ~0.1 nm, or slightly less than an atomic diameter.

Within the confines of an evacuated or "eutactic" assembler workspace, the location of every structural atom may be known to within the uncertainty created by thermal noise. Internal structural components may possess relatively high stiffness, so positional uncertainty caused by thermal noise may be small. In the classical analysis, positional variance $\Delta X^2 = kT / k_s$, where $k = 1.381 \times 10^{21}$ J/kelvin (*Boltzmann's constant*), T is temperature in kelvins, and k_s is component stiffness, typically 10 N/m for nanometer-scale diamondoid components, hence the standard deviation of component position is ~0.02 nm at room temperature, or ~0.01 nm at cryogenic (e.g., LN_2) temperatures. By employing sufficiently stiff mechanisms the position of every structural atom in the system can be known to within a fraction of an atomic diameter with high reliability and without the need for explicit positional sensing. Steric tool hindrance—in molecular manufacturing, the conical volume occupied by a tool tip or manipulator platform—is a technical design problem that may be overcome by using sufficiently stiff tools or multi-tip tools, working *in vacuo*. Additionally, while some synthetic methods might involve the manipulation of individual atoms by appropriate tools, such as the hydrogen abstraction tool which removes a single selected hydrogen atom from a diamondoid surface, other reactions will involve small clusters of atoms or larger molecular components.

What about friction and wear among nanomechanical components? Properly designed molecular machines lack wear mechanisms, although other damage mechanisms remain. Dry bearings with atomically precise surfaces can have negligible static friction and wear, and very low dynamic friction or drag. Nanomechanical components are better viewed as moving smoothly in a force field than as sliding subject to friction. For example, total drag power dissipated by an isolated molecular sleeve bearing of radius 2 nm spinning at 1 MHz is dominated by band-stiffness scattering, amounting to 0.000004 picowatts (pW), very small compared to the 1-1000 pW power draw that will be typical of micron-size medical nanorobots. The strong precise surfaces of nanomachines experience no change during a typical operational cycle, hence zero wear. Within the single-point failure model, the first step in a wear process (e.g., a dislocated atom) is regarded as fatal, hence cumulative wear plays no role in determining device lifetimes. In eutactic nanomachines, contaminants of all kinds are rigorously excluded. The equivalent of wear particles cannot appear until the device has failed. On this scale, repulsive fields provide "lubrication" and an oil molecule would be a contaminant object, not a lubricant.

The general conclusion is that molecular nanotechnology violates no physical laws. In 1998, it was generally accepted that molecular nanotechnology will be developed, although there remains some disagreement about how long it will take. To progress from today's limited capabilities to the complex nanorobots described later in this trilogy will clearly require a great deal of research and development effort. Still, the proper question is no longer "if" but "when." This Chapter presents a few of the many possible technical paths leading to a working molecular assembler that were being discussed in 1998. The reader is strongly urged to consult the most current literature for the latest results. Gimzewski and Joachim note that "bottom-up approaches to nanofabrication may one day compete with conventional top-down approaches in providing nanotechnologies for the next millenium. Top-down refers to increasing miniaturization through extension of existing microfabrication schemes, [whereas] the bottom-up scenario is one of ever-increasing complexity at the molecular level while maintaining control on an atom-by-atom basis".

Top-Down Approaches to Nanotechnology

Nanotechnology was first proposed by Nobel physicist Richard Feynman in December 1959, in a talk in which he also issued a seemingly "impossible" challenge to build a working electric motor no larger than a 1/64th-inch (400-micron) cube, backed by a $1000 prize to spur interest in the new field. Just 11 months later, engineer William McLellan had constructed a 250-microgram 2000-rpm motor out of 13 separate parts and collected his reward. (McLellan's entire motor is only as big as the period at the end of this sentence.) In 1995, a $250,000 Feynman Grand Prize was made available, this time sponsored by the Foresight Institute, for any engineer who could build the first programmable nanometer-scale robotic arm. How long will it take for history to repeat?

In his famous 1959 talk, Feynman proposed the prototypical top-down strategy for building complex nanomachinery—essentially a completely teleoperated machine shop, including mills, lathes, drills, presses, cutters, and the like, plus master-slave grippers to allow the human operator to move parts and materials around the workshop.

To build a nanomachine using Feynman's scheme, the operator first directs a macroscale machine shop to fabricate an exact copy of itself, but four times smaller in size. After this work is done, and all machines are verified to be working properly and as expected, the reduced-scale machine shop would be used to build a copy of itself, another factor of four smaller but a factor of 16 tinier than the original machine shop. This process of fabricating progressively smaller machine shops proceeds until a machine shop capable of manipulations at the nanoscale is produced. The final result is a nanomachine shop capable of reconstructing itself, or of producing any other useful nanoscale output product stream that is physically possible to manufacture, using molecular feedstock.

As Feynman describes the process:

"When I make my first set of slave hands at one-fourth scale, I am going to make ten sets. I make ten sets of hands, and I wire them to my original levers so they each do exactly the same thing at the same time in parallel. Now, when I am making my new devices one-quarter again as small, I let each one manufacture ten copies, so that I would have a hundred hands at the 1/16th size....If I made a billion little lathes, each 1/4000th the scale of a regular lathe, there are plenty of materials and space available because in the billion little ones there is less than two percent of the materials in one big lathe. It doesn't cost anything for materials, you see. So I want to build a billion tiny factories, models of each other, which are manufacturing simultaneously, drilling holes, stamping parts, and so on."

"As we go down in size, there are a number of interesting problems that arise. All things do not simply scale down in proportion. There is the problem that materials stick together by the molecular (van der Waals) attractions. There will be several problems of this nature that we will have to be ready to design for....[But] if we go down far enough, all of our devices can be mass produced so that they are absolutely perfect copies of one another. We cannot build two large machines so that the dimensions are exactly the same. But if your machine is only 100 atoms high, you only have to get it correct to one-half of one percent to make sure the other machine is exactly the same size—namely, 100 atoms high!"

Progress is being made on Feynman's top-down approach in a relatively new engineering field known as Micro Electro-Mechanical Systems or MEMS, originally an extension of chip-etch technology rather than micromanipulation. Conventional purely-electronic devices fabricated on silicon chips with ~0.2 micron features must be coupled to external systems to produce mechanical effects, but MEMS devices integrate mechanical components directly with electrical circuitry. Thus while microelectronic chips merely route electrons, microelectro-mechanical systems give the electronics immediate access to control applications, enabling single microdevices to interact directly with the physical world.

Generic microsystems research has proven to be a rewarding activity for two decades, with each successively smaller motor or machine widely publicized. The field gathered momentum after the first working micromotors were demonstrated in the late 1980s by groups at Berkeley and MIT. By 1990, tiny electrostatic motors with 100-micron rotors displayed operating speeds of ~250 Hz and operating lifetimes of ~10^6 revolutions. A new *Journal of Micromechanics and Microengineering* appeared in 1991, followed by the *Journal of Microelectromechanical Systems* in 1992 and the first MEMS Application Symposium in Tokyo in June 1996. By the mid-1990s, Sandia's Microelectronics Development Laboratory, one of the most prominent vendors of micromachines, could mass-produce planar micromechanical devices smaller than ~1 mm^2, incorporating motors with gears and cogs each no thicker than a human hair and turning at speeds of >4000 Hz. Wobble micromotors such as those fabricated in silicon by the University of Utah's micromachine lab demonstrated very little friction or abrasion, with measured operating lifetimes in excess of 300 million revolutions. (Diamond coatings could further reduce friction by an order of magnitude over silicon, and the wear rate of polycrystalline diamond is known to be ~10^4 times lower than for silicon.)

MEMS research has produced multi-micron-scale accelerometers, diverse microsensors (e.g., blood pressure microsensors attached to cardiac catheters), microscale cantilevers and jointed crank mechanisms, 5-micron barbs, flow microvalves and pressure microtransducers, micropistons and micropumps, microgear trains, microactuators and piezodriven micromotors, micromirrors and microshutters, Fresnel lens microarrays, microgyroscopes, a 2-mm long combustion chamber suitable for turbine use (making possible a 1 cm^3 gas-turbine generator that would deliver 50 W of power or 0.2 N of thrust), and multidevice microsystems that were available customized or off-the-shelf in mass quantities by 1998. Laser beams and ion beams were used to carve ~300-micron-diameter gears with 50-micron wide teeth in solid diamond. Microelectrodeposition and microcontact printing readily produced topologically complex 3-D microstructures with 1-10 micron feature sizes. Colloidal templating yielded 0.71-micron diameter hollow silica spheres with wall thicknesses from tens to hundreds of nanometers, 10-25 nm trenches, or 4-nm thick graphite sheet layers tiled on spherical surfaces ~0.2 microns in diameter. A CAD/CAM desktop micromanufacturing system invented at Lincoln Laboratory in Lexington, MA, machined three-dimensional structures, including spheres, in silicon, using computer-controlled laser/chemical etching of cubic-micron voxels at the rate of ~20,000 voxels/sec.

By 1998, smaller (~100 nm) accelerometers were being developed by such companies as Analog Devices in the U.S. and Daimler-Benz in Germany, and electronic control systems such as microswitches and microrelays were available commercially from several companies including Integrated Micromachines of California. Micromachining was a well-developed engineering discipline. Westinghouse Science and Technology Center used MEMS sensors to build a miniaturized mass spectrometer. Micro-opto-mechanical systems were under development. Chemists were proposing and building Integrated Chemical Synthesizers consisting of millimeter- to micron-sized chemical reactors, designed for specific applications, along with associated devices for moving reactant and product streams (pumps), mixing reactants (reaction chambers), analyzing streams, and separating products (separation chambers), constructed conventionally using silicon substrates and photolithographic techniques to form the desired structures but also utilizing the high selectivity of biological molecules such as enzymes and antibodies to fabricate sensors. MEMS had also led to the development of microgrippers that could manipulate individual 2.7-micron polystyrene spheres, dried red blood cells of similar size, and various protozoa. Market Intelligence Research Corp. of Mountain View, CA, estimated a total worldwide budget for micromachine research and development of $995 million in 1992 and over $3 billion by 1998. Systems Planning Corporation of Arlington, VA, projected that the microsystems industry could be worth more than $14 billion annually by 2001.

Some MEMS research followed surprising directions. For example, in the mid-1990s MEMS engineers at the University of Tokyo Department of Mechanical Engineering fabricated an 8-hinge, 8-plate wing microstructure capable of independent flight through the air. The structure was fabricated by traditional silicon micromachining techniques and was then released from the wafer surface, with the initially flat sections lifted into flight position using a micromanipulation system with glass probes several microns in diameter. Each wing was ~500 microns square; applying 300 VAC at 10 KHz caused resonant vibration and the flapping amplitude of the wings exceeded 30°. In another 4-

wing configuration, the flapping resonance frequency was ~150 Hz and the device freely flew up and down a thin glass pole when placed in an alternating electromagnetic field. Another bi-wing design made of magnetized nickel coated silicon measuring 2000 × 10,000 microns successfully flew without power supply cables and guides in the manner of a butterfly or mosquito, when an alternating magnetic field strength of >400 oersted and 12 Hz was applied; a similar device with 320 × 800 micron wings also flew in a 300 Hz magnetic field. "If we release the silicon mosquito from the silicon chip, it flies off and we cannot find it again," complained researcher H. Miura. "It's very small, like dust." MEMS "gnat robots" have also been described.

Using electron-beam (e-beam) lithography, in 1997 researchers at the Cornell University Nanofabrication Facility built what they believed were the world's smallest mechanical devices, including a 4-micron-wide Fabry-Perot interferometer and, just for fun, the world's smallest guitar, carved out of crystalline silicon and no larger than a single human cell. The "nanoguitar" was 10 microns long and 2 microns wide, with six strings each ~50 nm (~200 atoms) wide that would probably resonate at >10-100 MHz if plucked. Presentation of the nanoguitar enthralled the public, surprising at least one Cornell researcher (in another lab) who remarked that "anyone with a sufficiently advanced e-beam lithography system could make it. I make 50 nanometer silicon objects routinely. Doesn't seem like a big deal to me, but evidently most people have no idea what we're capable of these days."

Direct-write e-beam lithography uses a tightly-focused beam of electrons, steered by electromagnetic deflectors, to trace out patterns on a wafer resist surface. The electrons chemically alter the resist, which is then etched away, leaving the desired pattern. By 1998, e-beam spot sizes had been focused to less than 1 nm, and 5-nm beam widths were routine, although forward scattering of the electrons imposed a limit of resist exposure resolution of ~10 nm for e-beams and ~5 nm for focused ion beams. E-beam writing times were very slow—for example, 80 hours were required to carve a single photolithography mask for a 1G DRAM, though much of this time was required for proximity calculations by the AI (artificial intelligence) software [E.A. Rietman, personal communication, 1999] (a better approach is to use a neural net). By 1998, an electron interferometer containing atomically smooth mirrors spaced a few atomic layers apart had been fabricated, and in 1999, careful electron-energy-loss spectroscopy experiments revealed that silicon dioxide must be at least 4 atoms (~0.7 nm) thick to act as a conventional electrical insulator.

Atom lithography, employing a low-energy neutral atomic beam (usually Na or Cr) cooled and collimated by optical molasses, then concentrated at the nodes of the standing wave formed by the superposition of a laser beam and its reflection, was predicted also eventually to allow surface deposition of 5-10 nm features. The first directional atomic beam (of sodium) with a ~0.002 radian beamspread and ~femtogram pulses (~26 million atoms, or ~$(100\ nm)^3$, per pulse) was demonstrated in early 1999, with one commentator noting that "the longer dream...is atomic holography...[that] could combine beams of atoms to build a 3-D solid object". At the same time, another group achieved a continuous stream of rubidium atoms lasting up to ~0.1 sec in duration, with a beam radius potentially as small as ~1 nm. Current-carrying wires have been used to guide cold lithium atoms "just as optical fibers guide light"; the first curved focusing atom mirror also was demonstrated in early 1999. By 1998, the speculative possibility of quantum-mechanical micromanipulation also had been described.

Other approaches offered comparable results. For instance,engineers at the University of Minnesota NanoStructure Laboratory used nanoimprint lithography (NIL) in 1996 to etch out patterns of lines, grooves and circles as small as 25 nanometers in a polymer, and by 1998 NIL had been demonstrated in nanoscale quantum-wire, quantum-dot, and ring transistors with feature sizes below 10 nm. Nanochannel array glasses containing 33-nm capillaries arranged in a 2-D hexagonal close packing configuration at a number density of ~$300/micron^2$ had been available since 1992.

Thus by 1998, top-down MEMS techniques could in theory already make nanoscale (though not atomically precise) parts. Might these techniques also produce assemblies of such parts, creating complex machines? Very simple mobile robots of ~1 cm^3 volume were commonplace, so for a more challenging demonstration of MEMS' ability to manufacture complete working microrobots, in 1994 Japanese researchers at Nippondenso Co., Ltd. fabricated a 1/1000th-scale working electric car. As small as a grain of rice, the microcar was a 1/1000-scale replica of the Toyota Motor Corp's first automobile, the 1936 Model AA sedan. The tiny vehicle incorporated 24 parts, including tires, wheels, axles, headlights and taillights, bumpers, a spare tire, and hubcaps carrying the company name inscribed in microscopic letters, all manually assembled using a mechanical micromanipulator of the type generally used for cell handling in biological research. In part because of this handcrafting, each microcar cost more to build than a full-size modern luxury automobile.

The Nippondenso microcar was 4800 microns long, 1800 microns wide, and 1800 microns high, consisting of a chassis, a shell body, and a 5-part electromagnetic step motor measuring 700 microns in diameter with a ~0.07-tesla magnet penetrated by an axle 150 microns thick and 1900 microns long. Power was supplied through thin (18 micron) copper wires, carrying 20 mA at 3 volts. The motor developed a peak torque of 1.3×10^6 N-m (mean 7×10^7 N-m) at a mean frequency of ~100 Hz (peak frequency ~700 Hz), propelling the car forward across a level surface at a top speed of 10 cm/sec. Some internal wear of the rotating parts was visible after ~2000 sec of continuous operation; the addition of ~0.1 microgram of lubricant to the wheel microbearings caused the mechanism to seize due to lubricant viscosity. The microcar body was a 30-micron thick 20-milligram shell, fabricated with features as small as ~2 microns using modeling and casting, N/C machine cutting, mold etching, submicron diamond-powder polishing, and nickel and gold plating processes. Measured average roughness of machined and final polished surfaces was 130 nm and 26 nm, respectively. The shell captured all features as small as 2 mm on the original full-size automobile body. Each tire was 690 microns in diameter and 170 microns wide. The license plate was 10 microns thick, 380 microns wide and 190 microns high. Nippondenso subsequently used similar manufacturing techniques to build a prototype of a capsule intended to crawl through tiny pipes in a power plant or chemical plant like an inchworm, hunting for cracks. In 1999, three Japanese electronics companies announced the creation of a 0.42-gram, 5-mm long "ant size" robot reportedly able to lift 0.8-gram loads and to move at ~2 mm/sec, as part of the government's ongoing Micro Machine Project.

By 1998, micromanipulators with high placement precision had been demonstrated in various laboratories around the world. For example, the Spider-II micromanipulation robot employed bimorphic piezoactuators within a 260-micron cubic work volume and a three-axis gripper placement accuracy of 8 nm. A microassembly robot using stick-and-slip actuators had 5-nm resolution over a 200-nm scanning range with a maximum speed of 4 mm/sec and 0.4% repeatability at 10 KHz, delivering a maximum driving force of 155 mN and rigidity of 6.3 N/micron.

In terms of system size, micromachined MEMS devices (typically $\sim 10^{-13}$ m^3 in volume) lie exactly intermediate between the world ($\sim 10^{-4}$ m^3) and the nanoscale world ($\sim 10^{-22}$ m^3). The medical nanorobots described in this book generally will have dimensions, precisions, and component part sizes ~1000-fold smaller than the Nippondenso microcar, which itself was, in turn, ~1000-fold smaller than a macroscale automobile. It is difficult to see how MEMS techniques involving statistical materials deposition and removal could fabricate parts to better than ~10 nm feature sizes or tolerances, or could position parts during assembly to much better than ~10 nm spatial resolution, nor does it appear likely that MEMS techniques, on their own, can manufacture or position any structure to atomic precision. However, this does not rule out the possible use of MEMS techniques to fabricate crude useful parts for subsequent "finishing" to atomic precision by other techniques, or to facilitate the assembly of nanoscale components (of atomic precision or otherwise) that have be.

Bottom-up Pathways to Molecular Manufacturing

To manufacture machines of any kind generally requires two primary capabilities—fabrication of parts, and assembly of parts. As will be shown below, by 1998 at least a primitive parts fabrication and parts assembly capability had been demonstrated at the molecular level using three different enabling technologies: biotechnology, supramolecular chemistry, and scanning probes. This Section reviews the most preliminary steps and basic approaches to molecular manufacturing. The primary focus here is on bridging technologies that may lie between what can already be done and what is desired to be done—specifically, the manufacture of mechanical nanorobots. The reader is again cautioned that in 1998 this field was experiencing an explosion of interest and pathbreaking research efforts, so the current literature should be consulted for the latest results.

Biotechnology

The first broad category of enabling technology for molecular manufacturing in the mechanical tradition is biotechnology. Molecular biologists study and modify systems of molecular machines, and genetic engineers reprogram these systems, sometimes to build novel molecular objects having complex functions. By 1998, biotechnologists could make almost any DNA sequence or poly-peptide chain desired (though not extremely long aperiodic ones), assemble "devices" such as artificial chromosomes and viruses, and were actively pursuing gene therapy, all of which are examples of molecular engineering. While earlier writers had suggested that biomolecules could be used as mechanical components, Drexler evidently was the first to point out, in 1981, that complex devices resembling biomolecular motors, actuators, bearings, and structural components could be combined to build versatile molecular machine systems analogous to machine systems in the macroscopic world. "Development of the ability to design protein molecules," Drexler wrote, "will open a path to the fabrication of devices to complex atomic specifications, thus sidestepping obstacles facing conventional microtechnology. This path will involve construction of molecular machinery able to position reactive groups to atomic precision."

Perhaps the best-known biological example of such molecular machinery is the ribosome, the only freely programmable nanoscale assembler already in existence. In nature, ~8000 nm^3 ribosomes act as general-purpose factories building diverse varieties of proteins by bonding amino acids together in precise sequences under instructions provided by a strand of messenger RNA (mRNA) copied from the host DNA, powered by the decomposition of adenosine triphosphate (ATP) to adenosine diphosphate (ADP). Each ribosome is a compact ribonucleoprotein particle consisting of two subunits, with each subunit consisting of several proteins associated with a long RNA molecule (rRNA). Biotechnology can employ the ribosomal machinery of bacteria to produce novel proteins, which proteins then might serve as components of larger molecular structures.

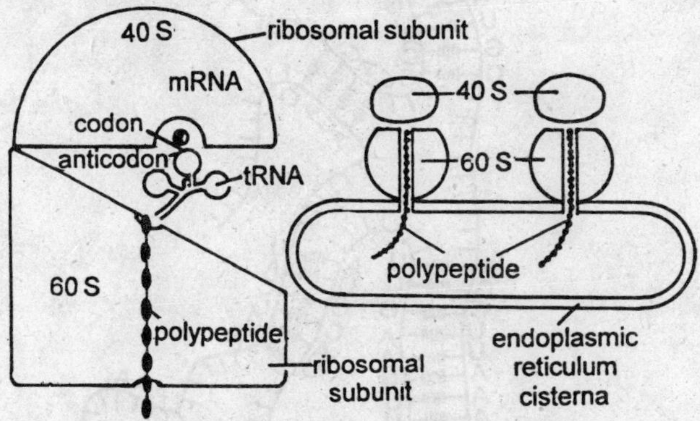

Fig. 5.1. Structure of ribosome.

Of course, such biological mechanisms have been disparaged as "slipshod devices working in spite of haphazard design." Dan Heidel, a University of Washington biomimetics engineer, agrees that biological systems are sloppy but observes that they are designed to tolerate it, and offers high praise for the lowly ribosome:

"The average ribosome sits in an aqueous environment surrounded by thousands of cosolutes at concentrations just a hair's breadth from precipitating out of solution. The ribosome itself consists of three RNA molecules that spontaneously fold into floppy spaghetti noodle piles of chemically active groups surrounded by >50 proteins of similarly dubious structure. The other solutes are bombarding this motley assemblage at tens to hundreds of miles an hour and the solvating water molecules are a constantly shifting and unpredictable network of hydrogen bonds and polar ionic charges constantly impinging upon the ribosome. The ribosome must recognize one particular cognate transfer RNA (tRNA) for the codon in its A-site out of ~60 other nearly identical tRNAs. This particular tRNA must be discriminated from all its tRNA brethren along with whatever other solutes can fit into the A-site pocket using only three base pairs for discrimination. In fact, most codons use only two of the three base pairs for recognition so some tRNAs are recognized by as few as 4 hydrogen bonds. All of

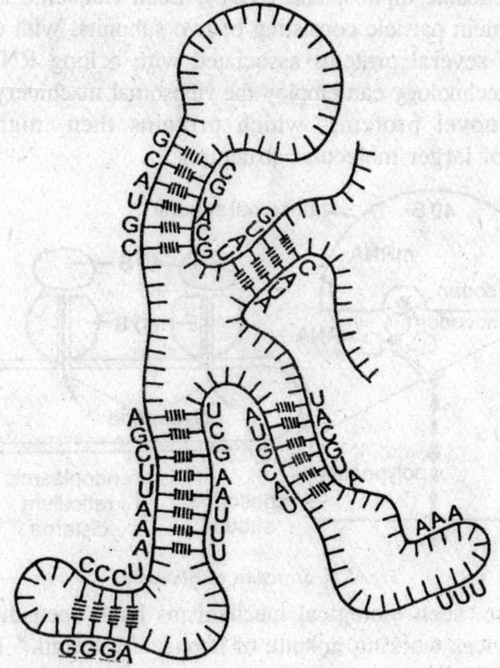

Fig. 5.2. Clover leaf model of tRNA.

this requires sub-Angstrom precision in the active sites but since the ribosome is a massive complex of subunits held together by only ionic, hydrogen, hydrophobic and Van der Waals bonds, it's like getting submillimeter accuracy from points on a bunch of taped-together water balloons being rolled down a staircase."

"This ridiculous machine can assemble proteins at a 20 Hz subunit-incorporation frequency with a 10^3 error rate. Ribosomes are even more impressive when one considers that the number of amino acids in each ribosome's own protein subunits combined with the 10^3 protein assembly error rate ensures that each ribosome has several errors in it and that each cell almost certainly has no two ribosomes exactly alike. I dare any engineer on this planet to rationally design an assembler at any scale that is able to operate at a comparable level of performance under similarly adverse operating conditions."

Thus ribosomes can manufacture, with >99.9% fidelity to arbitrary specifications, linear strings of amino acids of great length which may then fold up to produce three-dimensional protein structures performing a broad range of highly differentiated functions, providing good analogs for working nanomachines capable of forming self-assembling systems. Much of the great diversity of protein activity flows from the selectivity with which they bind to specific receptor sites on other molecules. Some of these binding mechanisms might be copied to construct molecular robotic arms, while other devices could be made by adapting naturally-occurring proteins, or by synthesizing new, custom-built ones. Protein engineering might be used to create proteins with desired qualities as building blocks for constructing the first primitive assemblers.

The field of protein engineering has its own journal of the same name, and has already achieved remarkable results in the synthesis of novel structures. Drexler cites the creation of the first de novo structure by DeGrado, the development of synthetic branched protein-like structures that depart substantially from protein models, and the engineering of a branched, nonbiological protein with enzymatic activity. Since then, numerous alpha-helical peptides have been designed *de novo*, a 20-residue artificial peptide exhibiting designed beta-sheet secondary structure has been created, and various peptide analogs have also been synthesized. By 1998, advances in computational techniques allowed the design of precise sidechain packing in proteins with naturally occurring backbone structures, and the study of de novo backbone structures had begun. The catalytic task space was being labouriously investigated at the molecular level—for example, one study of comparative enzyme structures revealed that changing as few as four amino acids converted an oleate 12-desaturase to an oleate hydroxylase, and as few as six substitutions could convert an oleate hydroxylase into an oleate desaturase. To further expand the toolkit, Mills and Fahy proposed exploiting the degeneracy of the genetic code (wherein amino acids are specified by up to 6 different codons) by assigning some codons to unnatural amino acids, allowing the creation of artificial proteins with unprecedented structural or catalytic properties. With 61 usable codons available, there is room for up to 41 novel amino acids in addition to the natural 20 amino acids; unnatural amino acids have been introduced into beta-lactamase and T4 lysozyme enzymes in a site-directed fashion, using artificial tRNAs and mRNAs in an in vitro translation system, and into an engineered peptide as a reporter group. Additionally, artificial novel base pairs can be incorporated enzymatically into DNA and other schemes are possible. By incorporating one additional base pair, the number of possible codons is expanded from $4^3 = 64$ codons to $6^3 = 216$ codons, potentially allowing up to 193 novel amino acids to be incorporated into artificial protein structures—or at least 48 new amino acids assuming that present levels of redundancy are retained and the existing 4-letter code is kept as a subset for "upward compatibility". If the genetic alphabet is increased to eight bases ($8^3 = 512$ codons), the number of available codons for novel amino acid-like molecules increases to as many as 489.

First Letter	Second Letter: U	C	A	G	Third Letter
U	UUU(Phe) UUC (Phe) UUA (Leu) UUG (Leu)	UCU (Ser) UCC (Ser) UCA (Ser) UCG (Ser)	UAU (Tyr) UAC (Tyr) UAA (Ochre) UAG (Amber)	UGU (Cys) UGC (CYS) UGA ? UGG (Tryp)	U C A G
C	CUU (Leu) CUC (Leu) CUA (Leu) CUG (Leu)	CCU (Pro) CCC (Pro) CCA (Pro) CCG (Pro)	CAU (His) CAS (His) CAA (Glun) CAG (Glun)	CGU (Arq) CGC (Arq) CGA (Arq) CGG (Arq)	 C A G
A	AUU (Heu) AUC (Heu) AUA (Aeu) AUG (Met)	ACU (Thr) ACC (Thr) ACA (Thr) ACG (Thr)	AAU (ASpN) AAC (ASPN) AAA (LYS) AAG (Lys)	AGU (Ser) AGC (Ser) AGA (Arg) AGG (Arg)	U C A G
	GUU (Aal) GUC (Val) GUA (VaL) GUG (Val)	GCU (Ala) GCC (Ala) GCA (Ala) GCG (Ala)	GAU (Asp) GAC (Asp) GAA (Glu) GAG (Glu)	GGU (Glu) GGC (Gly) GGA (Gly) GGG (Gly)	U C A G

Fig. 5.3. Genetic code (64 codons).

One of the greatest challenges in protein engineering (where one objective is to create functional, atomically-precise 3-D aperiodic structures) has been the difficulty of protein fold design, because individual amino acids have no strong, natural complementarity. In 1998, Duan and Kollman successfully folded a solvated 36-residue (~12,000-atom) protein fragment (i.e., a peptide) by molecular dynamics simulation into a structure that resembles an intermediate native state. During a 1-microsec simulation, the chain folded during 150 nanosec into a compact structure resembling the intermediate native state, as known by NMR, then unfolded and refolded again for a shorter period. In 1998, whole-protein folding to the final stable native state had not yet been computed deterministically using molecular dynamics—the real protein will fold and refold hundreds of times before it stumbles into the stable conformation with the lowest free energy—but research pursuing this objective was active and ongoing. Structural changes that occur during protein function (e.g., enzymatic action) were also being avidly studied.

A different protein-based approach to near-atomically-precise 3-D structures avoids the folding problem by making use of semi-rigid protein nanoshells of deterministic size and shape to guide the ordered assembly of inorganic particles. For example, the capsid of the cowpea chlorotic mottle virus, a protein container with a 28-nm external diameter and an 18-nm internal cavity diameter, is composed of 180 identical coat protein subunits arranged on an icosahedral lattice. Each subunit presents at least nine basic residues (arginine and lysine) to the interior of the cavity, creating a positively charged interior surface which provides an interface for inorganic crystal nucleation and growth, while the outer capsid surface is not highly charged. In one experiment, the empty capsid shell was found to act as a spatially selective nucleation catalyst in paratungstate mineralization, in addition to its role as a size- and shape-constrained reaction vessel. Noted the authors: "The range of viral morphology and size allow great flexibility for adapting this methodology to control the size and shape of the entrapped material, which is limited only by access to the protein interior and could include inorganic and organic species. The electrostatic environment within the cavity could be altered by site-directed mutagenesis to induce additional specific interactions."

It may also be possible to borrow various existing protein devices and apply them to new uses. The bacteriophage DNA injection system is self-assembling and could be engineered for other applications. Vogel

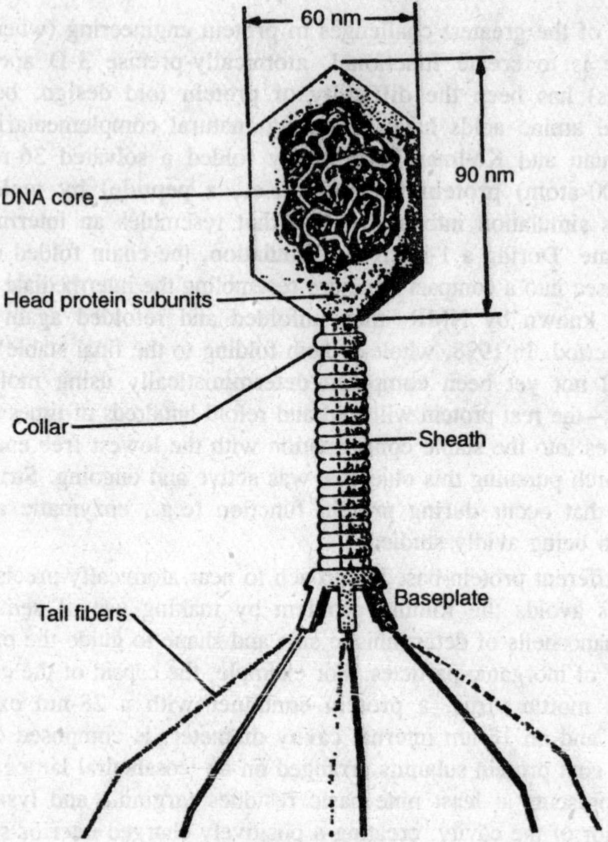

Fig. 5.4. Bacteriophage.

has attached kinesin motors to flat surfaces in straight grooves; when ATP was added, the motors were activated, passing 25-nm wide microtubules hand over hand down the line in the manner of a ciliary array. Montemagno genetically engineered the 12-nm wide ATPase molecular rotary motor so that one end would adhere to a metal surface and the other end would provide an attachment site for 1-micron fluorescent streptavidin-coated bead payloads. When ATP was added, each bead individually attached to a biomotor in the motor array began twirling at ~10 Hz, generating a >100 pN force. Bead rotation was maintained continuously for more than 2 hours, indicating the high reliability of these ~100% efficient self-assembling biomotors. Montemagno explains that his laboratory's "long-term goal is the integration of the F_1ATPase biological motor with nanoelectromechanical systems (NEMS) to create a new class of hybrid nanomechanical devices." Other natural nanomotors such as the self-assembling bacterial flagellar motor and the hexagonal "packaging RNA" or pRNA pumping mechanism that packs DNA into the capsid shell of the bacteriophage Phi 29 are also being intensively studied.

Yet another material which allows moderately well-defined three-dimensional nanoassemblies is DNA. The ideas behind DNA nanotechnology have been around since 1980, but activity in this field accelerated in the 1990s after numerous experimental difficulties were surmounted. Horn and Urdea reported branched and forked DNA polymers. Niemeyer and Smith exploited DNA specificity to generate regular protein arrays, and both suggested using self-assembled DNA as an early material for molecular nanotechnology. Shi and Bergstrom attached DNA single strands to rigid organic linkers, showing that cyclical forms of various sizes could be formed with these molecules. Mirkin's group attached DNA molecules to 13-nm colloidal gold particles with ~6 nm particle spacings, with the goal of assembling nanoparticles into macroscopic materials; Alivisatos, Schultz and colleagues used DNA to organize 1.4-nm gold nanocrystals in arrays with 2-10 nm spacings. In either case, the specificity of DNA pairing should allow the construction of complex geometries since the use of colloidal particles can potentially add structural elements (which are more rigid than any polymer strand) to the set of building blocks for nanometer-scale structures—although to take full advantage of this capability would require particles which are atomically precise and which possess several chemically distinct anchoring points on each particle. Damha has synthesized V-shaped and Y-shaped branched RNA molecules, branching RNA dendrimers with "forked" and "lariat" shaped RNA intermediates, and even trihelical DNA. Henderson and coworkers have designed a simple DNA decamer that can form an extended linear staggered quadruplex array reaching lengths of >1000 nm, and have made branched oligonucleotides that template the synthesis of their branched "G-wires"; depending on the ratio of linear to branched building blocks, extensive DNA arrays with differing connectivities but irregular interstices can be created.

In 1998 the most intensive work on three-dimensional engineered DNA structures was taking place in Nadrian Seeman's laboratory in the New York University Department of Chemistry. Seeman originally conceived the idea of rigid 3-D DNA structures in the early 1980s, while examining DNA strands that had arranged themselves into unusual

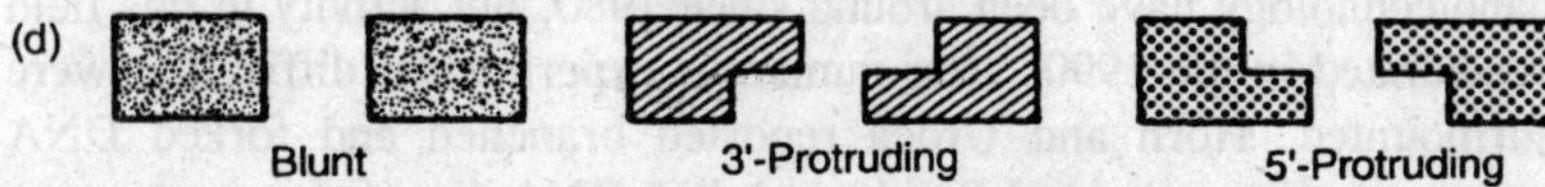

Fig. 5.5. Restriction endonuclease.

four-armed Holliday junctions. Seeman recognized that DNA had many advantages as a construction material for nanomechanical structures. First, each double-strand DNA with a single-strand overhang has a "sticky end," so the intermolecular interaction between two strands with sticky ends is readily programmed (due to base-pair specificity) and reliably predicted, and the local structure at the interface is known (sticky ends associate to form B-DNA). Second, arbitrary sequences are readily manufactured using conventional biotechnological techniques. Third, DNA can be manipulated and modified by a large variety of enzymes, including DNA ligase, restriction endonucleases, kinases and exonucleases. Fourth, DNA is a stiff polymer in 1-3 turn lengths and has an external code that can be read by proteins and nucleic acids.

During the 1980s, Seeman worked to develop strands of DNA that would zip themselves up into more and more complex shapes. Seeman made junctions with five and six arms, then squares, stick-figure cubes comprised of 480 nucleotides, and a truncated octahedron containing 2550 nucleotides and a molecular weight of ~790,000 daltons. The cubes were synthesized in solution, but Seeman switched to a solid-support-based methodology in 1992, greatly improving control by allowing construction of one edge at a time and isolating the growing objects from one another, allowing massively parallel construction of objects with far greater control of the synthesis sequence. By the mid-1990s, most Platonic (tetrahedron, cube, octahedron, dodecahedron, and icosahedron), Archimedean (e.g., truncated Platonics, semiregular prisms and prismoids, cuboctahedron, etc.), Catalan (linked rings and complex knots), and irregular polyhedra could be constructed as nanoscale DNA stick figures.

Seeman's DNA strands that formed the frame figures were strong enough to serve as girders in a molecular framework, but the junctions

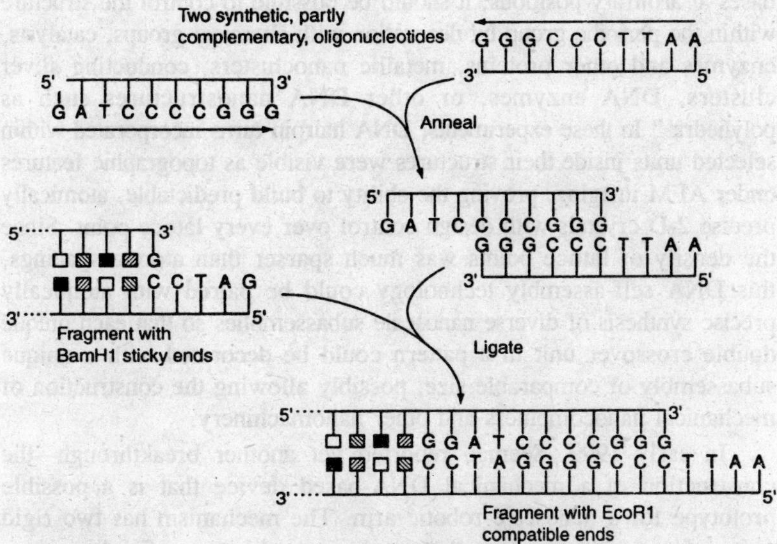

Fig. 5.6. Oligonucleotides.

were too floppy. In 1993 Seeman discovered the more rigid anti-parallel DNA "double crossover" motif, which in 1996 he used to design and build a stiff double junction to keep his structures from sagging. The next goal was to bring together a large number of stick figures to form large arrays or cage-shaped DNA crystals that could then be used as frameworks for the assembly of other molecules into pre-established patterns. These DNA molecules would serve as the scaffolding upon which new materials having precise molecular structure could be assembled.

In 1998, Erik Winfree and colleagues in Seeman's laboratory reported the design and construction of two-dimensional DNA crystals using self-assembling double crossover molecules. Repeating array units were approximately 2 nm × 4 nm × 16 nm in size, and examination of the array with an Atomic Force Microscope (AFM) revealed domains of up to 500,000 interconnected units, showing that the self-assembly process could be very reliable under ideal conditions. While the work described the construction of two- and four-unit lattices, the number of component tiles that could be used in the repeat unit did not appear to be limited to such small numbers, suggesting that complex patterns could be assembled into periodic arrays, which might also serve as templates for nanomechanical assembly. Noted the authors: "Because oligonucleotide synthesis can readily incorporate modified bases at arbitrary positions, it should be possible to control the structure within the periodic group by decoration with chemical groups, catalysts, enzymes and other proteins, metallic nanoclusters, conducting silver clusters, DNA enzymes, or other DNA nanostructures such as polyhedra." In these experiments, DNA hairpin turns incorporated within selected units inside their structures were visible as topographic features under AFM imaging, proving the ability to build predictable, atomically precise 2-D crystals with design control over every lattice point. Since the density of lattice points was much sparser than atomic spacings, this DNA self-assembly technology could be paired with atomically precise synthesis of diverse nanoscale subassemblies so that each unique double crossover unit in a pattern could be decorated with a unique subassembly of comparable size, possibly allowing the construction of mechanical nanocomputers and other nanomachinery.

In early 1999, Seeman reported yet another breakthrough—the construction of a mechanical DNA-based device that is a possible prototype for a nanoscale robotic arm. The mechanism has two rigid arms a few nanometers long that can be rotated between fixed positions by introducing a positively charged cobalt compound into the solution surrounding the molecules, causing the bridge region to be converted from the normal B-DNA structure to the unusual Z-DNA structure. The free ends of the arms shift position by ~ 2.0 nm during the structural conversion. Explained Seeman: "Using synthetic DNA as a building material, we have constructed a controllable molecular mechanical system. In the long-term, the work will have implications for the development of nanoscale robots and for molecular manufacturing."

DNA can also serve as an assembly jig in solution phase. Bruce Smith and colleagues are devising a method for the assembly and covalent linkage of proteins into specific orientations and arrangements as determined by the hybridization of DNA attached to the proteins, called DNA-Guided Assembly of Proteins (DGAP). In this method, multiple DNA sequences would be attached to specific positions on the surface of each protein, and complementary sequences would bind, forcing protein building blocks (possibly including biomolecular motors, structural protein fibers, antibodies, enzymes, or other existing functional proteins) together in specific desired combinations and configurations, which would then be stabilized by covalent interprotein linkages. This technique could also be applied to nonprotein components that can be functionalized at multiple sites with site-specific DNA sequences, although proteins, at least initially, may be more convenient building blocks due to their size, their surface chemistry, the wide variety of functions and mechanical properties they can confer on the resulting assemblies, and the many existing techniques for introducing designed or artificially evolved modifications into natural proteins of known structure. (In 1998, custom DNA and peptide sequences could be ordered online.) Methods for covalently attaching functional proteins to a DNA backbone in a specified manner at ~8.5 nm (25 base-pair) intervals, addressable protein targeting in macro-molecular assembly, and "protein stitchery" were being explored by others. Drexler notes that evolution has not maximized the stability of natural proteins, and that substantially greater stability may be engineered by various means (e.g., increasing folding stability by >100 Kcal/mole.

Molecular and Supramolecular Chemistry

A second broad category of enabling technology for molecular manufacturing includes molecular chemistry—the conscious design of completely artificial, nonbiological chemical structures that could potentially serve as molecular "parts" or which have specific devicelike functionality. For example, catalysts may very loosely "be thought of as rudimentary assemblers that are slightly programmable through changes in the reaction milieu (i.e., changes in pH, temperature, etc.)". The number of natural molecular "parts" is immense—perhaps a few hundred thousand bioorganic compounds have been separated, purified and identified—but by 15 February 1999 chemists had already registered 19,245,458 well-characterized artificial molecules or other "substances" in the CAS Registry. Synthesis of natural molecules containing ~100 atoms (e.g., a ~1 nm^3 molecular "part"), using methods of classical organic synthesis often requiring ~1 synthesis step per atom with ~90% yield per step, was state-of-the-art in 1998.

"What is exciting about modern nanotechnology," says Nobel chemist Roald Hoffmann, "is (a) the marriage of chemical synthetic talent with a direction provided by 'device-driven' ingenuity coming from engineering, and (b) a certain kind of courage provided by those incentives, to make arrays of atoms and molecules that ordinary, no, extraordinary chemists just wouldn't have thought of trying. Now they're pushed to do so. And of course they will. They can do anything."

Molecular chemists study how small numbers of atoms combine covalently with each other to produce molecules with a wide variety of different properties. Supramolecular chemists are primarily concerned with the much weaker noncovalent forces that can arise between molecules—such as hydrogen bonds and van der Waals interactions—that can be sufficient to bind collections of molecules tightly enough to form functional nanometer-sized structures. *Supramolecular chemistry* studies "chemistry beyond the molecule" or "the chemistry of the noncovalent bond". Its subject matter includes atomically precise supermolecules (well-defined discrete oligomolecular species resulting from the intermolecular association of a few components, such as a receptor and its substrate) and polymerlike supramolecular assemblies (polymolecular entities that result from the spontaneous association of a large undefined number of components).

Self-assembly is a key concept in supramolecular chemistry as it is in nature, allowing the manufacture of large numbers of compound objects simultaneously and in parallel, rather than sequentially. Molecular self-assembly is a strategy for nanofabrication that involves designing molecules and supra-molecular entities so that complementarity causes them to aggregate into desired structures. Self-assembly of atomically precise supermolecules thus demands well-defined adhesion between selected molecules, which may require steric (shape and size) complementarity, interactional complementarity, large contact areas, multiple interaction sites, and strong overall binding.

G.M. Whitesides explains that self-assembly has a number of advantages as a strategy. First, it carries out many of the most difficult steps in nanofabrication—those involving the smallest atomic-level modifications of structure—using the very highly developed techniques of synthetic chemistry. Second, it draws from the enormous wealth of examples in biology for inspiration; self-assembly is one of the most important strategies used in biology for the development of complex, functional structures, such as the well-studied examples of complex biomachine self-assembly of whole bacteriophage virions and flagellar rotor motors from their smaller molecular "parts". Third, self-assembly can incorporate biological structures directly as components in the final systems. Fourth, self-assembly requires target structures to be among the most thermodynamically stable available to the system, thus tends to produce structures that are relatively defect-free and self-healing. As Lehn points out: "By increasing the size of its entities, nanochemistry works its way upward towards microlithography and microphysical engineering, which, by further and further miniaturization, strive to produce ever smaller elements". Hierarchical self-assembly of many billions of micron-scale spherical components into a periodic honeycomb-structured photonic crystal optical device ~30 microns thick and ~1 cm^2 in area was demonstrated in 1998.

There is a wide range of different molecular systems that can self-assemble, including those which form ordered monomolecular structures by the coordination of molecules to surfaces, called *self-assembled monolayers* (SAMs), self-assembling thin films and Langmuir-Blodgett films, and self-organizing nanostructures. In these systems, a single layer of molecules affixed to a surface allows both thickness and composition in the vertical axis to be adjusted to 0.1 nm by controlling the structure of the molecules comprising the monolayer, although control of in-plane dimensions to < 100 nm is very difficult. Fluidic self-assembly of microscale parts and the dynamics of Brownian self-assembly have also been described, and the theory of designable self-assembling molecular machine structures is beginning to be addressed.

Another approach using solid-phase peptide synthesis and a massively convergent self-assembly process was employed by M.R. Ghadiri and colleagues, who designed and synthesized a number of self-assembling peptide-based nanotubes using cyclic peptides with an even number of alternating D- and L-amino acids for the building blocks of the nanotubes. The alternating stereo-chemistry of the cyclic peptides allowed all the side chains of the amino acids to be pointing outwards which would not be possible in an ordinary all L-cyclic peptide. In this conformation, the amide backbone can H-bond in a direction perpendicular to the plane of the cyclic peptide. The stacking of two cyclic peptides forms an H-bonding network resembling an antiparallel beta-sheet as is commonly found in natural proteins. The H-bonding lattice quickly propagates perpendicular to the plane of the cyclic peptide, forming a tubular microcrystalline structure with 0.75-nm pores, or, in another experiment, 1.3-nm pores. Another group of nanotubes was designed with a highly hydrophobic outer surface and a hydrophilic inner pore. These nanotubes were easily inserted into a lipid bilayer and have been shown to be highly efficient ion channels; nanotubes with slightly larger pores transport small molecules such as glucose as well. Stable flat nanodisks with diameters continuously (chemically) adjustable from 30-3000 nm have been self-assembled in surfactant solutions.

Self-assembled crystalline solids such as zeolites incompletely fill space, leaving substantial voids that may be occupied by solvent molecules or other guest molecules, producing solid-state host-guest complexes known as clathrates (from the Latin *clathratus*, meaning "enclosed by the bars of a grating"). MacNicol and coworkers reported the first rationally designed clathrate host in 1978, but the first true de novo design of an organic clathrate was in 1991 and involved the formation of an extensive three-dimensional diamondlike porous lattice built from a single Tinkertoy-type subunit containing four tetrahedrally arrayed pyridone groups that acted as connectors to assemble the units together in a well-defined geometry. However, the crystal was held together only by weak hydrogen bonds and the links had a rotational degree of freedom, rendering the exact crystal molecular arrangement unpredictable. Engineered nanoporous molecular crystals can provide molecular-scale voids with controlled sizes, shapes, and embedded chemical environments at resolutions of 0.3-4.0 nm. Other self-assembling crystal structures have been described including the crystal engineering of diamondoid networks, template-directed colloidal crystallization or colloidal epitaxy, organic templating to form crystalline zeolite-type structures with ordering lengths <3 nm, 3-D polymer channels with chemically functionalizable channel linings, and micromolding combined with templating and cooperative assembly of block copolymers to produce hierarchical ordering over discrete and tunable characteristic length scales of ~ 10 nm, ~ 100 nm, and ~ 1000 nm in a single body.

Dendrimers, also known as arborols or fractal polymers, are a well-known set of self-assembling structures that possibly could be used as tools to assist in the assembly of early nanomachines. Dendrimers are large regularly-branching macro-molecules resembling fractal patterns, made by an iterative process in which small linear molecules are allowed to bind to each other at a certain number of sites along their length, building up branches upon branches with each iteration working outward from a core molecule having at least two chemically reactive arms, somewhat mimicking tree growth. Different core molecules or building-block chains produce macromolecules with different shapes, and the outer surface can be terminated with chemical groups of specific functionality, producing hundreds of differently-surfaced branched molecules that have already found use in medical research. Growth is regulated, so size can be accurately controlled—dendrimers are typically a few nanometers wide but have been constructed with masses exceeding a million protons ($\sim 10^5$ atoms) and with diameters larger than 30 nm. In 1999, convergent assembly of specified 6-mer dendrimers, via trimerization of precursor dimer "parts," was demonstrated.

Autocatalysis or *self-replication* is yet another approach to the chemical self-assembly of subunits. In 1989, J. Rebek reported the synthesis of super-molecules that could generate copies of themselves when placed in a sea of simpler molecular parts, with each component part consisting of up to a few dozen atoms. Molecular self-replication has continued to be studied by others. Ghadiri reported a self-replicating peptide in 1996, and by 1998 his laboratory had moved on to studies of more complex autocatalytic cycle networks. It is unknown how complex such assembly cycles may be made, but Ghadiri was investigating a self-replicating 256-component molecular ecosystem incorporating ~32,000 possible binary interactions, in 1998.

By 1998, chemists had already synthesized a vast variety of simple molecular parts of which only a tiny sampling can be mentioned here. For instance, carbon rings have been linked edge-on or by corners to create propeller-shaped molecules called propellanes or rotanes, or stacked with short hydrocarbon links to make gear-looking molecules such as cyclophane and superphane. Hydrocarbon molecular polyhedra in the shape of triangular (e.g., triangulane), cubic (e.g., cubane, first made by P. Eaton in 1964), pentagonal and hexagonal prisms, collectively known as *prismanes*, have been created, along with more unusual geometrical forms such as paddlane, housane, basketane, churchane, pagodane, and bivalvane (one carbon atom at each vertex). Cryptands, spherands, cryptaspherands and carcerands, corannulene baskets, and calixarenes are bowl-shaped molecules whose benzene ring walls hold their cavities rigid. The bowl rim of a calixarene can be lined with chemical groups to determine which guests it will accept, or two bowls can be bonded together, making a hollow nanoscale reaction vessel, and various molecular cages and capsules with precise sizes and characteristics have been assembled. A variety of nonchiral and chiral square molecules have been synthesized, as well as molecular "bottlebrushes"; DNA-polymer constant-force springs; the ferric wheel; molecular barrels and molecular tennis balls; various helical, boxlike, and other hollow organic structures; molecular "turnstiles" and "molecular tweezers"; simple and complex trefoil knot molecules; and stable carbyne molecular rods with chain lengths up to ~300 carbon atoms having alternating single and triple covalent bonds. Chemists are justifiably awed by the creation of structures like the ferric wheel; "for me," observed Nobel chemist Roald Hoffmann, "this molecule provides a spiritual high....The molecule is beautiful because its symmetry reaches directly into the soul, [playing] a note on a Platonic ideal."

Molecular gear systems first became a popular object of chemical study during the 1980s. For example, G. Yamamoto described "compounds that exist in conformations which are regarded as static meshed gears with two-toothed and three-toothed wheels and some of them behave as dynamic gears." H. Iwamura prepared one system that formed a chain of beveled molecular gears with ~GHz rotation rates, and another "doubly connected bevel gear system...[Transfer] of information from one end of the molecular system to the other end could take place in large molecules via cooperativity of the torsional motions of the chain." Kurt Mislow gave numerous examples of molecular gear systems that he had synthesized using the methods of traditional solution chemistry, which "resemble to an astonishing degree the coupled rotations of macroscopic mechanical gears." He added: "It is possible to imagine a role for these and similar mechanical devices, molecules with tiny gears, motors, levers, etc., in the nanotechnology of the future."

In 1998, Gimzewski and coworkers synthesized and observed a 1.5-nm diameter gearlike single molecule (hexatertbutyl decacyclene) rotating within a supramolecular bearing (a void in a 2-D lattice) at room temperature on the surface of a Cu{100} lattice. The authors suggest that added nonthermal noise (e.g., temperature differences produced by tunnel current heating of the rotor) could be rectified by asymmetries in the rotor/neighbour potential energy curve, allowing the rotor to turn unidirectionally. Says Gimzewski: "Our wheel is frictionless in the conventional sense of the word, and it doesn't wear out." Gimzewski's experimental work confirms Drexler's calculations[10] that van der Waals bearings can operate in molecular systems, that multiatomic bearing surfaces with higher load capacities than sigma bond bearings can be built, that these bearings can run with no lubricants, and that such bearings have sufficiently low energy barriers to allow turning by thermal vibrations alone.

By 1998, limited control of the mechanical action of molecular parts had been demonstrated. For example, in 1994 T. Ross Kelly's group created a "paddlewheel" molecule (a spinning propeller-shaped wheel) with a built-in chemically-controlled brake. In solution, the wheel spins freely, but when Hg^{++} ions are added, a blocking ligand physically rotates into a new position (tripping the brake) and stopping the rotation; removal of the mercury causes the wheel to spinning. In 1997, Kelly's group used similar techniques to synthesize the first molecular ratchet, a propeller with three benzene ring "blades" that serve as gear teeth. A row of four rings (the pawl) sits between two of the blades, such that the propeller cannot turn without pushing it aside; because of a twist in the pawl, it is easier to turn clockwise than counterclockwise under thermal agitation, providing mechanical rectification without net motion or energy extraction. Shinkai and colleagues fabricated a pair of molecular tongs for grasping metal ions that uses two crown ether rings as the jaws, linked by two benzene rings joined through two double-bonded nitrogen atoms. Irradiation with ultraviolet light induces photoisomerization of the double bond between the nitrogen atoms, closing the jaws; heat reverses the process, opening the jaws. In related systems, collectively called "butterfly molecules," irradiation with visible light or pH changes can also induce opening. Seeman's chemically-operated mechanical DNA system provides similar control of molecular mechanical movement.

Noncovalent self-assembly of molecular parts has also been demonstrated. For example, Fraser Stoddart and colleagues have extensively investigated the rotaxanes, molecules in which one part is threaded through a hole or loop in another, and kept from unthreading by end-groups, like a ring trapped on a barbell. In one system, a ring-shaped molecule slides freely along a shaft-like chain molecule, moving back and forth between stations at either end with a frequency of ~500 Hz, making an oscillating "molecular shuttle". This shuttling behaviour can be controlled by a series of different chemical, electrochemical, or photochemical external stimuli. Using a different approach, G. Wenz threaded ~120 molecular beads onto a single, long, poly-iminooligomethylene polymer chain, making a "molecular necklace." Other groups are pursuing related research regarding [n]rotaxanes and pseudorotaxanes (mechanically-threaded molecules), while still others are using the necklace technique as a means, for example, of self-assembling ~2 nm wide cyclodextrin nanotubes.

Another interesting molecular system that can be made entirely by self-assembly is the catenanes, which have two or more closed rings joined like the links of a chain. The rings are mechanically linked—there are no covalent bonds between separate links. The first [2]catenane (2 linked rings) was constructed by Edel Wasserman of AT&T Bell Laboratories in 1960, using simple hydrocarbon rings; the first [3]catenane was synthesized in 1977. In 1994, Fraser Stoddart and David Amabilino constructed the first [5]catenane, having ~372 atoms, dubbed "olympiadane" because of its resemblance to the Olympic rings, and in 1997 the same group announced the first deterministically-ordered 7-ring heptacatenane; another group later reported the spontaneous self-assembly of a 10-component catenane. By 1998, much research on larger polymeric chains of linked rings (e.g., oligocatenanes), supramolecular "daisy chains", fullerene-containing catenanes, and supramolecular weaving was in progress. The smallest [2]catenane constructed to date has dimensions 0.4 nm × 0.6 nm. Self-assembled mechanically-interlocked 2-dimensional and 3-dimensional "infinite" arrays of single molecular ring-like species are known. Interestingly, in 1998 it was discovered that many viral coats (which also self-assemble) appear to be a 2-dimensional "chain-mail" weave of mechanically-interlocked protein rings—for instance, the spherical bacteriophage HK97 capsid shell consists of exactly 72 interlinked protein rings, specifically 60 hexamers and 12 pentamers.

Complex molecular parts also can be built up from simpler molecular parts by covalent bonding as low-symmetry shape-invariant molecular object polymers, or via processes variously known as modular chemistry, heterosupramolecular chemistry, chemical assembly, or structure-directed synthesis including the well-known Diels-Alder reactions which may be employed repetitively to fabricate a series of atomically-precise molecular rods, rings, and nanocages (e.g., "beltenes" and "collarenes") composed of polyacenequinone units with nanometer-scale dimensions. Such efforts have been described as steps toward "molecular LEGO", "molecular Meccano", "molecular Tinkertoys", or "molecular building blocks". Some structure-directed assembly has been achieved with simple DNA forms. Other simple techniques for synthesizing nanowires, nanorods, nanotubes, and nanocages, or for selecting polymers for length, are well-known.

For instance, Jean-Marie Lehn has created rectangular grid complexes by mixing rodlike ligands that make use of different numbers of binding sites—one example is a 4 × 5 array consisting of nine ligands and 20 metal ions, nanoscale grids which he claims might one day find use "as components within a futuristic information storage and processing nanotechnology". C.M. Drain has described a 21-component $(5\ nm)^2$ square array of nine porphyrins tethered together by 12 palladium ions shaped like a four-pane window. The structure is built from three different kinds of porphyrins— (1) an X-shaped unit that coordinates to four metal ions, forming the center of the array, (2) a T-shaped unit that coordinates to three metals and forms each side of the array, and (3) an L-shaped unit that coordinates to only two metals and forms each corner of the array. When these components are placed in solution in the correct ratio, they form the square array within a half hour at room temperature with about 90% yield. The same porphyrin units can be combined in different ratios and induced to form wires or tapes. The properties of such arrays can be fine-tuned by choosing the appropriate metal ion linker and functionalized porphyrin unit.

Jeffrey Moore has investigated a three-dimensional nanoscaffolding comprised of a molecular lattice that could serve as a framework for catalysts, photosynthetic molecules, or more complex molecular devices. Moore's original objective was to synthesize modular building blocks with physical and chemical properties that would dictate a "programmed assembly" protocol for "nanoarchitectures". The characteristics of each modular unit shaped the weak intermolecular attractions—electrostatic, van der Waals, and hydrogen-bonding forces—between it and its neighbours so that units would array themselves in a larger structure with the desired spacing and geometry. Moore's group first linked phenylacetylene subunits into oligomers of various lengths which bend and twist like wire sculptures until their ends meet to form closed polyhedral molecules—the basic scaffold components (of which there were at least half a dozen shapes). These building blocks then self-assembled or folded into geometries at least partly controlled by designed hydrogen-bonding interactions. The molecular polyhedra each possessed several phenylene groups, "chemical handles" upon which other molecular groups could be hung. Moore then focused on ~10-nm dendrimeric forms and later began attaching photosensitive antenna molecules to these structures. By 1997, Moore had begun investigating norbornadiene on Si{100} in an effort to develop ways to covalently attach molecules onto surfaces with subnanometer spatial precision, and had suggested a packing model for interpenetrated diamondoid structures. Robson's group has also investigated possible scaffolding molecules.

Josef Michl is attempting to build a molecular construction kit using molecular rods and connectors, pursuing an explicit vision of "molecular Tinkertoys" by working with simple molecular structures that form stiff, flexible rods. Michl has assembled rods from a mixture of carbon-boron molecules (e.g., 10-vertex or 12-vertex carboranes) and carbon-hydrogen molecules, providing fine control over the total rod length. The rods are built up from more primitive molecular parts, such as propellane (a strained form of C_5H_6) and cubane (a strained form of C_8H_8), to make "staffanes" which are a series of cage units in a linear series. (Strained molecules are constructed with bonds that are forced out of their normal angles, as for example 90° in the case of cubane, compared with carbon's normal orientation of 109.5°.) Michl has fabricated rods whose lengths vary from 0.5 nm to 2.5 nm, in precise 0.1-nm steps. There are other ways of making rodlike molecules, but Michl's are highly inert, do not absorb visible or UV light, are stable up to at least 200°C, and do not react with the oxygen in air even at high temperatures.

Related work is underway in Michl's laboratory on connectors to join the rods together. Metal atoms would be the simplest solution, offering the useful quality of strong joints that can be easily disassembled. Different metals give different binding geometries—square, octahedral, and so on. Many of the connectors are metal-containing species; some are symmetrically trisubstituted or hexasubstituted benzenes or tetrasubstituted cyclobutadiene complexes. Michl must build the right chemical groups onto the rod termini to make them adhere to the connectors as desired. In one example, tiny crosses were built by attaching carboxylate groups to one end of the staffanes, allowing them to bind to a connector made of two rhodium atoms. The groups at the other end of the rods were converted to an ester group, which doesn't bind to rhodium, and so the crosses self-assembled correctly. Michl has also fabricated "star connectors" in which the rods are built into the molecules as covalently bonded arms of a star. For instance, one 3-arm star connector uses three large carboranes coupled to a central benzene ring, allowing a pedestal to be attached vertically to the benzene ring via a ruthenium "sandwich" bond. Additional devices could be installed on this molecular scaffolding, such as optically active molecules or even molecular mechanical "windmills". Nanoscale square and hexagonal planar grids have been fabricated. Extensive ab initio quantum mechanical calculations accompany the experimental work and aid in the interpretation of the results. Michl's ultimate goal is "the production of thin layers of solids of completely controlled aperiodic structure consisting of an inert covalent scaffolding carrying selected active groups."

G. Leach and colleagues used computers to simulate various diamond and graphite nanostruts using molecular dynamics calculations, analyzing solid rectangular struts with varying aspect ratios, cross-sections, terminating atoms (for diamond), potential energy functions, and temperatures. The most dramatic differences were seen between struts with 100:1 aspect ratios and struts with lower (10:1 or 1:1) aspect ratios—the diamond strut with a 100:1 aspect ratio and a unit cell ($\sim$0.4 nm)^2 cross-section begins to curl (with end-to-end distance decreasing 1.2 nm after 20 picosec) due to thermal vibration (both at

150 K and 300 K), while a unit strut with an aspect ratio of 10:1 or a 100:1 strut with a $(1\ \text{nm})^2$ cross-section showed end-to-end length fluctuations of only ~0.1 nm. The authors concluded that support struts with a cross-section of at least $(1\ \text{nm})^2$ would be sufficient to keep both end-to-end distance and transverse fluctuations below ~0.1 nm in struts with 10:1 aspect ratios. Other work by the same group claims that many nanomachine designs may actually perform better than molecular dynamics results might suggest, or may require fewer atoms for the same positional stability. The elastic and wear characteristics of diamond are being studied computationally by others.

No discussion of molecular "parts" could be complete without mention of the fullerenes, first discovered in 1985 and vigorously investigated ever since. Fullerenes are one of the three known *allotropic* forms of carbon. In graphite, the most common allotrope, carbon atoms are arranged in hexagonal rings and strongly bonded into parallel planar sheets, with much weaker bonding between the sheets, giving graphite its excellent lubricating properties. In diamond, the second allotropic form, carbon atoms are arranged in a symmetric, tetrahedral structure, giving immense strength. Exposed to air, both diamond surfaces and graphite edges are quickly coated or "passivated" with hydrogen or other atoms that tie up the dangling bonds. In fullerenes, the third allotropic form, carbon atoms form large and hollow cage-like structures, often roughly spherical or tubular, made up of closed curved one-atom-thick sheets of carbon atoms arranged in a number of five-, six-, and higher-membered rings. Interestingly, the fullerenes need no passivating hydrogens or other atoms to satisfy their chemical bonding requirements at the surface—in this sense, "fullerenes are the first and only stable forms of pure, finite carbon".

A new journal, *Fullerene Science and Technology*, appeared in 1993, and by the mid-1990s ~3000 papers on fullerenes and their properties had been published and 149 fullerene-related patents had been issued in the U.S. alone (through 1996). By 1998, an incredible variety of fullerenes of many shapes and sizes had been synthesized.

The original fullerene, C_{60}, consists of exactly 60 carbon atoms arranged in a soccerball structure (lone carbon atom at each vertex) with 20 hexagons and 12 pentagons (with each pentagon entirely surrounded by hexagons). In theory, small molecules such as H_2 or CO, and probably CH_4, would fit inside C_{60}. A slightly larger variant, C_{70}, is the same except that an extra belt of carbon atoms has been inserted around the equator. Smaller and larger variants have been observed; as cages get bigger, the corners (where the 12 pentagons needed for closure reside) get sharper. The topologically smallest possible fullerene is a dodecahedron consisting of 12 pentagons and 20 carbon atoms, but the ring isomer of C_{20} (a hoop of single carbons) is apparently energetically favoured over the bowl or spherical (fullerene) isomers; in 1998, C_{28} was the smallest fullerene that had been observed experimentally, and C_{36} could be made in significant quantities by arc-discharge. The approximate diameter (measured to atomic centers) of a spheroidal fullerene of formula C_n is $D_{ball} \sim 1.1 (n/60)^{1/2}$ nanometers. In 1998, purified C_{60} cost $27.50/gm (99.5% pure) or $60/gm (99.9%), C_{70} $250/gm (98%), and C_{84} $3,750/gm (90%) from Dynamic Enterprises Ltd. of London.

Insertion of many additional equatorial belts of carbon atoms (incorporated as hexagons) into a spheroidal fullerene results in a long cylinder with graphite-like ("graphene") walls and spherical endcaps, making a class of fullerenes known as the capsular fullerenes or single-walled carbon nanotubes (SWNT) that come in many sizes and chiral forms. The molecular form of carbon nanotubes has been likened to rolled chicken wire. SWNTs may be synthesized by vapourizing graphite spiked with 1% catalyst from the nickel, cobalt and iron group above 3000°C, then allowing the vapour to slowly condense. Tubes form because the metal atoms interact with dangling bonds at the end of a tube, favouring tube extension over hemispherical capping by newly arriving carbon vapour atoms. (For any given mass number, ball-shaped fullerenes are energetically favoured over tube-shaped fullerenes during high-temperature synthesis in the absence of catalyst.) SWNTs typically self-assemble in 1.1 nm wide tubes, although smaller single-walled nanotubes and larger single-walled and multi-walled nested nanotubes exist. SWNT carbon atoms are bonded in virtually flawless hexagonal arrays. Simulations show that isolated flaws migrate to the ends of the tube and are eliminated, a phenomenon known as "self-healing". Single-molecule nanotubes of $C_{1,000,000}$ or larger, with lengths >100,000 times longer than their widths (e.g., ~1 mm long), had been synthesized by 1998. While graphite is a very brittle material, the graphene walls of a carbon nanotube are quite resilient. Computer simulations and experiments confirm that nanotubes kink when bent, then snap back when released. Nanotube diameter is $D_{tube} = 0.078 (n^2 + nm + m^2)^{1/2}$ nanometers, where (n, m) is the "rollup vector" defined by the number of steps required for a repeat pattern along two crystallographic cylinder wall axes. The cohesive energy per atom required to curve a flat graphene sheet into a cylinder is $E_{rollup} = (13 \text{ zJ-nm}^2 / D_{tube}^2)$. (1 zeptojoule (zJ) = 10^{21} joules.) As an example, a (10,10) nanotube has D_{tube} = 1.36 nm and E_{rollup} = 7 zJ/atom. Young's modulus for individual SWNTs, assuming a hollow (open-ring) cross-section, has been estimated as high as $\sim 5.5 \times 10^{12}$ N/m^2; elastic bending modulus (measured as 1×10^{12} to 0.1×10^{12} N/m^2) decreases sharply with increasing diameter (from 8-40 nm). In 1998, purified carbon single-walled nanotubes were available via the Internet for $1400/gm from Tubes@Rice, or for $200/gm at lesser purity from CarboLex.

From spatial geometry, it is known that hexagonal tiling produces flat sections (like cylinder walls), and inserting pentagons into an hexagonal array produces positive curvature (like spheres), making endcaps. But 7-sided heptagons can also be inserted into hexagonal arrays to induce negative curvature, thus permitting concave surface deformations such as saddle-shaped fullerenes or possibly helical tubular structures. Indeed, note Colbert and Smalley, "the use of hexagonal, pentagonal, and heptagonal substructures are sufficient to produce caged carbon structures of any topology". Nature makes use of this same geometrical principle in the radiolarians, tiny protozoans with fullerene-like siliceous skeletons, and in the conical nucleoprotein core particle of the HIV-1 virus. R. Smalley observes that a graphene sheet has the highest tensile strength of any known 2-dimensional network, and the packing density of atoms in the sheet (e.g., atoms/m^2) is higher than any other network made of any atoms in the periodic table and higher than the packing density in any 2-dimensional slice through any 3-dimensional object—even diamond which has the highest known 3-D packing density. Hence the graphene sheet is effectively impermeable under normal chemical conditions.

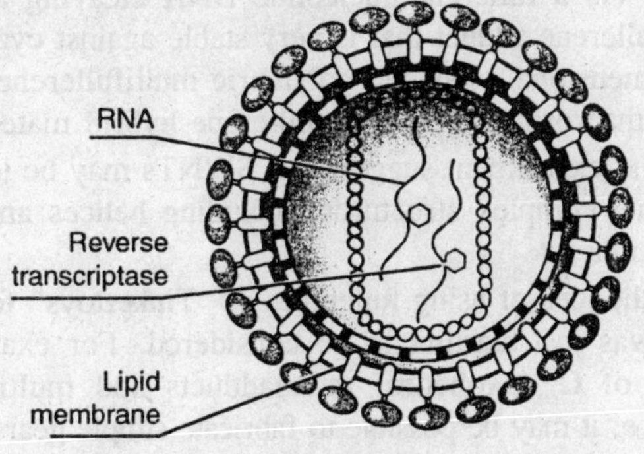

Fig. 5.7. HIV virus.

Carbon nanotubes have also been observed with a variety of endcap shapes and with tubes that reduced or increased their diameter for some distance before terminating. Toroidal fullerenes can be produced experimentally; conelike, spindlelike, and helical fullerene objects have also been examined. By 1998, the mechanical properties of fullerenes and carbon nanotubes were being extensively investigated and there were experimental demonstrations of:

1. A limited ability to cut fullerene nanotubes to specific desired lengths;
2. Dissolving derivatized SWNTs in common organic solvents;
3. The synthesis of self-oriented nanotube arrays; and
4. The trapping of individual C_{60} molecules in a perforated Langmuir-Blodgett film "workpiece holder" (though an even simpler means for producing 2-nm hole arrays with 78 nm hole spacings was later reported).

Nanotubes were also being investigated as nanoscale sensor components.

Given an ability to synthesize a wide variety of fullerene shapes and sizes, the next task is to find ways to join them together in specific desired molecular architectures. Two hydrogens were readily added to fullerene carbon atoms, making specific isomers of $C_{60}H_2$ and $C_{70}H_2$, and in the 1990s fullerene chemistry began to be explored. A little more than a decade after the discovery of C_{60}, gram quantities of buckyballs became widely available for chemical experimentation. Progress in covalent fullerene chemistry exploded. By the late 1990s, fullerene surfaces could be regioselectively functionalized in many interesting ways, including: (i) a fullerene dimer, (ii) a fullerene-polyester polymer, (iii) a fullerene dendrimer, (iv) a fullerene-rotaxane and catenane, (v) a fullerene-nucleotide DNA cleaving agent, (vi) a Diels-Alder fullerene adduct that is very stable against cycloreversion, and (vii) 3-dimensional extended polymeric multifullerene forms such as acetylenic macrorings and DNA/fullerene hybrid materials.

Theory and experiment suggest that SWNTs may be joined at 30° angles to create complex structures, including helices and three-way nanotube junctions.

In 1998, the idea of using fullerenes as "Tinkertoys" for molecular construction was just starting to be considered. For example, given the synthesis of C_{60} exohedral monoadducts and multiple adducts described above, it may be possible to fabricate simple gears by bonding rigid ligands onto the external surfaces of carbon nanotubes in the manner of gear teeth, although the chemical synthesis of nanotubes with precisely positioned teeth will not be easy. Given the success in

1996 of IBM scientists in positioning individual organic molecules (each having a total of 173 atoms and a 1.5-nm diameter) at room temperature by purely mechanical means, it might also be possible to align and maintain these molecular gear teeth in atomically precise meshed positions. If this could be done, would such devices actually work in the manner of macroscale gears?

J. Han and colleagues at NASA/Ames performed a detailed 2000-atom molecular dynamics simulation to investigate the properties of molecular gears fashioned from carbon nanotubes with teeth added via a benzyne reaction known to occur with C_{60}. Computationally, one gear is powered by forcing the atoms near the end of the nanotube to rotate, and a second gear is allowed to rotate by keeping the atoms near the end of its nanotube constrained to a cylinder (i.e., the ends of the shaft were constrained to not elongate but were allowed to move within a plane transverse to the tube symmetry axis). The meshing aromatic gear teeth transfer angular momentum from the powered gear to the driven gear. Each gear is made of a 1.1-nm diameter (14, 0) nanotube with seven benzyne teeth. The spacing between two nanotubes is 1.8 nm and the smallest distance between a tooth atom and a tube atom is ~0.4 nm. The results show that the gears can operate up to 70 GHz *in vacuo* at room temperature without overheating or slipping. As rotational speed rises above 150 GHz, the gears overheat and begin to slip with tooth tilting up to 20°, but no bond or tooth breaking occurs up to at least 3000 K and slipping gears can always be returned to proper operation by lowering the temperature or the rotation rate. A related nanotube gear system (58 sprockets, 290-464 atoms) simulated by Robertson and colleagues at the Naval Research Laboratory (NRL) showed similar overheating at 500 GHz but was stable when accelerated to only 20 GHz.

The NASA group also simulated several other types of nanotube-based gear systems. For instance, a rack and pinion system was designed using a gear made from a (14, 0) nanotube with teeth separated by two hexagon rings and a shaft made from a (9, 9) tube with teeth separated by three rings. Gear and shaft are 1.94 nm apart, with tooth face normal to the radial direction of nanotube (14, 0) for the gear, but in the axis direction of nanotube (9, 9) for the shaft. The gear could receive power, driving the shaft, or vice versa, and worked well for shaft translational velocities up to ~100 m/sec. (Most devices described in Drexler typically move at 1 m/sec.) Since shaft mass is almost twice gear mass, it takes more power for the gear to drive the shaft. In another simulation involving a large 1.4-nm gear coupled to a smaller 0.8-nm gear, the large gear drives the smaller gear smoothly, but if power is instead applied to the smaller gear at a sufficiently large acceleration, then the smaller gear does not drive the larger one but instead "bounces back and forth several times, like elastic collisions of a small ball between two boards."

Another research group at Oak Ridge National Laboratory (ORNL) used classical molecular dynamics to investigate the properties of molecular bearings consisting of an inner and an outer carbon nanotube. The graphite bearings ranged in size from inner shafts between 0.4-1.6 nm in diameter up to 12 nm long, and outer cylinders between 1.0-2.3 nm in diameter up to 4 nm long. The original simulations found excessive vibrational motion, but subsequent work using a more complete quantum approach found that under certain conditions the nanobearing is "frictionless" and undergoes superrotation, a classical dynamical behaviour reminiscent of superfluidity. The regime of superrotary motion is somewhat restricted when the nanobearing is under load, which suggests that a very careful design is required to ensure optimum performance.

The ORNL group simulated a fullerene motor consisting of two concentric graphite cylinders (shaft and sleeve) with one positive and one negative electric charge attached to the shaft. Rotational motion of the shaft was induced by applying one, or sometimes two, oscillating laser fields. The shaft cycled between periods of undesirable rotational pendulum-like behaviour and good unidirectional motor-like behaviour. The NASA/Ames group simulated a pulsed-laser-powered carbon nanotube gear-motor system which rotated consistently in one direction, although that direction could be either clockwise or counterclockwise.

G. Leach and colleagues also simulated various carbon nanotubes treated as nanostruts using molecular dynamics calculations, as described earlier for diamondoid block struts. The conclusions were similar though less severe. Nanotubes with 100:1 aspect ratio and 0.8-nm diameter at 300 K experienced periodic oscillations in length of ~0.4 nm at ~122 GHz, although multiple modes clearly were being excited; nanotubes of lesser aspect varied by at most ~0.04 nm.

Scanning Probe Technology

The third general pathway leading to molecular manufacturing involves a technology known as scanning probe microscopes (SPMs). The first of the SPMs was the Scanning Tunneling Microscope (STM) developed in the late 1970s and early 1980s by Gerd Karl Binnig and

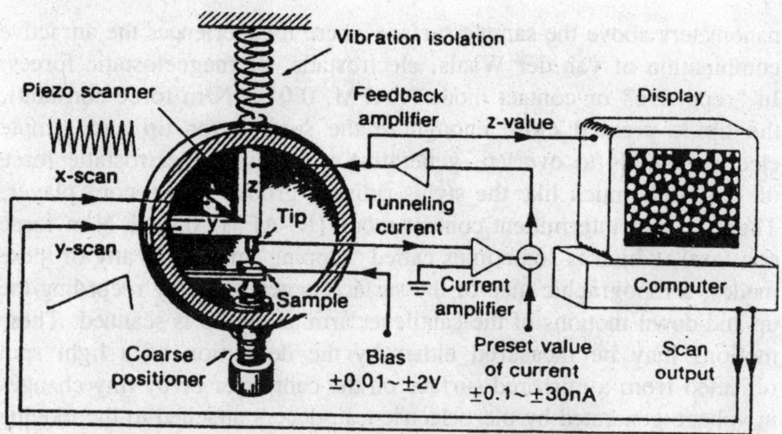

Fig. 5.8. Scanning tunneling microscope (STM).

Heinrich Rohrer at an IBM research lab in Zurich, Switzerland, earning these scientists, along with Ernst Ruska, the 1986 Nobel in Physics. The STM was initially used as an imaging device, capable of resolving individual atoms by recording the quantum tunneling current that occurs when an extremely sharp conductive probe tip (usually tungsten, nickel, gold, or PtIr) is brought to within about one atomic diameter of an atom, and then adjusting the position of the tip to maintain a constant current as the tip is scanned over a bumpy atomic surface. A height change as small as 0.1 nm can cause tunneling current to double. The tip is connected to an arm that is moved in three dimensions by stiff ceramic piezoelectric transducers that provide subnanometer positional control. If the tip is atomically sharp, then the tunneling current is effectively confined to a region within ~0.1 nm of the point on the surface directly beneath the tip, thus the record of tip adjustments generates an atomic-scale topographic map of the surface. STM tips can scan samples at ~KHz frequencies, although slower scans are used for very rough surfaces. In some modern STMs (e.g., the DI Nanoscope), the sample is moved while the tip is held stationary.

A major limitation of the STM was that it only worked with conducting materials such as metals or semiconductors, but not with insulators or biological structures such as DNA. To remedy this situation, in 1986 Binnig, Quate and Gerber developed the Atomic Force Microscope (AFM) which is sensitive directly to the forces between the tip and the sample, rather than a tunneling current. An AFM can operate in at least three modes. In "attractive" or non-contact mode (NC-AFM, 0.01-1 N/m force constant), the tip is held some tens of nanometers above the sample surface where it experiences the attractive combination of van der Waals, electrostatic, or magnetostatic forces. In "repulsive" or contact mode (CAFM, 0.01-1 N/m force constant), the tip is pressed close enough to the surface for tip and sample electron clouds to overlap, generating a repulsive electrostatic force of ~10 nN, much like the stylus riding a groove in a record player. There is also intermittent-contact mode (IC-AFM, 0.01-1 N/m force constant), which is sometimes called "tapping" mode. In any of these modes, a topographic map of the surface is generated by recording the up-and-down motions of the cantilever arm as the tip is scanned. These motions may be measured either by the deflection of a light spot reflected from a mirrored surface on the cantilever or by tiny changes in voltage generated by piezoelectric transducers attached to the moving cantilever arm. Typical AFM cantilevers have lengths of 100-400 microns, widths of 20-50 microns, and thicknesses between 0.4 to several microns. AFM tips may be positioned with ~0.01 nm precision, compressive loads as small as 1-10 pN are routinely measured, and the tips may be operated even in liquids.

S. Vetter notes that STM technology has also improved, reaching resolutions of ~0.001 nm in the z direction (vertical) and ~0.01 nm in the xy plane, well beyond atomic resolution. The STM remains the instrument with the best resolution. The conducting surface limitation has been overcome in some cases by coating the target with an extremely thin conducting layer, developing tips with multiple electrodes, and coating a conducting substrate with a sample so thin as to allow enough conduction even if the sample is characterized as an insulator in bulk.

By 1998, the growing family of SPMs included at least forty types of instruments and techniques that relied on interactions between a scanned surface and a nearby probe. Different instruments measured different forces and thus could be used to characterize different properties of the surface. For example, friction force microscopes (FFMs), magnetic force microscopes (MFMs), shear force microscopes (ShFMs), scanning capacitance microscopes (SCMs), scanning conducting ion microscopes, chemical force microscopes, and electrostatic force microscopes (EFMs) measured frictional drag or other binding forces. Magnetic resonance force microscopes (MRFM) used a field generated from a small magnet mounted on the tip of the cantilever arm to probe nuclear magnetic moments in a small region on the surface of the sample, imaging atom types and even detecting the spin of a single electron. By the mid-1990s, AFMs were already a $100 million/year industry and SPMs generally were an off-the-shelf technology costing up to $50,000-$500,000 for complete systems, with the whole industry worth up to ~$0.5 billion annually. (Low-performance "homebrew," student, and science-fair STMs have been built for as little as $50.)

How might SPMs be used for molecular manufacturing? In the crudest approach, SPMs might be employed as nanoscale milling machines to carve out "nanoparts" from appropriate substrates. For example, Kim and Lieber used an AFM to perform nanomachining operations on a molybdenum trioxide crystal—applying a 100 nN load at the tip, they milled a triangular-shaped 50-nm planar chunk from the crystal, then slid the part 200 nm across the work surface by pushing it with the AFM tip. Sheehan and Lieber milled a 50-nm rectangular part and two other parts with concavities complementary to the rectangle by mechanically etching a MoO_3 layer deposited atop an MoS_2 surface by pressing down with an AFM tip at a 50 nN load. Like a numerical-controlled machine tool, the AFM motions could be programmed to perform a series of steps, like automatically carving the chemical formula "MoO_3" into the crystal. Nanoparts with features as narrow as 10 nm can be nanomilled in a limited set of materials. In theory, large arrays of independently-controlled SPM tips could be operated in parallel to carve out great numbers of similar objects simultaneously, though such objects would of course not be atomically identical without further "finishing." AFM tips have also been used to bulldoze nanoscale lines and rectangular <10-nm features into nonconducting photoresist, perform "dip-pen" AFM nanolithography, form grooves, and fabricate a single-electron transistor via an STM nano-oxidation process. A carbon nanotube affixed to an AFM tip was used as a pencil to write 10-nm-wide features onto silicon substrates at ~0.5 mm/sec: An electric field issues from the nanotube, removing hydrogen atoms from a hydrogen monolayer atop a silicon base; the exposed silicon surface oxidized, producing narrow SiO_2 tracks.

Perhaps more significantly, SPMs have been employed in an increasingly sophisticated manner to manipulate individual atoms and small clusters of atoms. The possibility of modifying surfaces with scanning probe tips was evident from the earliest years of STM research, since inadvertent contact between tips and surfaces routinely caused such modifications, and the possibility of transferring material between tip and sample had already been discussed. Suggestions for controlled surface modification soon appeared, and in 1985 Becker and

Golovchencko used voltage pulses on an STM tip to pluck a single germanium atom from the {111} surface of a sample. In 1988, J. Foster and colleagues at IBM Almaden pinned small organic molecules to a graphite surface by applying a small electrical pulse through an STM. Additional pulses enlarged or erased this molecular feature, and often split the organic molecule into smaller pieces, though not in a controlled manner. Other workers used an STM to dump small clusters of gold atoms on a platinum surface, clusters of copper atoms on a gold surface, and molecules of tungsten carbide from a supply flowing over a surface. Others reported positioning carbon monoxide molecules and platinum atoms on platinum surfaces, moving clusters and single atoms of silicon across a silicon surface at room temperature, and fabricating via STM a "molecular corral" or "quantum corral" a ring of atoms so small that the enclosed electrons were forced to exhibit quantum behaviour. In 1999, Tomanek and Kral proposed an atomic fountain pen in which a carbon nanotube filled with atoms is electrically induced to release these atoms, one by one every 15 microsec, onto the work surface.

More precise control was achieved by Eigler and Schweizer at IBM Almaden in 1989, when they used an STM to position 35 individual xenon atoms on a nickel surface to spell out the corporate logo "IBM". To accomplish this, a bias voltage was applied to weaken the adsorption of each atom to its nickel substrate, then each atom was dragged, one by one, to the desired locations at a speed of ~0.4 nm/sec to build a meaningful pattern. It took 22 hours to make the entire logo, or ~38 min/atom. The experiment had to be carried out at 4 K (i.e., liquid helium temperatures) because the arrangement would have been unstable at room temperature. Numerous similar examples of "atomic graffiti" followed the IBM experiment, including atoms patterned in the shape of a world map; the word "atom" and "nanoworld", spelled out with atoms, in Japanese; the word "Peace," Einstein's famous equation "$E = mc^2$," and even a reproduction of the well-known Einstein portrait with the scientist's tongue sticking out; a sketch of a "molecule man" using 28 individual CO molecules on a platinum surface; gold nanoparticles spelling out "USC" and "Zyvex"; and sulfur atoms removed from a surface to make letters 2 nm high. In 1996, Gimzewski's group used an STM tip to manipulate individual buckyballs along terraces on a grooved copper plate, making a "molecular abacus."

Some variant of an SPM would appear to be the most promising tool for direct molecular manipulation. Once a tip lies within ~1 nm of a surface, the potential barrier can be lowered sufficiently for atoms to be induced from the surface by field evapouration; an applied tip voltage ionizes the atom which can then be guided around by the tip. Depending on voltage and separation, atoms, atom clusters, or molecules can be pried out, pushed in, or nudged around on a surface. By 1998, techniques for direct atomic manipulation proceeded at room temperature and ~1000 times faster than the original IBM logo had been assembled. For example, the Nanomanipulator, an interactive haptic control system created by a group at the University of North Carolina at Chapel Hill, allows near-real-time manipulation of individual gold atoms across a surface using a hand-held master-slave controller that drives an STM probe while the position of the atom is displayed on a monitoring screen visible to the user. Partially funded by NSF, Silicon Graphics, and TopoMetrix, the Nanomanipulator is a fully-integrated system that enables investigators to "feel" the interatomic forces as the user pushes atoms around on a surface. In one demonstration, a TV-watching user herded a gold atom, slipping and sliding, into a slot in the planar workplace in about 1 minute; when force feedback was turned off, movements tended to run wild. IBM Almaden has also experimented with removing individual atoms from metal surfaces using an STM hooked into a Virtual Reality Dataglove apparatus.

To manufacture atomically precise parts, it also will be necessary to manipulate covalent bonds at the probe tip. STMs have broken and created chemical bonds; ~1 volt pulses were used to pull atoms out of crystals, binding them to the tip, and then to reinsert them back into the crystal. In 1995, the first demonstration of catalysis on a nanometer scale was reported by scientists at the Molecular Design Institute at LBL. They used an AFM modified to function like an ultrafine-point pen for catalytic calligraphy to change the chemical composition of a material surface one molecule at a time. A surface was prepared as a self-assembled monolayer (SAM) of alkylazide molecules capped with a crown of three nitrogen atoms, then platinum-coated chromium was deposited onto an AFM silicon tip just a few atoms wide. The SAM was soaked with a hydrogen-containing solvent, then scanned by the AFM over a 100 micron^2 area, with the platinum catalyzing a covalent bonding reaction in which hydrogen was added to the azides, transforming them into amines as revealed by selective fluorescent tags.

Tip technology is an important and fast-growing subspecialty of SPM research. In 1990, Drexler and Foster suggested the use of custom-made, synthetic proteins to be mounted on the point of an SPM tip. Modified antibodies or specially designed proteins could serve as simple, first-generation "grippers" for binding and manipulating specific molecules, bringing them into position to react with other molecules in a precise and selective way. Functionalized SPM tips have been created to exploit antibody-antigen recognition and in the context of chemical force microscopy. A major convenience for molecular manipulation is that the same instrument that does the chemistry and the molecular orienting can also be used to inspect the results.

In 1996, Smalley's group at Rice University attached a single nanotube to the pyramidal silicon tip of an AFM and showed that it was quite robust and could image the bottom of deep trenches inaccessible to conventional tips. By 1998, progress had been made in functionalized carbon nanotubes for use on AFM tips. Most notably, Wong and colleagues prepared nanotube tips by oxidation in air at 700°C, burning off all but 2% of the original material and leaving the ends covered with carboxyl (COOH) groups whose chemistry is rich and well understood. Four different kinds of tips were created: (1) the original carboxyl tip, which is acidic; (2) an amine-terminated tip (made by forming an amide bond to one of the amine groups in ethylenediamine ($H_2NCH_2CH_2NH_2$)), which is basic; (3) a hydrocarbon-terminated tip (made by forming an amide bond to benzylamine ($C_6H_5CH_2NH_2$)), which is hydrophobic; and (4) a biotin-terminated tip (made by forming an amide bond to a biotin derivative), which shows specific binding to streptavidin. AFM contact forces between tips and selected substrates were shown to be sensitive to pH and to the chemical details of the substrate in ways consistent with the tips' intended chemistry.

Wong's tips had three closely related advantages over previous techniques. First, when a functional group is attached to the apex of an Si_3N_4 or SiO_2 tip, these groups usually adhere to the sides of the tip as well; upon using the tip as a tool, there is a constant hazard that contact with the sides of the tip will alter the workpiece in unwanted places. Unlike these tips built with bulk techniques, the ends of Wong's tips are very different from their sides—the carboxyl groups are attached only on the ends of the nanotube, not on the sides. Second, Wong's tips have lateral dimensions set by the nanoscale dimensions of nanotubes, not by top-down fabrication techniques. Note the authors: "...the small effective radius of nanotube tips significantly improves resolution beyond what can be achieved using commercial silicon tips....we have recently demonstrated that lateral resolution of <3 nm can be achieved by using COOH-terminated single-walled nanotubes tips". Third, single-walled nanotube tips are much closer to yielding truly single atom tips with controlled chemistry than any other alternative. J. Soreff notes that a (10, 10) nanotube with a 1.4-nm diameter has just 20 atoms at an open end. Even given a statistical distribution of tubes with varying numbers of carboxyl groups attached to their ends, one should be able to build a ligand which covers the whole end of the tube, yielding a method for ensuring that just one molecule of known structure and orientation was present at the end of the probe.

Theoretical studies have employed molecular dynamics simulations and other techniques to investigate the behaviour of mechanosynthetic tool tips. For example, Drexler proposed using an acetylene radical to perform selective abstractions of hydrogen from a diamond surface, and this proposal was further studied in Musgrave's ab initio quantum chemistry analysis and Sinnott's molecular dynamics modeling of the reaction at room temperature on a diamond {111} surface; other H-abstraction studies have been done. One possible structure for a hydrogen abstraction tool might be similar to an anthracene whose end had been modified; prior to use, the structure must be activated by removal of the terminal hydrogen. Brenner and colleagues describe further work modeling the parallel abstraction of several hydrogen atoms from the diamond {111} surface. Such reactions are similar to those involved in the well-studied chemical vapour deposition of diamond, but adding only positional control and applied mechanical force; more detailed theoretical studies of diamond mechanosynthesis reactions have begun.

R. Merkle has proposed a simple mechanosynthetic system capable of fabricating a large class of useful stiff hydrocarbons. Merkle's system includes positionally-controlled hydrogen abstraction and deposition tools (to remove or add hydrogen atoms to a workpiece), carbene and dimer deposition tools (to add one or two carbons to a workpiece), a silicon radical tool (to remove a carbon), and four other specialized tools. Each proposed tool has a molecular handle structure that could be permanently affixed to a suitably functionalized AFM tip (or other appropriate mechanosynthetic nanomanipulator). As each operation is performed on the butadiyne (a.k.a diacetylene, C_4H_2) hydrocarbon feedstock in an inert (e.g., vacuum or noble gas) environment, one or more tools are brought to the workpiece at the necessary position and angle, and any necessary force is applied via the AFM. Interestingly, except for a small number of "vitamin parts" involving transition metals, Merkle's system can in theory synthesize all of its own mechanosynthetic tools, thus establishing net positive production of system components and illustrating a limited subset-level "quantitative parts closure" as first defined in 1982 by Freitas and Gilbreath in connection with *self-replicating machine systems*. Merkle acknowledges that several of the proposed reactions involve the simultaneous coordinated action of two, three, or even four positionally controlled tools, but suggests that two 6-degree-of-freedom manipulators plus some combination of appropriate jigs and fixtures should be sufficient in a more parsimonious system. However, steric hindrance of multiple tools at one site also was not explicitly addressed. Another potential problem to be overcome is loss of positional registration as the various functionalized tool heads are changed out at the AFM tip; such tips might be paired with a second positionally-invariant metrology tip on each tool head, allowing precise calibration of workspace location before proceeding back to the workpiece.

Others have suggested fabricating nanoparts by building up structures by stacking layers of individual atoms, perhaps constrained by sacrificial joiners and scaffolding, and honed by differential etching operations.

Assuming that "*nanoparts*" can be fabricated to atomic precision, can SPMs also be used to assemble these nanoparts into working nanomachines? Some preliminary work has been accomplished. For instance, arrays of nanoscale holes have been synthesized that could serve as "workpiece holders." In one experiment, C_{60} buckyballs were scattered about on a prepared surface, then an AFM pushed the individual spheres into their own little holes, securely seating them pending further processing. The C_{60}-impregnated perforated film was so durable that the buckyballs remained trapped after several months of storage. Similarly, Jung and colleagues in Gimzewski's group synthesized 4-legged porphyrin-based molecular "nanoparts" that were specifically designed for easy positioning on a copper surface. Pushing with an STM at room temperature, the authors displaced the ~1.5-nm wide nanoparts in predefined directions and rotated the 4-legged molecules at will, arranging, for example, the parts into a precise hexagonal configuration on the copper surface. The STM has been used to induce, and to view, the rotation of an individual oxygen molecule trapped on a platinum surface, as it is intentionally and stably rotated into any one of three distinct orientations. The forces required to slide, or alternatively to roll, a carbon nanotube across a graphitic surface have been directly measured by AFM.

The first known instance of a crude but purely mechanical "nanopart" assembly operation was performed in 1995 by Paul E. Sheehan and Charles M. Lieber. Two nanoscale parts with rectangular slots and a third 50-nm rectangular sliding "latch" member were milled from a MoO_3 crystal using an AFM, then the rectangular "latch" member was slid repeatedly from one slot to the other using the same AFM, making a three-nanopart reversible mechanical latch. Noted the authors: "The lateral force needed to break the latch, 41 nN, was large considering the small latch contact area, which suggests that relatively robust assemblies can be created with such devices. Such a reversible latch could serve as the basis for mechanical logic gates....Our results...demonstrate the ability to machine complex shapes and to reversibly assemble these pieces into interlocking structures."

In 1998, manipulation of nanoscale parts in the vertical dimension (e.g., normal to the surface plane) had only just begun. The first three-dimensional structure built out of single "nanoparts" was demonstrated in 1997 by Requicha's group at the USC Laboratory for Molecular Robotics. First, an AFM was used to push 5-nm gold nanoparticles in the third dimension up and over surface protrusions or obstructions—for example, a 5-nm gold particle was shoved up onto a 2-nm-high protrusion, then slid back off again onto the surface. In another series of experiments conducted in air at room temperature with 15-30 nm gold nanoparticles deposited on silicon previously coated with a silane layer, Requicha's group used an AFM to build a simple 3-D pyramidal structure by pushing one gold "nanopart" on top of two others, then off again; the group also used the AFM to rotate and translate a dimer unit formed by two linked "nanoparts".

Three-dimensional nanoassembly might perhaps be more readily achieved if nanoparts could assist in their own assembly process. R. Merkle suggests that an SPM could be used to position DNA-tagged molecular building blocks. A building block tagged with a specific single-stranded DNA should preferentially adhere to a surface covered with the complementary DNA. If the building block has a second single-stranded DNA tag which is complementary to single-stranded DNA on the tip of an SPM, and if it is initially attached to the SPM tip by this tag, then it could be positioned by the SPM and attached to the surface. The relative strength of attachments could be adjusted by changing the number of base pairs in the complementary region between the two strands. A related suggestion, due to G. Fahy, is that SPM tips for assembly could be designed with more than one binding site.

The existence of at least two binding sites could allow precise orientation of a nanopart, permitting the nanopart to be presented to a desired target site (or another nanopart, possibly attached to another manipulator) at exactly the location and in exactly the orientation desired, assuming that each nanopart can bind to each tip in only one possible orientation. A convenient post-assembly nanopart release mechanism is also required.

In 1997, the Avouris group at IBM Yorktown Heights demonstrated that the tip of an atomic force microscope (AFM) could be used to control the shape and position of individual multiwalled carbon nanotubes dispersed on a surface. Nanotubes could be bent, straightened, translated, rotated, and (under certain conditions) cut.

In 1998, Zyvex LLC (a developmental engineering company whose goal is to create a molecular assembler capable of manufacturing atomically precise structures) demonstrated the ability to manipulate carbon nanotubes in three dimensions inside a scanning electron microscope (SEM) having ~6 nm resolution at near-video scan rates. Zyvex's custom piezoelectric vacuum manipulator achieved positional resolutions comparable to SPMs along with the ability to manipulate objects along one rotational and three linear degrees of freedom with 0.1 nm spatial resolution. This prototypical device could probe, select and handle limited classes of nanometer-scale objects such as carbon nanotubes under real-time SEM inspection. Carbon nanotubes were attached to commercial atomic force microscope (AFM) tips either by van der Waals forces alone or by "nanosoldering" by a concentrated electron beam from the SEM. Forces applied to the nanotubes could be measured from the cantilevers' deflections (spring constants 0.01-100 N/m). Sometimes nanotubes were transferred from tip to tip. The manipulator could function both as a research tool for investigating properties of carbon nanotubes and other nanoscale objects without surface restrictions, and as a rudimentary building device for larger nanotube assemblies. Zyvex believed that this capability to select and manipulate nanoscale components and to examine directly their suitability as construction materials during various phases of the construction process would play an important role in enabling the technology of assembling mechanical and electronic devices from prefabricated components. Most impressively, the Zyvex system allowed the simultaneous coordinated operation of three independently controlled AFM tips within the same workspace, at different orientations. (A four-tip system was expected to be operational by mid-1999.) Two-handed coordinated micromanipulation under the view of an SEM had previously been demonstrated by others.

Tuzun and colleagues calculated the requirements for coaxial docking of two nanotubes of different diameters to form a molecular bearing. They looked at bearings formed from nanotubes 11 rings long, with 10 carbons per ring in the shaft and either 30 or 34 carbons per ring in the sleeve. For the computer simulations, Tuzun placed the sleeve of the bearing along the z axis and gave the shaft a small initial velocity towards it. A perfectly aligned shaft falls straight into a potential well from the van der Waals attraction to the sleeve; displaced or misoriented shafts can bounce off the edge of the sleeve. The docking envelopes for the molecular dynamics calculations and for the rigid body calculations had essentially the same shape, but with slightly different sizes. For the atomistic calculation, a shaft aligned parallel to the sleeve (with a 30 carbon ring sleeve) can be displaced by 0.26 nm before it fails to dock, while in the rigid body calculation the displacement can only be 0.16 nm. The authors note that "if an end of the shaft points closely enough to the center of the sleeve, it will fall into the nonbonded potential energy well and the two nanotubes will dock." These calculations are important because they tell us what tolerances can be accepted during the manual assembly process. Note the authors: "A question just as important as how to design or operate nanomachines is how to assemble them". Interestingly, one of Zyvex's experiments produced a multiwalled carbon nanotube that thinned in three steps along its length, suggesting the most likely explanation that the inner shells had been pulled from the outer shells. This phenomenon has also been observed by others and is known as the "sword and sheath" failure the exact inverse of the purposeful nanotube insertion operation investigated by Tuzun and colleagues.

In order to produce large numbers of nanoparts and nanoassemblies, massively parallel SPM arrays and microscale SPMs would be most convenient. Force sensing devices such as piezoelectric, piezoresistive, and capacitive micro-cantilevers made it possible to construct microscale AFMs on chips without an external deflection sensor. In 1995, Itoh and colleagues at the University of Tokyo fabricated an experimental piezoelectric $ZnO_2 on SiO_2$ microcantilever array of ten tips on a single silicon chip. Each cantilever tip lay ~70 microns from its neighbour, and measured 150 microns long, 50 microns wide and 3.5 microns thick, or ~26,000 micron^3/device, and each of the devices could be operated independently in the z-axis (e.g., vertically) up to near their mechanical resonance frequencies of 145-147 KHz at an actuation sensitivity of ~20 nm/volt—for instance, 0.3-nm resolution at 125 KHz.

Parallel probe scanning and lithography has been achieved by Quate's group at Stanford, which has progressed from simple piezoresistive microcantilever arrays with 5 tips spaced 100 microns apart and 0.04-nm resolution at 1 KHz but only one z-axis actuator for the whole array, to arrays with integrated sensors and actuators that allow parallel imaging and lithography with feedback and independent control of each of up to 16 tips, with scanning speeds up to 3 mm/sec using a piezoresistive sensor. By 1998, Quate's group had demonstrated arrays of 50-100 independently-controllable AFM probe tips mounted in 2-D patterns with 60 KHz resonances, including a 10 × 10 cantilevered tip array fabricated in closely spaced rows using throughwafer interconnects on a single chip.

MacDonald's group at the Cornell Nanofabrication Facility has pursued similar goals. In 1991, the team fabricated their first submicron stylus, driven in the xy plane using interdigitating MEMS comb drives, including the first opposable tip pair. By 1993, they had produced a 25-tip array on one xyz actuator, and by 1995 a complete working microSTM (including xy comb drives) measuring 200 microns on an edge and a microAFM measuring 2 mm on an edge including a 1-mm long cantilever with a 20-nm diameter integrated tip on a 6-micron high by 1-micron diameter support shaft. MacDonald's group demonstrated tip arrays with 5 micron spacings, exploiting the same process used to make the working microSTM. With the same technology tips or small arrays of tips could be spaced 25-50 microns apart and integrated with individual z-axis microactuators, so that one xy-axis manipulator could support many tips with each tip having a separate z actuator. By 1997, the group had built and tested an array of microSTMs on the surface of an ordinary silicon chip, with each tip on a cantilever 150 microns long with 3-D sensing and control. The largest prototype array has 144 probes, arranged in a square consisting of 12 rows of 12 probes each, with individual probe needles about 200 microns apart. Further development was to focus on increasing the range of movement and on fitting more and smaller probes into the same space.

Molecular Components and Molecular Assemblers

Assuming that various elementary molecular parts fabrication and assembly operations can be performed for certain limited classes of structures, then what structures would be interesting and useful to build? This Section briefly describes a small set of possibly useful molecular mechanical components and then concludes with a discussion of the concept of the molecular assembler.

Molecular Assemblers

Drexler has proposed the molecular assembler, a device resembling an industrial robot which would be capable of holding and positioning reactive moieties in order to control the precise location at which chemical reactions take place. This general approach would allow the construction of large atomically precise objects by a sequence of precisely controlled chemical reactions. Much like the ribosome in biology, an assembler would build various classes of useful molecular structures following a sequence of instructions. During this process, the assembler would provide three-dimensional positional and full orientational control over the molecular component (analogous to the individual amino acid in the ribosome model) that is being added to a growing complex molecular structure (analogous to the growing polypeptide in the ribosomal model). In one approach, a molecular assembler may be capable of forming any one of several different kinds of chemical bonds (e.g., by changing tool tips), not just a single kind such as the peptide bond that the ribosome makes. In bonding atoms or molecules to one another, the assembler would provide any needed energy (especially if the reaction happens not to be energetically favoured) through physical force, thus performing mechanosynthesis (as opposed to the traditional means of chemical synthesis in solution). In another approach, a molecular assembler might be capable only of noncovalent assembly operations, wherein nanoparts are fabricated by other means and then presented to the assembler, which assembles the nanoparts into working nanomachines.

The first simple molecular assembler will almost certainly be a macroscale device, perhaps a modified SPM system as was being pursued by Zyvex in 1998. Multiple SPM heads could be equipped with a small number of nanoscale tool tips. In one scenario, nanoparts fabricated using bulk chemistry techniques would be inspected and selected by the SPM, then assembled one by one into working nanomachines (e.g., the desired useful nanoscale products). Such assembly operations will be very slow, because the placement of each new component may require simultaneous rotations and translations of large macroscale SPM components. Assembly time scales roughly linearly with assembler size because smaller assembler components moving at a given velocity need to travel less distance to accomplish a given physical operation, hence consuming less time and energy per physical operation. An important early developmental goal thus will be to design and fabricate nanoscale molecular assemblers.

At its most basic level, the simplest possible nanoscale molecular assembler may be comprised of one or more nanoscale manipulators. For example, the diamondoid telescoping manipulator arm has about 4 million atoms excluding the base and control and power structures; doubling the size to account for support structures gives a molecular weight per arm of ~100 megadaltons and a total molecular volume of ~140,000 nm^3. Designed for high strength and stiffness, this robot arm should be able to hold a molecular fragment stiffly enough to make it react with a chosen end of a carbon-carbon double bond with an error rate of only 10^{-15}. The robot arm could guide chemical reactions with high reliability at room temperature *in vacuo*, with little need for sensing the positions of the molecules with which it is working. Alternatively, with appropriate tool tips, such an arm could grasp and manipulate individual prefabricated nanoparts in solution phase. Assembly of nanoparts without fabrication may require only a very small tool set.

The manipulator arm must be driven by a detailed sequence of control signals, just as the ribosome needs mRNA to guide its actions. However, such detailed control signals can be provided by external acoustic, electrical, or chemical signals that are received by the robot arm via an onboard sensor or power transducer, using a simple "broadcast architecture", a technique which can also be used to import power. Such transducers may be extremely small, on the order of $(\sim 10\ nm)^3$ each. Interestingly, the biological cell may be regarded as an example of a broadcast architecture: the nucleus, located external to the cytoplasm, broadcasts mRNA chemical signals to millions of spatially diverse cytoplasmic ribosomes, thereby remotely controlling the construction of cellular proteins.

Thus a two-armed mechanical molecular assembler that receives power and instructions from some external agency may have a total molecular volume of ~300,000 nm^3 (a cube ~67 nm on an edge), containing ~16 million atoms with a molecular weight of ~200 megadaltons—very roughly the mass and scale of a medium-size virus particle, such as an adenovirus. For comparison, the average enzyme (a biochemical crimping tool) weighs about ~0.1 megadalton, while a typical ribosome (a primitive "protein assembler") weighs ~4.2

megadaltons. A simple mechanical assembler in the 10-100 megadalton range cannot be ruled out. However, such a device might have a very small set of tool tips and an extremely limited manufacturing repertoire.

Multiple nanoscale assemblers each capable of independent simultaneous actuation may require control signals more conveniently provided by an onboard nanocomputer. This programmable nanocomputer must be able to accept stored instructions which are sequentially executed to direct the manipulator arm to place the correct moiety or nanopart in the desired position and orientation, thus giving precise control over the timing and locations of chemical reactions or assembly operations. The mechanical nanocomputer analyzed by Drexler requires ~16 nm^3 per logic gate and ~40 nm^3 per data register. If such components could be used to construct the equivalent of the most primitive 4-bit Intel 4004 microprocessor, then the nanocomputer could process ~10^5 bits/sec (~25,000 ops/sec) at a ~1 KHz clock speed within a mechanism volume of ~36,000 nm^3, again neglecting power supply, I/O linkages, and the like. An additional ~160,000 nm^3 of rod logic registers adds ~1 kilobyte of onboard RAM memory or ~40 kilobytes of internal tape memory (conservatively assuming a tape storage density comparable to linear DNA). (A memory tape containing all bits needed for complete self-description may be considerably longer, even allowing for substantial data compression.) Doubling the total volume to account for support structure and other overhead gives a minimum mechanical nanocomputer molecular volume of ~400,000 nm^3, roughly 70 million atoms with a molecular weight of ~800 megadaltons. Thus the smallest nanocomputer-driven nanoscale molecular assembler with two manipulator arms may have a total molecular volume of ~700,000 nm^3 (a cube ~88 nm on an edge), containing ~86 million atoms with a molecular weight of ~1 gigadalton.

By 1998, only a small amount of research targeted at actual assembler design had begun. Following Drexler's original discussions, during 1991-1998 R. Merkle authored or coauthored a continuing series of papers discussing various operational aspects and specific components of assembler design, including mechanosynthetic positional control, general design considerations for assemblers, the broadcast architecture, convergent assembly, binding sites, positioning devices, mechanosynthetic path sets, designs for a neon pump and a fine motion controller, and possible assembler casings. J.S. Hall has considered high-level designs for a nanoscale parts-fabrication and parts-assembly nanorobot. Zyvex, founded in 1996 by James von Ehr, has set itself the task of building the first programmable nanoassembler in a 5-10 year time frame (e.g., by ~2006). W. Goddard and colleagues have proposed a series of molecular dynamics simulations of simple assemblers, although by 1998 these studies evidently had not yet begun. According to Goddard's original proposal:

"Ultimately we need a programmable synthetic system to make a real device. Even though we may not have tools for all the chemical steps and may not have designs for all the pumps, engines, and transmissions needed, we propose to study the dynamics of simplified prototype assemblers. In these studies we anticipate having (1) a reservoir or supplies area for providing the various building units (atoms and fragments) required, (2) a work area in which we construct the nanomachine device (initially we will consider assembling the structure on top of a diamond surface), and (3) a molecular scale nanohand which will extract the atoms from (1) and carry them to (2). We will then use extensions of our massive molecular dynamics program to operate the system: moving the tip from reservoir to work area, moving it to contact the appropriate surface site, moving it to regenerate the active tip, and then moving it back to add new atoms and molecules. This will include proper temperature effects, molecular vibrations, energy release upon the various chemical steps, etc. The ground rules here are that a realistic force field be used and that all pieces be treated at the atomic level (but some might be semi-rigid). This will use the force field developed for nanosynthesis. The purpose of these simulations is to examine issues of vibration caused by chemical forces as the tool picks up and delivers atoms to the growing surface. Also we want to consider the effect of energy release in the chemical steps on the thermal fluctuations in these systems (which may cause displacements and vibrations)."

Building mega-atom or giga-atom nanoproducts one at a time would be incredibly expensive and time consuming. For example, imagine that we wish to construct a simple medical nanorobot such as the 1-micron spherical *respirocyte*, which consists of ~18 billion atoms (dry structure). A factory employing a coordinated team of 100 macroscale SPM assemblers, each able to place ~1 atom/sec-SPM on a convergently-assembled workpiece achieves a pitiful manufacturing throughput rate of ~2 *respirocytes* per decade. Nanoscale assemblers with appendages ~10^6 times smaller than macroscale SPM assemblers might plausibly achieve net assembly rates of $\sim 10^6$ atoms/sec-nanoassembler, but even a production line employing ~ 300 such nanoassemblers (which the abovementioned 100-SPM factory team could build in ~ 1 year, assuming $\sim 10^7$ atoms/nanoassembler) can only manufacture ~ 1 respirocyte nanorobot per minute. At that rate, it would take ~ 2 million years to build the first ~ 1 cm^3 therapeutic dosage containing $\sim 10^{12}$ respirocyte nanorobots.

The necessary solution to this mass-production bottleneck is to employ any of several massively parallel approaches to manufacturing, including such techniques as self-assembly, convergent assembly, or, most usefully, *self-replication*. The basic advantage of self-replication is readily illustrated. Consider a single $\sim 10^8$ atom "seed" nanoassembler (having an onboard nanocomputer) that has been painstakingly built in $\sim 10^6$ sec using the abovementioned 100-SPM macroscale assembler team, working at ~ 1 atom/sec-SPM. The seed nanoassembler is first programmed to build a copy of itself, which it accomplishes (working at $\sim 10^6$ atoms/sec-nanoassembler) in ~ 100 sec. These two nanoassemblers then each build a copy of themselves in another ~ 100 sec; now there are four nanoassemblers. After ~ 48 generations, requiring a total of ~ 80 minutes to complete, there are $\sim 3 \times 10^{14}$ nanoassemblers. These nanoassemblers are then reprogrammed for the manufacture of 18-billion-atom *respirocytes* and are fed the appropriate, presumably different, feedstock. This vastly expanded nanoassembler manufacturing system can now produce $\sim 10^{12}$ respirocytes (~ 1 cm^3 therapeutic dose) per minute.

The design of machines able to make copies of themselves was first described by von Neumann. Many variations on this theme have been reviewed by Freitas and Gilbreath and Sipper, and design for self-replication in the context of nanoscale assemblers has been considered by Drexler, Merkle, and Hall. As Drexler notes: "It may seem somehow paradoxical that a machine can contain all the instructions needed to make a copy of itself, including those selfsame complex instructions, but this is easily resolved. In the simplest approach, the machine reads the instructions twice: first as commands to be obeyed, and then as data to be copied. Adding more data does not increase the complexity of the data-copying process, hence the set of instructions can be made as complex as is necessary to specify the rest of the system. By the same token, the instructions transmitted in a replication cycle can specify the construction of an indefinitely large number of other artifacts."

The estimated information content of self-replicating systems—the length of the instruction tape—can be surprisingly small. Von Neumann's original analysis concluded that perhaps twelve different kinds of units of unknown complexity could be required as building materials, and Haldane inferred that as many as $\sim 10^5$ individual parts might be needed to make a replicator. This inference was refuted just three years later with the arrival of the first in a stream of very simple but ingenious designs for mechanical self-replicating machines that were assembled and operated in the late 1950s. In the first example, a pair of joined ratchet-like blocks, when placed in a "sea" of left and right blocks and then physically agitated, replicated itself from this well-ordered input substrate, making more block-pairs until all the single blocks were used up, a process similar to chemical autocatalysis. More complex congeries of blocks, including clever four-block, eight-block, and 12-block replicators that could replicate themselves in a sea of blocks were presented by Penrose. Jacobson demonstrated a 3-unit replicator consisting of toy train engines circulating on HO model railroad tracks, and Morowitz designed a simple two-unit device with one unit comprised of about a dozen components including switches, batteries and electromagnets, that could assemble copies of itself from parts floating on the water surface of a bathtub. In 1998, Lohn and colleagues gave two designs for simple self-replicating systems that could be constructed out of wood, batteries and electromagnets, with explicit analogies to nanoscale fabrication. Several important conclusions may be drawn from these examples:

First, replication is fundamentally so simple a task that machines capable of displaying this behaviour predate most of the modern electronic computer era.

Second, assembly is an inherently simpler operation than fabrication. A complex part may embody hundreds or thousands of prior fabrication and assembly operations, yet may be installed within an assemblage in a single step. Hence nanopart assembly may be an easier candidate for early implementation in first-generation self-replicating molecular manufacturing systems than is molecular fabrication.

Third, and most important, the simplest nanoreplicator may require the ability to assemble, but not to (atomically) fabricate, in order to replicate itself. A nanomachine capable of assembly alone can replicate itself only from a very limited set of well-ordered input materials of relatively high complexity, but such a machine can be extremely simple both in structure and function. Certainly a nanomachine capable of both assembly and fabrication can replicate itself from a more diverse and disordered set of more elementary input materials but only at the expense of far greater internal structural and functional complexity.

At least two distinct models of replication have been identified in replicating systems design. The first model may be called the "unit replication" or organismic model, in which the replicator is an independent unit which employs the surrounding substrate to directly produce an identical copy of itself; both the original and the copy remain fertile and may replicate again, thus exponentiating their numbers. The second model may be called the "unit growth" or factory model, in which a population of specialist devices, each one individually incapable of self-replication, can collectively fabricate and assemble all necessary components comprising all specialist devices within the system, hence the factory is capable of expanding its size (or of manufacturing duplicate factory systems) indefinitely in an appropriate environment. The factory model is sometimes called "bootstrapping," which is any production of more productive capacity, from less. T. McKendree likens the situation to an ant colony: "There necessarily is the line of reproducing queens, making a sufficient number of ants feasible; [however,] most of the queen's products are 'workers' that perform useful functions, but themselves are incapable of reproduction."

It cannot be emphasized too strongly that mechanical medical nanodevices will not self-replicate inside the human body, nor will they have any need for self-replication themselves. Machines that perform medical tasks are fundamentally different from machines that manufacture other machines. As R. Merkle explains: "While self-replicating systems are the key to low cost [manufacture], there is no need (and little desire) to have such systems function in the outside world. Instead, in an artificial and controlled environment they can manufacture simpler and more rugged systems that can then be transferred to their final destination. Medical devices designed to operate in the human body don't have to self-replicate: we can manufacture them in a controlled environment and then inject them into the patient as needed. The resulting medical device will be simpler, smaller, more efficient and more precisely designed for the task at hand than a device designed to perform the same function and self-replicate. This conclusion should hold generally: optimize [product] device design for the desired function, manufacture the [product] device in an environment optimized for manufacturing, then transport the [product] device from the manufacturing environment to the environment for which it was designed. A single device able to do everything would be harder to design and less efficient."

In an effort to stimulate scientific and engineering interest in constructing the first nanoassembler, in November 1995 the Foresight Institute created the $250,000 *Feynman Grand Prize*, with funding contributed by Zyvex founder James von Ehr and St. Louis venture capitalist Marc Arnold. The prize will be awarded to the individual or group that first achieves both of two significant nanotechnology breakthroughs—first, the design and construction of a functional nanometer-scale robotic arm, and second, the design and construction of a functional 8-bit adder computing nanodevice. According to Grand Prize rules, the robotic nanomanipulator must fit entirely inside a 100-nm cube, carry out actions directed by input signals of specified types, be able to move to a directed sequence of positions anywhere within a 50-nm cube, complete all directed actions with a positioning accuracy of 0.1 nanometer or better, and perform at least 1,000 accurate, nanometer-scale positioning motions per second for at least 60 consecutive seconds. The adder must fit entirely within a 50-nm cube, and must be capable of adding accurately any pair of 8-bit binary numbers, discarding overflow, accepting input signals of specified types, and producing its output as a pattern of raised nanometer-scale bumps on anatomically precise and level surface. (J.S. Hall notes that a conventional 8-bit adder may be constructed using a total of 94 AND, OR, and NOT gates; using XOR gates, the total may be reduced to 37 gates.) Both devices may accept inputs from acoustic, electrical, optical, diffusive chemical, or mechanical means, although any mechanical driving mechanism used for input must be limited to a single linkage that either slides or rotates on a single axis. To demonstrate the capacity for mass production, at least 32 copies of each device must be provided for analysis and destructive testing by judges.

Once nanoassemblers are available, the design of far more complex nanomachines each containing tens or hundreds of billions of precisely arranged atoms will require new molecular CAD tools and techniques, including automated hierarchical design decomposition of objects (that must be built up from an array of nanoparts) and a shape description language. Nanosystem design compilers, conceptually similar to silicon compilers that generate a complex pattern of transistors and conductors from an abstract specification of the properties of a digital circuit, will also be required for efficient nanorobot design. Given a fully specified design that meets all constraints necessary to physically permit assembly, all operations of the assembly process must be specified using an assembly process compiler. Explains Drexler: "The design of structures and assembly procedures by hierarchical decomposition directly generates a tree of assembly steps. If this tree has been chosen to generate parts of the appropriate sizes and numbers, then it can be mapped onto a manufacturing layout. The vast number of manipulator motions to be specified at the finest, earliest, and most-dispersed levels of the assembly process can be planned in parallel by identical software systems running almost independently on separate processors. The result of this parallel assembly process compilation is a set of instructions that, when executed by the transportation system and manipulator controllers of a manufacturing mechanism, will result in the assembly of an object corresponding to the initial design."

Since molecular manufacturing systems can be used to make more molecular manufacturing systems, it is believed that the capital costs of production can be quite low. Drexler claims that an analysis of inputs, outputs, and productivity suggests the total cost of production can be in the range familiar in agriculture and in the production of industrial chemicals—on the order of tens of cents per pound. Merkle has made similar assertions, for example: "Eventually (after amortization of possibly quite high development costs), the price of assemblers (and of the objects they build) should be no higher than the price of other complex structures made by self-replicating systems. Potatoes—which have a staggering design complexity involving tens of thousands of different genes and different proteins directed by many megabits of genetic information—cost well under a dollar per pound." Such analogies from conventional agriculture or industrial chemistry are not strictly applicable to the medical products and pharmaceutical areas, where structural precision, product repeatability, and intensive quality control are of supreme importance, and where most products are subject to rigorous government regulation—unlike potatoes, which vary widely in size, shape, and molecular constitution, and generally do not require stringent product liability insurance, multiphase clinical trials, or FDA approval.

Still, it is likely that molecular manufacturing will eventually allow reasonably inexpensive production of large batches of designed medical nanorobots. As a very crude analogy, even in 1995 a typical mail-order bioscience vendor could manufacture 15 micromoles (~9

$\times$ 10^{18} product items) of customized oligonucleotides for \$18/base up to at least 110 bases with a convenient 48-hour turnaround on orders; other online vendors offer similar or slightly higher prices up to 200-mer sequences with additional charges for necessary purifications. If a billion-base structure could be induced to self-assemble from 10 million different sets of 100-mer custom-built oligonucleotides (total ~20 billion atoms), forming a desired ~20-billion-atom nanomachine, then, optimistically assuming a ~100% yield, the cost would be ~\$2 $\times$ 10^9 per nanomachine or ~\$2000 per ~1 cm^3 of product volume which would include 10^{12} finished nanomachines. In 1998, many specialty biotechnology-related drug treatments were comparably expensive. For example, treatments for multiple sclerosis using two closely related ~22.5 kilodalton interferons were administered in dosages of ~1 milligram/week at a treatment cost of ~\$10,000/yr, which is equivalent to a treatment cost of ~\$2 $\times$ 10^9 (2 "nanodollars") per giga-atom. The famous HeLa (Henrietta Lacks) cell line is an example of a once tiny population of unique cancerous cells that were purposely replicated *in vitro* (~24-hour replication time) over many decades and are now widely available worldwide at very low cost (e.g., \$216 per 1-cm^3 ampule of the ATCC CCL-2 HeLa cell line). Swallowable therapeutic pills containing bacteria (i.e., natural biological nanomachines) are already widely available over the counter for gastrointestinal refloration, as for example Salivarex which "contains a minimum of ~2.9 billion beneficial bacteria per capsule", and Alkadophilus which "contains 1.5 billion organisms per capsule", both at a 1998 price of ~\$0.2 $\times$ 10^9 per bacterium.

Another useful analogic approach to cost estimation is suggested by the plunging price of computer power during the 20th century. The plotted data is from a compilation of 67 historical computers by Moravec through 1989, with an additional 27 data points for other computers spanning 1973-1999 added by the author. The chart shows that the cost of 1 TFLOP has fallen from \$2 $\times$ 10^{20} in 1908 (the mechanical Hollerith Tabulator) to just \$6 $\times$ 10^6 in 1998 (the SGI/Cray T3E-1200E), a 90-year price decline of almost 14 orders of magnitude—on average, a halving every ~2 years, a century-long example of Moore's Law. This trend may give us some confidence that even if the first individual 1 $micron^3$ nanorobot costs ~\$100,000 to assemble (much like the first Nippondenso microcar), eventually the unit price may fall at least 14 orders of magnitude to below ~\$$10^9$ per nanorobot or <\$1000 per ~1 cm^3 dosage (10^{12} nanorobots).

Recycling or remanufacturing each nanomachine ~1000 times before final discard would then imply a net ~\$1/cm^3 treatment cost. Of course, in 1998 all such estimates were mere guesswork, but the ultimate expectation of relatively low-cost nanomedical treatments was certainly rational, if not provable.

This Chapter has demonstrated that there are many different pathways leading toward molecular nanotechnology, each one of which provides incremental benefits to motivate further travel down the paths. The likelihood of all the paths being blocked is low, even if we have little confidence that we can predict which approach will succeed first. Progress along the many pathways will provide precursor products, including some early medical applications, but the ultimate achievement of molecular manufacturing will finally make nanomedicine feasible.

As Drexler concluded *Nanosystems*: "Each step along [these pathways] will present great practical challenges, but each step will also bring valuable new capabilities. The long-term rewards, measured in terms of scientific and technological capabilities, appear large." This author agrees. For the remainder of the present work, we shall assume that a molecular manufacturing technology will someday exist that can economically manufacture macroscopic batches of microscopic machines comprised of atomically-precise nanoscale components.

6

Molecular Transport and Sortation

Human Body Chemical Composition

The human body consists of ~7 × 10^{27} atoms arranged in a highly aperiodic physical structure. Although 41 chemical elements are commonly found in the body's construction, CHON comprises 99% of its atoms. Fully 87% of human body atoms are either hydrogen or oxygen.

Somatic atoms are generally present in combined form as molecules or ions, not individual atoms. The molecules of greatest nanomedical interest are incorporated into cells or circulate freely in blood plasma or the interstitial fluid. The gross molecular contents of the typical human cell, which is 99.5% water and salts, by molecule count, and contains ~5000 different types of molecules. There are 261 of the most common molecular and cellular constituents of human blood, and their normal concentrations in whole blood and plasma. This listing is far from complete. The human body is comprised of ~10^5 different molecular species, mostly proteins—a large but nonetheless finite molecular parts list. By 1998, at least ~10^4 of these proteins had been sequenced, ~10^3 had been spatially mapped, and ~7,000 structures (including proteins, peptides, viruses, protein/nucleic acid complexes, nucleic acids, and carbohydrates) had been registered in the Protein Data Bank which at that time was maintained at Brookhaven National Laboratory. Given the current accelerating pace of improving technology, it is likely that the sequences and 3-D or tertiary structures of all human proteins will have been determined by the second decade of the 21st century.

Transporting and sorting such a broad range of essential molecular species will be an important basic capability of many nanomedical systems. The three principal methods for distinguishing and conveying molecules that are most useful in nanomedicine are diffusion transport, membrane filtration, and receptor-based transpor. The Chapter ends with a brief discussion of binding site engineering.

Diffusion Transport

Fluidic transfer of material, known as convective-diffusive transport, can occur either by convection due to bulk flow or by diffusion due to Brownian motion. In convective transport, material is carried along fluid streamlines at the mean velocity of the fluid, with a velocity distribution such as that in Poiseuille flow. Bulk flow is customarily regarded as the most important physiological transport mechanism in the human circulation. Only for the smallest molecules, such as water or glucose, does the time required to diffuse across the width of a capillary roughly equal the time taken by a fluid element to flow the same distance (~0.02 sec). Larger molecules such as fibrinogen take ~100 times longer (~2 sec) to diffuse across one capillary width.

However, bulk flow in the body is usually laminar. Transported materials travel parallel to (and thus cannot reach) fluid/solid interfaces such as the surfaces of blood vessels or membranes. Wall interactions are made possible by diffusion, a random process in which particles can move transversely to fluid streamlines in response to molecular-scale collisions.

Additionally, the movement of micronscale devices within a bulk fluid flow is dominated by viscous, not inertial, forces. Molecular transport to and from such nanodevices is governed by diffusion, not by bulk flow.

Brownian Motion

A particle suspended in a fluid is subjected to continuous collisions, from all directions, with the surrounding molecules. If the velocities of all molecules were the same all the time, the particle would experience no net movement. However, molecules do not have a single velocity at a given temperature, but rather have a distribution of velocities of varying degrees of probability. Thus from time to time, a suspended particle receives a finite momentum of unpredictable direction and magnitude. The velocity vector of the particle changes continuously,

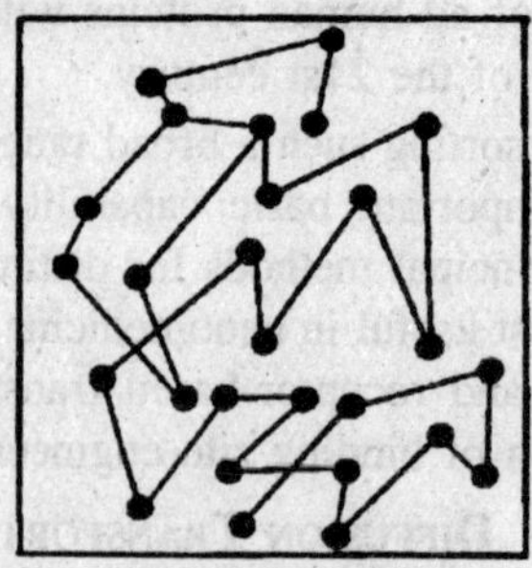

Fig. 6.1. Brownian movement.

resulting in an observable random zigzag movement called *Brownian motion*.

Einstein approximated the RMS (root mean square) displacement of a particle of radius R suspended in a fluid of absolute viscosity η and temperature T, after an observation period τ, as:

$$\Delta x = (kT\,\tau / 3\,\pi\eta\,R)^{1/2} \text{ (meters)} \qquad ...(1)$$

where k = 1.381 × 10^{-23} joule/kelvin (K) or 0.01381 zJ/K (Boltzmann's constant). Particles under bombardment also experience a rotational Brownian motion around randomly oriented axes, with the RMS angle of rotation:

$$\Delta\alpha = (kT\,\tau / 4\,\pi\eta\,R^3)^{1/2} \text{ (radians)} \qquad ...(2)$$

although for $\tau < \tau_{min} = M / 15\,\pi\eta\,R$, where M is particle mass, rotation is ballistic.

In human blood plasma, with η = 1.1 centipoise (1.1 × 10^{-3} kg/m-sec) and T = 310 K, a spherical 1-micron diameter nanodevice (R = 0.5 micron) translates ~1 micron in 1 sec ($v_{brownian}$ ~ 10^{-6} m/sec) or ~8 microns (~the width of a capillary) after 77 sec ($v_{brownian}$ ~ 10^{7} m/sec), and rotates once in ~16 sec (τ_{min} = 2 × 10^{-8} sec). In the same environment, a rigid 10-nm particle (roughly the diameter of a globular protein) would translate ~8 microns in one second ($v_{brownian}$ ~ 10^{-5} m/sec) while rotating ~250 times, due to Brownian motion (τ_{min} = 2 × 10^{-12} sec).

The instantaneous thermal velocity over one mean free path, the average distance between collisions, is much higher than the net Brownian translational velocity would suggest. For a particle of mass M = 4/3 $\pi\rho$ R^3 with mean (working) density r, the mean thermal velocity is:

$$v_{thermal} = (3\,kT / M)^{1/2} \qquad ...(3)$$

At T = 310 K, a spherical 1-micron diameter nanodevice of normal density (e.g., taking $\rho \sim \rho_{H_2O}$ = 993.4 kg/m^3 to minimize ballasting requirements) has $v_{thermal} \sim 5 \times 10^{-3}$ m/sec; for a spherical 10-nm diameter protein with $\rho \sim$ 1500 kg/m^3, $v_{thermal} \sim$ 4 m/sec.

Passive Diffusive Intake

Medical nanodevices will frequently be called upon to absorb some particular material from the external aqueous operating environment. Molecular diffusion presents a fundamental limit to the speed at which this absorption can occur. (Once a block of solution has passed into the interior of a nanodevice, it may be divergently subdivided and transported at ~0.01-1 m/sec along internal pathways of characteristic dimension ~1 micron far faster than the <1 mm/sec bulk diffusion velocity across 1 micron distances.)

For a spherical nanodevice of radius R, the maximum diffusive intake current is:

$$J = 4\pi\ RDC \qquad ...(4)$$

where J is the number of molecules/sec presented to the entire surface of the device, assumed to be 100% absorbed, D (m^2/sec) is the translational Brownian diffusion coefficient for the molecule to be absorbed, and C (molecules/m^3) is the steady-state concentration of the molecule far from the device.[337] (Blood concentrations in gm/cm^3 are converted to molecules/m^3 by multiplying by ($10^6 \times N_A$/MW), where $N_A = 6.023 \times 10^{23}$ molecules/mole (Avogadro's number) and MW = molecular weight in gm/mole or daltons.) For rigid spherical particles of radius R, where $R >> R_{H_2O}$, the EinsteinStokes equation gives:

$$D = kT/(6\ \pi\eta\ R) \qquad ...(5)$$

though this is only an approximation because D varies slightly with concentration, with departure from molecular sphericalness, and other factors.

Measured diffusion coefficients in water for various molecules of physiological interest, converted to 310 K. Diffusion coefficient data for ionic salts such as NaCl and KCl, which dissociate in water and diffuse as independent ions, are for solvated electrolytes. A 1-micron (diameter) spherical nanodevice suspended in arterial blood plasma at 310 K, with C = 7.3×10^{22} molecules/m^3 of oxygen and D = 2.0×10^{-9} m^2/sec, encounters a flow rate of J = 9.2×10^8 molecules/sec of O_2 impinging upon its surface. (The same calculation applied to serum glucose yields J = 1.3×10^{10} molecules/sec.) The characteristic time for change mediated by diffusion in a region of size L scales as ~L^2/

D. Across the diameter of an L = 1 micron nanodevice, small molecules such as glucose diffuse in ~0.001 sec, small proteins like hemoglobin in ~0.01 sec, and virus particles diffuse in ~0.1 sec. Diffusion coefficients of the same molecules in air at room temperature are a factor of ~60 higher, because η_{air} ~ 183 micropoise at 20°C.

In blood, the diffusivity of larger particles is significantly elevated because local fluid motions generated by individual red cell rotation lead to greater random excursions of the particles. The effective diffusivity $D_e = D + D_r$, where the rotation-induced increase in diffusivity $D_r \sim 0.25\ R_{rbc}^2\ '\gamma$, with red cell radius R_{rbc} ~ 2.8 microns (taken for convenience as a spherical volume equivalent) and a typical blood shear rate $'\gamma \sim 500\ sec^{-1}$, giving $D_r \sim 10^{-9}\ m^2/sec$ in normal whole blood. The elevation of diffusivity caused by red cell stirring is just 50% for O_2 molecules. However, for large proteins and viruses the effective diffusivity increases 10-100 times, and the effective diffusivity of particles the size of platelets is a factor of 10,000 higher than for Brownian molecular diffusion.

The diffusion current to the surface of a nanodevice can also be estimated for various nonspherical configurations. For instance, the diffusion current to both sides of an isolated thin disk of radius R is given by J = 8 R D C. The two-sided current to a square thin plate of area L^2 is $J = (8/\pi^{1/2})$ L D C. The steady-state diffusion current to an isolated cylinder of length L_c and radius R is approximated by $J = 2\pi L_c DC / (\ln(2L_c/R) - 1)$, for $L_c >> R$. The diffusion current through a circular hole of radius R in an impermeable wall separating regions of concentration c_1 and c_2 is $J = 4\ R\ D\ (c_1 - c_2)$.

Active Diffusive Intake

Foraging nanodevices operating in aqueous environments may only modestly exceed the maximum rates of passive diffusive intake by engaging in active physical movements designed to increase access to the desired molecules, as described quantitatively below.

Diffusive stirring

The first strategy for active diffusive intake is local stirring. For this, the nanodevice is equipped with suitable active appendages used to manipulate the fluid in its vicinity. Transport by stirring is characterized by a velocity v_a, the speed of the appendage, and by a length L_a, its distance of travel, which together define a characteristic stirring frequency $n_{stir} \sim v_a/L_a\ sec^{-1}$. Movement of molecules over a distance L_a by diffusion alone is scaled by a characteristic time $\sim L_a^2$ /D, which defines a characteristic diffusion frequency $n_{diff} \sim D / L_a^2$

sec^{-1}. Stirring will be more effective than diffusion only if $n_{stir} > n_{diff}$, that is, if $v_a > D / L_a$. For local stirring, L_a cannot be much larger than the size of the nanodevice itself. Assuming L_a = 1 micron and D = 10^{-9} m^2/sec for small molecules, then v_a > 1000 microns/sec, a faster motion than is exhibited by bacterial cells but quite modest for nanomechanical devices. With D = 10^{-11} m^2/sec for large proteins and virus particles, v_a > 10 microns/sec, well within the normal microbiological range.

The ratio of stirring time to diffusion time, or Sherwood number, is:

$$N_{Sh} = L_a v_2 / D \qquad ...(6)$$

provides a dimensionless measure of the effectiveness of stirring vs. diffusion. For bacteria absorbing small molecules, $N_{Sh} \sim 10^{-2}$. Micron-scale nanodevices with 1-micron appendages capable of 0.01-1 m/sec movement can achieve $N_{Sh} \sim$ 10-1000 for small to large molecules, hence could be considerably more effective stirrers.

In a classic paper, Berg and Purcell analyzed the viscous frictional energy cost of moving the stirring appendages so that the fluid surrounding a spherical object (e.g., a nanodevice) of radius R, out to some maximum stirring radius R_s, is maintained approximately uniform in concentration. The objective is to transfer fluid from a distant region of relatively high concentration to a place much closer to the nanodevice, thereby increasing the concentration gradient near the absorbing surface. To double the passive diffusion current by stirring, the minimum required power density is:

$$P_d = \frac{12\,\eta\,D^2}{R^4}\left(\frac{R_s + 2R}{R_s - 2R}\right)(\text{watt} / \text{m}^3) \qquad ...(7)$$

If η = 1.1 × 10^{-3} kg/m-sec, R = 0.5 micron, D = 10^{-9} m^2/sec for small molecules, and using a modest L_a = 1 micron stirring apparatus giving R_s = 3R, then $P_d \sim 3 \times 10^7$ watts/m^3. This greatly exceeds the 10^2-10^6 watts/m^3 power density commonly available to biological cells but lies well within the normal range for nanomechanical systems which typically operate at up to ~10^9 watts/m^3. For D ~ 10^{-11} m^2/sec for large molecules, $P_d \sim 3 \times 10^3$ watts/m^3, which is reasonable even by biological standards. The maximum possible gain from stirring is ~R_s/R, because the current is ultimately limited to what can diffuse into the stirred region.

Local heating due to stirring is minor. Given device volume V ~1 micron^3, P_d = 3 × 10^7 watts/m^3, mixing distance L_{mix} ~ 5 microns, and thermal conductivity K_t = 0.623 watts/m-K for water, then $\Delta T \sim (P_d\ V\ /\ L_{mix}\ K_t)$ = 10 microkelvins; taking heat capacity $C_V = 4.19 \times 10^6$ J/m³-K for water, thermal equilibration time $t_{EQ} \sim L_{mix}^2\ C_V\ /\ K_t$ = 0.2 millisec.

Diffusive swimming

The second strategy for active diffusive intake is by swimming. Again, the nanodevice is equipped with suitable active propulsion equipment which enables it to move so as to continuously encounter the highest possible concentration gradient near its surface. Consider a spherical motile nanorobot of radius R propelled at constant velocity v_{swim} through a fluid containing a desired molecule for which the surface of the device is essentially a perfect sink. Applying the Stokes velocity field flow around the sphere to the standard diffusion equation, a numerical solution by Berg and Purcell found that the fractional increase in the diffusion current due to swimming is proportional to v_{swim}^2 for $v_{swim} << D/R$, and to $v_{swim}^{1/3}$ for $v_{swim} >> D/R$.

Diffusive intake is doubled at a swimming speed v_{swim} = 2.5 D/R, which for 1-micron devices is ~5000 microns/sec when absorbing small molecules, ~50 microns/sec for large molecules. The viscous frictional energy cost to drive the nanodevice through the fluid, derived from Stokes' law, requires an onboard power density of:

$$P_d = 9\ \eta\ v_{swim}^2\ /\ 2\ R^2 \qquad \text{...(8)}$$

If $\eta = 1.1 \times 10^{-3}$ kg/m-sec, v_{swim} = 2.5 D/R, R = 0.5 micron, D = 10^{-9} m²/sec for small molecules, then $P_d \sim 5 \times 10^5$ watts/m³. For large molecules with D = 10^{-11} m²/sec, $P_d \sim 50$ watts/m³. Thus the energy cost of diffusive swimming appears modest for nanomechanical systems; gains in diffusion by swimming for nanodevices will be restricted primarily by the maximum safe velocity that may be employed *in vivo*.

In general, outswimming diffusion requires movement over a characteristic distance $L_s \sim D/v_{swim}$. For bacteria moving at ~30 micron/sec and absorbing small molecules, then $L_s \sim 30$ microns, roughly the sprint distance exhibited by flagellar microbes such as *E. coli*. For micron-scale nanodevices moving at ~1 cm/sec, $L_s \sim$ 1-100 nm for large to small molecules.

Diffusion Cascade Sortation

Nanodevices may also use diffusion to sort molecules. One of the remarkable features of diffusive sortation is that an input sample consisting of a complex mixture of many different molecular species can sometimes be completely resolved into pure fractions without having any direct knowledge of the precise shapes or electrochemical characteristics of the molecules being sorted. This can be a tremendous advantage for nanodevices operating in environments containing a large number of unknown substances. Another major advantage is the ability to readily distinguish isomeric (though not chiral) molecules. As one example of many possible, molecules suspended in water will diffuse into an adjacent region of pure water at different speeds, giving rise to dissimilar time-dependent concentration gradients which may be exploited for sortation by interrupting the process before complete diffusive equilibrium is reached.

For simplicity, assume we wish to separate two molecular species initially present in solution in equal concentrations ($c_1 = c_2$), but having unequal diffusion coefficients ($D_1 < D_2$). Consider a separation apparatus with two chambers. Chamber A contains input sample concentrate. Chamber B contains pure water. A dilating gate separates the two chambers. The gate is opened for a time Δt approximated by

$$\Delta t = \frac{(\Delta X)^2}{2\,D_2} - \frac{L^2}{2\,D_2} \qquad \ldots(9)$$

which relates the diffusion coefficient to the mean displacement ΔX, taken here as L, the length of Chamber B.

After Δt has elapsed, the gate is closed. (A gate with 10-nm sliding segments moving at 10 cm/sec closes in 0.1 microsec.) The faster-diffusing component D_2 approaches diffusive equilibrium in Chamber B, but the slower-diffusing component does not; it is present only in smaller amounts. This gives a separation factor $c_2/c_1 \sim D_2/D_1$ for each diffusion sortation unit. If n units are connected in series, with each unit receiving as input the output of the previous unit, the net concentration achieved by the entire cascade is $\sim (D_2/D_1)^n$. Such cascades are commonplace in gaseous diffusion isotope separation and other applications.

Each unit consists of 5 chambers of equal volume, 7 dilating gates, 3 flap valves, 3 pistons, and 2 sieves which pass only water (or smaller) molecules. Each chamber is roughly cubical with L ~ 35 nm along the inside edge; including full piston throws and drives, controls, interunit piping and other support structures, each unit measures ~125 nm × 100 nm × 80 nm or ~0.001 micron^3 with a mass of ~10^{-18} kg.

The following is a precise description of one complete cycle of operation for each unit:

1. The cycle begins with fluid to be sorted in Chamber A, Chambers B and W full of pure water with piston W all the way out, Chambers R and D empty with pistons R and D all the way in, and all valves and gates closed.
2. Gate AB is opened for a time Δt, then closed. For a small molecule such as urea (MW = 60 daltons), Δt = 1 microsec; for a large molecule such as the enzyme urease (MW = 482,700 daltons), Δt = 35 microsec.
3. Valves AI- and AI+, and gate AR, are opened. Piston R is drawn fully out, slowly to preserve laminar flow and to prevent mixing. Fluid in Chamber A is drawn into Chamber R. Fluid passing through the DO gate of the previous unit in the cascade enters Chamber A through valve AI-. Fluid passing through the RO valve of the subsequent unit in the cascade enters Chamber A through valve AI+. All valves and gates are closed. Chamber A is now ready for the next cycle.
4. Gates WB and BD are opened. Piston W is slowly pushed all the way in while piston D is slowly pulled all the way out. Concentrated solution in Chamber B is transferred into Chamber D as pure water in Chamber W is transferred into Chamber B, again preserving laminar flow. Both gates are closed; Chamber B is now ready for the next cycle.
5. Gates RW and DW are opened. Pistons R and D are slowly and simultaneously pushed halfway in while piston W is pulled all the way out. Forced at high pressure (~160 atm) through ~0.3 nm diameter sieve pores, half of the solvent water present in Chambers R and D is pushed into Chamber W, filling Chamber W with water. (This design allows for easy backflushing if sieve pores become clogged.) Both gates are closed; Chamber W is now ready for the next cycle.
6. Valve RO and gate DO are opened. Pistons R and D are slowly and simultaneously pushed the rest of the way in. Concentrated return fluid passes through valve RO and back to the AI+ input port of the previous unit in the cascade for further extraction. Concentrated diffusant fluid passes through gate DO and on to the AI- input port of the subsequent unit in the cascade for further purification. The valve and gate are closed; Chambers R and D are now empty and ready for the next cycle.

7. Return to Step (1). Adjacent units operate in counterphase while previous and subsequent units operate in synchrony, in a two-phase system.

Increasingly purified sample passes through a multi-unit sortation cascade as described above. For small molecules, a cascade of n ~1000 units (total device volume ~1 micron^3) completely resolves two mixed molecular species with $D_2/D_1 = 1.01$. As a crude approximation, $D \sim 1 / MW^{1/3}$ for small spherical particles, so this cascade separates small molecules differing by the mass of one hydrogen atom which should be sufficient for most purposes. Structural isomeric forms of the same molecule, such as a-alanine and b-alanine, often have slightly different diffusion coefficients, thus are also easily separable using a diffusion cascade. However, stereoisomeric (chiral) forms cannot be sorted by diffusion through an optically inactive solvent like water.

For large molecules, a 1 million-unit cascade (total device volume ~1000 micron^3) provides $D_2/D_1 \sim 1.00001$, sufficient to completely separate large molecules differing by the mass of a single carbon atom. The fidelity of such fine resolutions depends strongly upon the ability to hold constant the temperature of the chamber, since D varies directly with temperature. Device temperature stability will be determined by at least three factors: (1) the accuracy of onboard thermal sensors in measuring T ($\Delta T/T < 10^{-6}$), (2) the rapidity with which the temperature measurement can be taken (10^{-9} to 10^{-6} sec), and (3) the time that elapses between the temperature measurement and the end of the diffusive sortation process (which may be of the same order as the gate closing time, ~10^{-6} sec).

Most of the waste heat is generated in this device by forced water sieving. To remain within biocompatible thermogenic limits (~10^9 watts/m^3), each unit may be cycled once every ~3 millisec, a 0.8% duty cycle of a ~23 microsec sieving stroke. Subject to this restriction, each unit would consume ~1 picowatt in continuous operation. A unit presented with a ~0.1 M concentration of small molecules processes ~10^6 molecules/sec (e.g., ~1 gm/hour of glucose using 1 cm^3 of n = 1000-unit cascades), or ~10^4 molecules/sec for a unit presented with large molecules at ~0.001 M, circulating ~10^9 molecules/sec of water as working fluid while running at 340 cycles/sec. Additional chamber segments on each unit, combined with more complex diffusion circuits among the many units in a cascade, should permit the simultaneous complete fractionation of the input feedstock even if hundreds of distinct molecular species are present.

By 1998, diffusion-based separation had been demonstrated in microfluidic devices.

Nanocentrifugal Sortation

Nanoscale centrifuges offer yet another method for rapid molecular sortation, by biasing diffusive forces with a strong external field. The well-known effect of gravitational acceleration on spherical particles suspended in a fluid is described by Stokes' Law for Sedimentation:

$$v_t = 2\, g\, R^2 (\rho_{particle} - \rho_{fluid}) / 9\, \eta \qquad ...(10)$$

where v_t is terminal velocity, g is the acceleration of gravity (9.81 m/sec^2), R is particle radius, $\rho_{particle}$ and ρ_{fluid} are the particle and fluid densities (kg/m^3), and η is coefficient of viscosity of the fluid. Particles which are more dense than the suspending liquid tend to fall. Those which are less dense tend to rise ($\rho_{particle} / \rho_{fluid} \sim 0.8$ for lipids, up to ~ 1.5 for proteins, and ~ 1.6 for carbohydrates in water).

This separation process may be greatly enhanced by rapidly spinning the mixed-molecule sample in a nanocentrifuge device. For ideal solutions (e.g., obeying Raoult's law) at equilibrium:

$$c_2 / c_1 = \exp\left[\left(\frac{MW_{kg}\omega^2}{2R_g T}\right)\left(1 - \frac{\rho_{fluid}}{\rho_{particle}}\right)\left(r_2^2 - \eta^2\right)\right] \qquad ...(11)$$

where c_2 is the concentration at distance r_2 from the axis of a spinning centrifuge (molecules/m^3), c_1 is the concentration at distance r_1 (nearer the axis), MW_{kg} is the molecular weight of the desired molecule in kg/mole, w is the angular velocity of the vessel (rad/sec), T is temperature (K) and the universal gas constant R_g = 8.31 joule/mole-K. The approximate spinning time t_s required to reach equilibrium is

$$t_s = \ln(r_2 / r_1) / (\omega^2 S_d) \qquad ...(12)$$

where S_d is the sedimentation coefficient, usually given in units of 10^{-13} sec or svedbergs. Research ultracentrifuges have reached accelerations of $\sim 10^9$ g's.

Consider a cylindrical diamondoid vessel of density ρ_{vessel} = 3510 kg/m^3, radius r_c = 200 nm, height h = 100 nm, and wall thickness x_{wall} = 10 nm, securely attached to an axial drive shaft of radius r_a = 50 nm. A fluid sample containing desired molecules enters the vessel through a hollow conduit in the drive shaft, and the device is rapidly spun. If rim speed v_r = 1000 m/sec (max), then $\omega = v_r / r_c$ = 5×10^9 rad/sec ($\omega / 2\pi = 8 \times 10^8$ rev/sec). The maximum bursting force $F_b \sim 0.5\ \rho_{vessel}\ v_r^2 = 2 \times 10^9\ N/m^2$, well below the $50 \times 10^9\ N/m^2$ diamondoid tensile strength conservatively assumed by Drexler. Since S_d ranges from $0.1\text{-}200 \times 10^{-13}$ sec for most particles of nanomedical interest, minimum separation time using acceleration $a_r / g = v_r^2 / g\ r_c = 5 \times 10^{11}$ g's, when $r_2 = r_c$ and $r_1 = r_a$, is $t_s = 0.003\text{-}6.0 \times 10^{-6}$ sec. Fluid sample components migrate at ~0.1 m/sec.

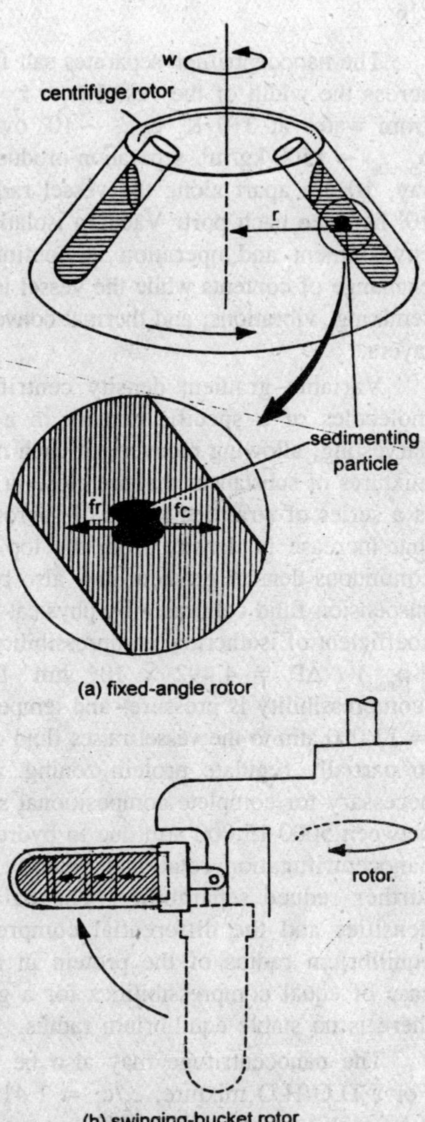

Fig. 6.2. Centrifuge.

Maximum centrifugation energy per particle $E_c = (MW_{kg} / N_A)\ a_r\ (r_c - r_a) \sim$ 10,000 zJ/molecule, or ~10 zJ/bond for proteins, well below the 180-1800 zJ/bond range for covalent chemical bonds. However, operating the nanocentrifuge at peak speed may disrupt the weakest noncovalent bonds (including hydrophobic, hydrogen, and van der Waals) which range from 4-50 zJ/bond. The nanocentrifuge has mass $\sim 10^{-17}$ kg, requires ~3 picojoules to spin up to speed (bearing drag consumes ~10 nanowatts of power, and fluid drag through the internal plumbing contributes another ~5 nanowatts); completes each separation cycle in $\sim 10^4$ revs ($\sim 10^{-5}$ sec), and processes ~300 micron³/sec which is $\sim 10^{13}$ small molecules/sec (at 1% input concentration) or $\sim 10^9$ large molecules/sec (at 0.1% input concentration).

The nanocentrifuge separates salt from seawater with $c_2/c_1 \sim 300$ across the width of the vessel ($r_c - r_a = 150$ nm); extracting glucose from water at 310 K, $c_2/c_1 \sim 10^5$ over 150 nm. For proteins with $\rho_{particle} \sim 1500$ kg/m^3, separation product removal ports may be spaced, say, 10 nm apart along the vessel radius while maintaining $c_2/c_1 \sim 10^3$ between each port. Vacuum isolation of the unit in an isothermal environment and operation in continuous-flow mode could permit exchange of contents while the vessel is still moving, sharply reducing remixing, vibrations, and thermal convection currents between product layers.

Variable gradient density centrifugation may be used to trap molecules of a specific density in a specific zone for subsequent harvesting, allowing recovery of each molecular species from complex mixtures of substances that are close in density. The traditional method is a series of stratified layers of sucrose or cesium chloride solutions that increase in density from the top to the bottom of the tube. A continuous density gradient may also be used, with the density of the suspension fluid calibrated by physical compression. For example, the coefficient of isothermal compressibility $\kappa = (\Delta V_1/V_1) / \Delta P_1 = (\Delta\rho_{fluid} / \rho_{fluid}) / \Delta P_1 = 4.492 \times 10^{-5}$ atm^{-1} for water at 1 atm and 310 K (compressibility is pressure- and temperature-dependent). Applying $P_1 = 12{,}000$ atm to the vessel raises fluid density to 1250 kg/m^3, sufficient to partially regulate protein zoning. A multistage cascade may be necessary for complete compositional separation. Protein denaturation between 5000-15,000 atm due to hydrogen bond disruption may limit nanocentrifugation rotational velocity. Protein compressibility may further reduce separability. The balance between the differential densities and the differential compressibilities will determine the equilibrium radius of the protein in the centrifuge; in the limiting case of equal compressibilities for a given target protein and water, there is no stable equilibrium radius.

The nanocentrifuge may also be useful in isotopic separations. For a D_2O/H_2O mixture, $c_2/c_1 = 1.415$ per pass through the device; $c_2/c_1 = 10^6$ is achieved in a 40-unit cascade. Tracer glycine containing one atom of C^{14} is separated from natural glycine using a 113-unit cascade, achieving $c_2/c_1 = 10^6$.

Membrane Filtration

Filtration through a permeable membrane is closely related to the process of diffusion, since in both cases random molecular motions help carry the process to completion. However, the presence of a

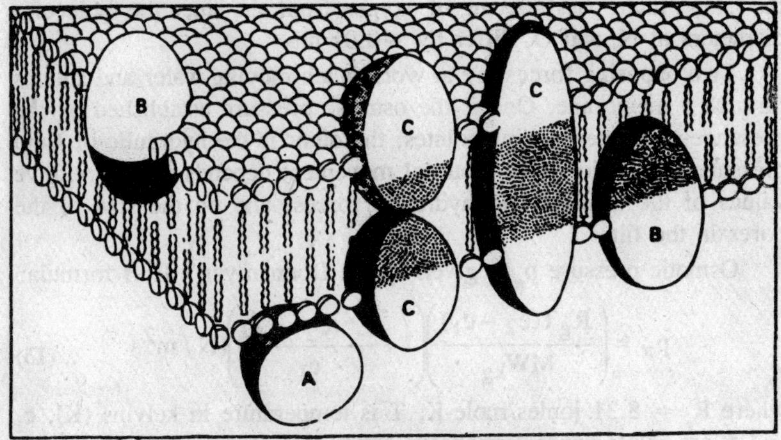

Fig. 6.3. Plasma membrane.

membrane adds a new measure of control that is not exploited in simple diffusive transport. This control may be either passive or active, unidirectional or bidirectional, as described below.

Simple Nanosieving

Nanometer-scale isoporous molecular sieves (with ovoid, square, or hexagonal holes) are common in almost every taxonomic group of eubacteria and archaeobacteria. Other well-known examples of nanoporous structures are the 6-nm pore arrays found in reverse osmosis and kidney dialysis membranes.

Likewise, it is possible for a nanodevice to sort molecules by simple sieving. In this process, a sample containing particles of various sizes suspended in water passes through a graduated series of filters perforated by progressively smaller holes of fixed size and shape. Between each filtration unit, the filtration residue consists almost exclusively of particles having a narrow range of sizes and shapes. For example, a series of n = 100 filtration units could reliably differentiate an input sample containing particles from 0.2-1.2 nm into 100 separate fractions, each fraction consisting predominantly of particles differing in mean diameter by ~0.01 nm. (In a practical system, several passes would be required to achieve complete discrimination.) A ~0.01 nm difference in molecular diameter corresponds to the mean contribution of ~1 additional carbon atom to the size of a small molecule (MW ~ 100 daltons), or to the mean contribution of ~100 additional carbon atoms to the size of a large molecule (MW ~ 100,000 daltons, ~17,000 atoms). A nanomembrane might even permit the (slow, multi-pass) sieving of oxygen from air, since the molecular diameters of N_2 and O_2 differ by ~0.01 nm.

Two opposing forces are at work when moving water and solutes through a membrane. One is the osmotic pressure established by the presence of nonpermeating solutes; the other is the hydraulic or fluid pressure. The velocity of material movement depends on the relative values of the osmotic and hydraulic forces, and on the size of the pores in the filter.

Osmotic pressure p_π is given by the Donnan-van't Hoff formula:

$$p_\pi = \left(\frac{R_g T(c_2 - c_1)}{MW_{kg}}\right)\left(1 + \frac{Z^2(c_2 - c_1)}{c_1}\right)(N/m^2) \qquad ...(13)$$

where R_g = 8.31 joules/mole-K, T is temperature in kelvins (K), c_2 and c_1 are solute concentrations on either side of the membrane in kg/m^3 ($c_2 > c_1$), MW_{kg} is the molecular weight of the solute in kg/mole, and the Z^2 term (dependent upon polymer-polymer interactions) is a correction factor for highly concentrated solutions which for some solvents and temperatures may equal zero; Z = net solute charge number and c_s is the concentration in kg/m^3 of a second solute, as for example when the first solute is a protein and the second solute is salt, as in human serum. Water at 310 K dissolves a maximum of c_2 = 370 kg/m^3 of sodium chloride (a 37.0% solution, by weight), and MW_{kg} = 0.05844 kg/mole for NaCl, so for salt water solutions the theoretical maximum $p_\pi \sim 1.6 \times 10^7$ N/m^2 ~ 160 atm. Natural bloodstream concentrations of salt produce p_π ~ 3 atm. Since osmotic pressure depends on the number of molecules present, the contribution from large molecules is usually negligible. For instance, total protein concentration in human blood serum is c_2 ~ 73 kg/m^3, MW_{kg} ~ 50 kg/mole (~50,000 daltons), so p_π ~ 0.04 atm.

In theory, extremely large hydraulic counterforces up to 10^5 atm may be applied in nanomechanical systems, for example by a piston, to overcome osmotic backpressure. As a practical matter, however, rapidly pushing small molecules at high pressure through nanoscale holes is an effective method for generating significant amounts of waste heat; a design compromise is required.

Consider a simple sorting apparatus in which a square piston is used to compress solvent fluid (say, water) trapped in a chamber $h_{chamber}$ in length and $L_{chamber}{}^2$ in cross-sectional area, forcing the fluid to filter through pores of radius r_{pore} (~ target molecule radius) covering a fraction α_H (~50%) of the surface of a square nanoscale sieve of thickness h_{sieve} and area $L_{sieve}{}^2$. Solute which is dissolved or suspended in the solvent is sorted based on molecular size; a sequence of sieving runs using sieves having progressively smaller pore radii produces an ordered sequence of molecular size fractions. (In small molecules, adding the mass of one hydrogen atom increases the mean linear dimension of the molecule by 0.11%.)

The first design constraint on this system relates to its maximum operating pressure. If ΔP is applied pressure (N/m²), then avoiding boiling the solvent water and denaturing proteins requires at least that $\Delta P_{max} < C_V \, \Delta T_{boil}$, where the heat capacity of water $C_V = 4.19 \times 10^6$ joules/m³-K and ΔT_{boil} = 373 K - 310 K = 63 K give $\Delta P_{max} <$ 2600 atm. The designs presented below operate at 6% of this maximum ($\Delta T \sim 3$ K) or less. There are two major design constraints on the duration of the power stroke, or t_p:

Molecular rotation constraint

Flow through the sieve must be slow enough to allow molecules to align with the holes. Assuming round pores, the total number of pores in the sieve is $N_{pore} = \alpha_H \, L_{sieve}{}^2/\pi \, r^2$. The volume processing rate (m³/sec) of the sieve $`V_{sieve} = `V_{chamber} = h_{chamber} \, L_{chamber}{}^2/t_p$, hence the molecule processing rate is $`N = `V_{chamber} \, c_{target}$ (molecules/sec) where c_{target} is the concentration of target molecules (molecules/m³). At any one time, each pore channel through the sieve can hold at most $N_{channel} = h_{sieve}/2 \, r_{pore}$ molecules in single file; during each power stroke, at most $N_{stack} = `N \, t_p/N_{pore}$ molecules pass through each pore. Hence the time available for molecular rotation $t_{rot} = t_p \, (N_{channel}/N_{stack})$, which assumes the layer of rotating target molecules in the vicinity of the pores approximates sieve thickness h_{sieve}, a reasonable assumption as long as the typical molecular diffusion time (across a distance h_{sieve}) $<< t_{rot}$. Taking $N_{rot} \sim 10$ as the mean number of molecular revolutions needed to ensure proper pore alignment with noncircular sieve holes (the most difficult case), then, from Eqn. 2, $\Delta\alpha = (kT \, t_{rot} / 4 \, \pi \, \eta \, r_{pore}{}^3)^{1/2} \geq 2 \, \pi \, N_{rot}$, where η is solvent viscosity (1.1×10^{-3} kg/m-sec for plasma) at T = 310 K. Solving for minimum t_p gives:

$$t_p \geq \frac{32\pi^4 \, N_{rot}^2 \, \eta \, c_{target} \, h_{chamber} \, r_{pore}}{kT \, \alpha_H \, h_{sieve}} \text{ (sec)} \qquad \text{...(14)}$$

Pressure/flow constraint

Flow through the sieve must be fast enough to establish a sufficient pressure to oppose osmotic backflows. From the Hagen-Poiseuille law, the volume processing rate through each pore is $`V_{pore} = \pi \, r_{pore}{}^4$

ΔP_{sieve} / 8 η h_{sieve} and the volume processing rate through the entire sieve is $'V_{sieve} = N_{pore}$ $'V_{pore}$ = $'V_{chamber}$; solving for maximum t_p gives:

$$t_p \leq \frac{8\eta \, h_{chamber} \, h_{sieve} \, L^2_{chamber}}{\alpha_H \, \Delta P_{sieve} \, r_{pore} \, L^2_{sieve}} \text{(sec)} \qquad ...(15)$$

Equating these two bracketing constraints, $h_{sieve} \geq 150$ nm for large molecules (r_{pore} ~ 5 nm) but $h_{sieve} \geq 1$ nm for small molecules (r_{pore} ~ 0.32 nm).

As a final constraint, power released by fluid flow through the chamber and sieve must not exceed safe thermogenic limits. Given the maximum safe power density for *in vivo* nanomachines as $D_n = 10^9$ watts/m^3, then:

$$D_{device} = \frac{P_{device}}{(h_{chamber} + h_{sieve}) \, L^2_{chamber}} \leq D_n \qquad ...(16)$$

where D_{device} = device power density (watts/m^3), total device power $P_{device} = P_{chamber} + P_{sieve}$ (watts), chamber fluid flow power $P_{chamber} = \pi \, L_{sieve}^4 \, \Delta P_{chamber}^2$ / 128 $h_{chamber}$ η, sieve fluid flow power $P_{sieve} = \alpha_H \, L_{sieve}^2 \, r_{pore}^2 \, \Delta P_{sieve}^2$ / 8 h_{sieve} η, and $\Delta P_{chamber} = 16 \, \alpha_H \, h_{chamber} \, r_{pore}^2 \, \Delta P_{sieve}$ / $\pi \, L_{sieve}^2 \, h_{sieve}$. For small molecules such as NaCl or glucose, $\Delta P_{sieve} \geq 160$ atm to overcome maximum osmotic backpressure; for large molecules (e.g., ~50,000 dalton proteins), we assume $\Delta P_{sieve} \geq 1$ atm to ensure sieving. The following designs are not optimized but illustrate the tradeoffs involved.

For small molecules (r_{pore} ~ 0.32 nm), an exemplar ~1 micron^3 device has $h_{chamber}$ = 1 micron, $L_{chamber} = L_{sieve}$ = 0.6 micron, h_{sieve} = 1.5 microns, t_p = 0.016 sec, ΔP_{sieve} = 160 atm, $\Delta P_{chamber}$ = 0.0001 atm. Piston velocity ~ 60 micron/sec and 'V = 2.5 × 10^{-17} m^3/sec for a 0.1 M solution of target molecules, yielding a processing rate of 1.5 × 10^9 molecules/sec (1.5 × 10^{-16} kg/sec); the device processes its own mass every ~7 sec or every ~430 power strokes. Device power P_{device} = 400 pW and power density D_{device} = 4 × 10^8 watts/m^3.

For large molecules (r_{pore} ~ 5.0 nm), an exemplar ~1 micron^3 device has $h_{chamber}$ = 1 micron, $L_{chamber} = L_{sieve}$ = 0.9 micron, h_{sieve} = 0.15 microns, t_p = 0.010 sec, ΔP_{sieve} = 1 atm, $\Delta P_{chamber}$ = 0.0004 atm. Piston velocity ~ 100 micron/sec and 'V = 9.8 × 10^{-17} m^3/sec for a 0.001 M solution of target molecules, yielding a processing rate of 5.9 × 10^7 molecules/sec (4.9 × 10^{-15} kg/sec); the device processes its own mass every ~0.2 sec or every ~20 power strokes. Device power P_{device} = 80 pW and power density D_{device} = 8 × 10^7 watts/m^3.

Sieve pores can become clogged by particles of radius $R \sim r_{pore}$ if the applied hydraulic pressure ΔP_{clog} exceeds the thermal energy of the trapped particles, or:

$$\Delta P_{clog} > 9kT / 8\pi R^3 \ (N/m^2) \qquad ...(17)$$

By this criterion, $\Delta P_{clog} > 500$ atm for small molecules and $\Delta P_{clog} > 0.1$ atm for large molecules. From the values of ΔP_{sieve} given above, clogging is unlikely for small molecules but is possible at the highest concentrations of large molecules. In 310 K water, large molecules diffuse ~3 nm and small molecules diffuse ~17 nm in ~10^{-7} sec, just far enough to clear the hole, so a ~10 MHz sawtooth pressure profile imposed on the power stroke may ensure sufficient backflushing action to avoid serious blockages. To reduce the possibility of clogging due to surface force adhesion, as a design criterion the work of adhesion should be reduced to $W_{adhesion} < \Delta P_{sieve} \ r_{pore} \sim 5 \times 10^{-3}$ J/m^2 for small molecules and $\sim 0.5 \times 10^{-3}$ J/m^2 for large molecules likely to come into contact with sieve pore surfaces. Clogging due to long-term random polymerizations can be minimized by periodically exchanging the entire contents of the input chamber with fresh solution, by operating the device at reduced power density, or by periodically replacing the sieve.

Dynamic Pore Sizing

A more efficient nanosieve system can be designed if pore size and shape can be actively modified during device operation, as for example by exchanging filters (from a membrane library stocking various pore sizes) each half cycle. Better, if pores can be reliably dilated or constricted in place during a period of time $\Delta t << t_p$, then filtration cascades can be more rapidly reconfigured to match changing input feedstock characteristics or to extract varying selections of desired molecules at will. Additionally, fully differentiating sieving cascades can be collapsed into a single unit, providing more compact devices especially useful in chemical sensor systems requiring preconcentration of sample. Control of pore shape should also provide finer discrimination among molecules of similar size but different shape, such as some isomers of nonchiral molecules.

Two or more overlapping surfaces containing regular arrays of perforations of fixed size and shape can conveniently generate a wide variety of pore geometries. Control of pore geometry is achieved by sliding or rotating one surface relative to the other surface by a small increment.

Circular dilating apertures can also be constructed using a matched set of overlapping segments, which may be driven either radially or tangentially to enlarge or contract the hole like an irising camera diaphragm, consistent with Akey's model of the nuclear pore complex present on the cell nucleus surface. Diaphragming mechanisms may be vertically staggered to maximize areal hole density in filtration surfaces (at the cost of increased vertical rugosity). Filters constructed of hydrogen-passivated diamondoid can have pores with <0.1 nm feature sizes, although H-free fullerene materials might avoid any possibility of dehydrogenation shearing.

Methods of positioning surfaces to accuracies of ~0.01 atomic diameter (!0.001 nm) have already been discussed. Assuming pore sizing blades require ~25 nm^2 of diamondoid contact surface per pore and each blade travels 25 nm at 0.01 meter/sec during one cycle, sliding friction dissipates ~0.01 zJ/pore, or $\sim 4 \times 10^{-18}$ watts/pore during each 2.5 microsec resizing cycle. Since fluid friction approaches kT for nanometer-size holes changing size in $\sim 10^{-9}$ sec, maximum blade speed is ~1 m/sec and the fastest resizing cycle is $\sim 10^{-8}$ sec.

A single sieving unit with controllable pores can be moderately efficient. Consider a design similar to that already described, except for a separate chamber and piston on either side of the filter block. Suppose that the particles desired to be extracted are of radius r, and the next smallest possible pore size is $r - \Delta r$. The device operates in two phases. In the first phase, the sample is placed in the first chamber, pore size is set equal to r, then the first piston forces the fluid through the membrane. Particles larger than r remain behind and are flushed from the first chamber. The pores then contract to $r - \Delta r$, and the second piston pushes the remaining filtrate back into the first chamber. After this second phase, particles of radius $\sim(r \pm \Delta r)$ remain in the second chamber at significantly higher concentration and may be removed for further use. Analogous double-sieve editing paradigms are commonplace in biological systems.

Other designs might work equally well, such as a 3-chamber flowthrough design using a variable pore membrane with pores of size r between the first and second chamber, a membrane with pores of size $r - \Delta r$ between the second and third chamber, and a piston at either end (one pushing, one pulling), thus concentrating molecules of size $r \pm \Delta r$ in the central chamber. M. Krummenacker suggests fixed chambers with a moving sieve operated as a dragnet. Filtration processes may be most useful in performing complete separations of complex mixtures. But they are inefficient in the sense that the energy expended to orient molecules passing through pores is wasted if the molecules are allowed to randomize on the other side; a eutactic mill-like molecule handling system might preserve this order and greatly improve energy efficiency.

Gated Channels

Besides controlling nanopore size and shape, individual molecular transport channels can be gated either mechanically (e.g., ligand gating) or electrically (e.g., voltage gating). Either method might usefully be employed to control molecular transport through the surfaces of medical nanodevices in a process that could very loosely be described as molecular transistor gating.

A good example of mechanical gating in biology is the nicotinic acetylcholine receptor channel, probably the best understood ligand-gated channel. Nerve impulses are communicated across neuromuscular junctions and autonomic ganglia via neurotransmitters such as acetylcholine. STM images confirm that the receptor itself is cylindrical, a bundle of 5 rod-shaped polypeptide subunits arranged like barrel staves with outside diameter ~6.5 nm. The receptor protrudes 6 nm on the synaptic side of the membrane and 2 nm on the cytoplasmic side. The water-filled channel pore lies along the symmetry axis, lined by 5 α-helices, with a 2.2 nm wide mouth on the synaptic surface, a 0.65 nm waist where the structure dives through the cell membrane, and a 2 nm wide cytosolic exit.

Normally, the channel is closed and no ions may pass. In this closed state, the channel is occluded at the waist by a ridge of large residues forming a tight hydrophobic ring. Each subunit has a bulky leucine at the bend in the α-helix, a critical position. When two acetylcholine molecules bind to the receptor, these helices allosterically tilt, shifting the position of the ridges. The pore becomes open because it is now lined with small polar residues rather than by large hydrophobic ones. This conformational change allows 2.5×10^7 Na^+ ions/sec to flow through the channel, about 10% of the diffusion-limited rate. (Anions like Cl^- cannot enter the pore because they are repelled by rings of negatively charged residues positioned at either end of the receptor.)

Acetylcholine binding opens the gate in less than 100 microsec under physiologic conditions. Subsequent rapid destruction of acetylcholine by acetylcholinesterase, an enzyme tethered to the membrane surface by a covalently attached glycolipid group, closes

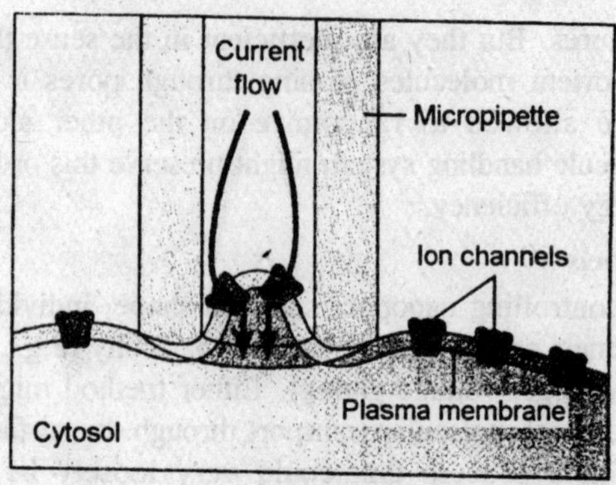

Fig. 6.4. Ion channels

the gate in ~1 millisec. Much faster gating action (~10^{-8} - 10^{-6} sec) could be achieved by nanodevices operating variable-scale nanopores in response to sensor data or other control signals. Such signals could drive the insertion or retraction of diamondoid rods, wedges, or trapdoors across the channel lumen to regulate the transmission of molecules having specific sizes, shapes, and charge distributions.

Transport channels through nanodevice surfaces may also be gated electrically. In contrast to the acetylcholine receptor, which is relatively nondiscriminating and allows both inorganic and organic cations to pass, the voltage-gated calcium channel has a highly discriminating mechanism with a Ca^{++}:Na^{+} permeation ratio on the order of 1000:1. (The high specificity of the voltage-gated Ca^{++} channel is a consequence of a single-file pore mechanism involving a pair of specific Ca^{++} ion binding sites. Selectivity is assured if either of the two sites is occupied by Ca^{++}, as monovalent ions do not bind strongly enough to the free site or generate sufficient electrostatic repulsion to push the first Ca^{++} ion through the channel.) Potassium channels are 100 times more permeable to K^{+} than to Na^{+}, and sodium channels favour the passage of Na^{+} over K^{+} by a factor of 12. All three of these voltage-gated channels are important in the generation and conduction of neural action potentials.

A nerve impulse is an electrical signal produced by the flow of ions across the plasma membrane of a neuron. Neuron interiors have high concentrations of K^{+} and low concentrations of Na^{+}. The resting potential of a neuron is 60 mV. An action potential may be generated when the membrane potential is slightly depolarized to 40 mV. This opens the Na^{+} voltage-gated channels, rapidly accelerating depolarization to a peak of +30 mV in ~1 millisec. Then Na^{+}

channels close and K^+ channels open, allowing K^+ ions to exit the cell, restoring the 60 mV resting potential. Only ~1 ion of every ~10^6 Na^+ and K^+ ions present in the local extracellular medium and the axoplasm participate in each such nerve impulse.

The sodium channel is a single polypeptide chain with four repeating units. Each repeating unit folds into six transmembrane alpha helices, including one that is positively charged called the S4 helix. The S4 helix is the voltage sensor that triggers the opening of the gate. Three positively charged residues on each S4 helix are paired at the resting membrane potential with negative charges on other transmembrane helices in a staircase geometry. The initial small depolarization event produces a spiral motion of each S4 accompanied by the net movement of one or two charges to the extracellular side of the membrane, essentially turning this left-handed hydrogen-bonded "molecular screw" through a ~60° rotation. This outward 0.5-nm translation of the four S4 segments opens the sodium gate by removing a steric barrier to ion flow. The energy cost of moving ~6 electrical charges (~10^{-18} coul) from the cytosolic to the extracellular side of the membrane against a ~100 mV potential (thus opening the gate in ~75 microsec) is ~100 zJ. Quantum tunneling activation of sodium channels, taking 1-1000 microsec, has been analyzed by Chancey.

Artificial ion-gated polymer membranes were reported in 1982, protein engineering of switchable pore-forming proteins is well-known, and "intelligent gels" are being developed that can change size and molecular porosity in response to chemical, electrical or thermal stimuli. In 1998, however, electroporation was a more commonly used method in biological research and a useful technique for "transfecting" cells in genetic studies. Electroporation employs a brief intense pulse of electricity to provide a force that opens cellular pores, enabling the insertion of macromolecules like DNA into cells of interest; laser pulses reduce cell loss to 10% by using a square-wave pulse to effect rapid and reversible pore formation. Artificial pH-gated 200-nm diameter nanopores and natural pH-gating of virion pores in the cowpea chlorotic mottle virus were demonstrated in 1998.

The first true voltage-gated nanomembrane was fabricated by Charles Martin and colleagues in 1995. This membrane consists of cylindrical gold nanotubules with inside diameters as small as 1.6 nm. When the tubules are positively charged, cations are excluded and only negative ions are transported through the membrane. When the membrane receives a negative voltage, only positive ions are transported through the tubules. Nanodevices may combine voltage gating with pore size and electrosteric constraints to achieve precision transport control with moderate molecular specificity at diffusion-limited throughput rates. In 1997, an ion channel switch biosensor with sub-picomolar sensitivity and quantitative detection time of ~600 sec was demonstrated by an Australian research group.

Receptor-Based Transport

The most efficient of all modes of molecular transport involves receptor sites capable of recognizing and selectively binding specific molecular species. Many receptors reliably bind only a single molecular type; others, such as sugar transporters, can recognize and transport several related sugar molecule types. In nanomechanical systems, artificial binding sites of almost arbitrary size, shape, and electronic charge may be created and employed in the construction of a variety of highly efficient molecular sortation and transport devices, described below.

Transporter Pumps

While gated channels enable ions to flow rapidly through membranes in a thermodynamically downhill direction, active pumps use a source of free energy to force an uphill transport of ions or molecules. In biology the energy supply is usually ATP or photons of light; medical nanomachines can make use of vastly more diverse energy sources. Actually, the term "pump" may be something of a misnomer because the action is highly specific—only one or a very small number of molecular species are selectively transported.

Molecular pumps generally operate in a four-phase sequence: (1) recognition (and binding) by the transporter of the target molecule from a variety of molecules presented to the pump in the input substrate; (2) translocation of the target molecule through the membrane, inside the transporter mechanism; (3) release of the molecule by the transporter mechanism; and (4) return of the transporter to its original condition, so that it is ready to accept another target molecule. Such molecular transporters that rely on protein conformational changes are ubiquitous in biological systems.

The minimum energy required to pump molecules is the change in free energy ΔG in transporting the species from one environment having concentration c_1 to a second environment having concentration c_2, given by:

$$\Delta G = kT \ln(c_2 / c_1) + \frac{Z_e F \Delta V}{N_A} \qquad \text{...(18)}$$

where k = 0.01381 zJ/K (*Boltzmann constant*), T = 310 K, Z_e is the number of charges per molecule transported (i.e., the valency), F = 9.65×10^4 coul/mole (*Faraday constant*), ΔV is the potential in volts across the membrane, and N_A is Avogadro's number. So for example, transport of an uncharged molecule across a $c_2/c_1 = 10^3$ gradient (typical in biology) costs ~30 zJ. An extremely aggressive $c_2/c_1 = 10^6$ concentration gradient costs ~60 zJ/molecule, plus another ~30 zJ/ion if we are moving Ca^{++} ions against a 100 mV potential. An artificial nanopump of dimension ~10 nm moving at a conservative ~1 cm/sec velocity operates at MHz frequencies, transporting ~10^6 molecules/sec for a continuous power consumption of ~0.03 pW at $c_2/c_1 = 10^3$. Such a pump has a mass ~10^{-21} kg.

Transporter pumps need not be limited to the movement of a single molecular species in a single direction, which biochemists call a uniport transport mechanism. Numerous well-known biological systems are capable of moving two molecules simultaneously in one direction (*symport mechanisms*), two molecules sequentially in opposite directions (*antiport mechanisms*), and charged molecules in one direction only, thus building up an electrical charge on one side of the membrane (electrogenic mechanisms). Such pumps exist in nature for numerous ions, amino acids, sugars, and other small biomolecules. Active drug efflux systems and multidrug resistance are made possible by the

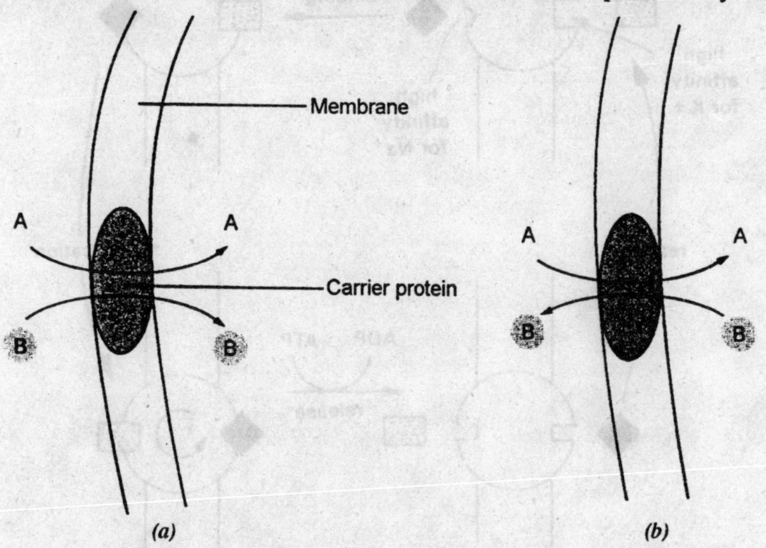

Fig. 6.5. Showing (a) symport; (b) antiport

expression of bacterial genes coding for molecular pumps that are constantly evolving new specificities to increasing numbers of microbicidal drugs.

Three Na^+ and two K^+ ions are transported per 10 millisec cycle, requiring the hydrolysis of one ATP molecule to ADP to drive the conformational changes. (More than one-third of the ATP consumed by a resting animal is used to pump these two ions.) Hydrolysis of ATP liberates ~80 zJ/molecule of free energy, so the antiporter is transporting Na^+ and K^+ at a cost of 16 zJ/ion at a 0.5 KHz frequency. Pump site density is ~1000/micron^2 of cell membrane in neural C fibers. (The Na^+ - K^+ pump can also be operated in reverse to synthesize ATP from ADP by exposing the mechanism to steep ionic gradients.) By contrast, artificial nanomechanical antiporter and symporter devices will operate at MHz frequencies. They should be able to transport much larger molecules, and may also be fully reversible.

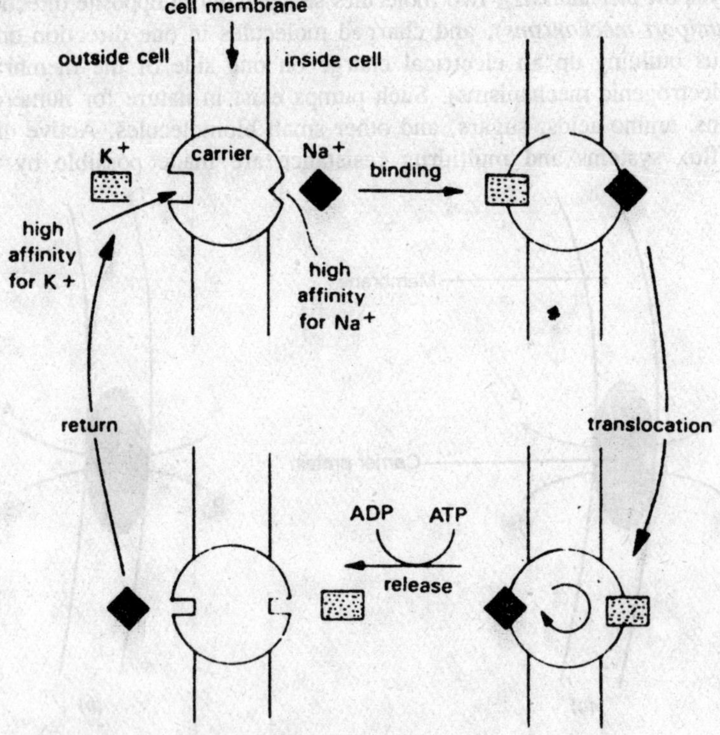

Fig. 6.6. Na⁺ and K⁺ pump.

Sorting Rotors

Drexler's molecular sorting rotor is a related class of nanomechanical device capable of selectively binding molecules from solution and then transporting these bound molecules against concentration gradients. The archetypal sorting rotor is a disk with 12 binding site "pockets" along the rim exposed alternately to the external solution and interior chamber by axial rotation of the disk. (Other designs may have more, or fewer, pockets.) Each pocket selectively binds a specific molecule when exposed to the solution. Once the binding site rotates to expose it to the interior chamber, the bound molecules are forcibly ejected by rods thrust outward by the cam surface (or using some other means by which receptor affinity can be adjusted during the inbound transport process). In the case of protein molecules, the debinding geometry must be carefully designed to avoid denaturation during ejection. Also, the rotor implicitly assumes that target molecules remain in the liquid or gaseous state after importation. M. Krummenacker observes that most bloodborne molecular species will precipitate as solids unless they are well-solvated; thus nanomedical sorting systems may require internal solvent or in some cases should be made completely eutactic. The discovery of positionally disordered water molecules resident inside protein hydrophobic cavities suggests that good rotor designs may also need to include solvent drainage channels.

Molecular sorting rotors can be designed from about 10^5 atoms (including housing and pro rata share of the drive system), measuring roughly 7 nm × 14 nm × 14 nm in size with a mass of 2×10^{-21} kg. Rotors turn at ~86,000 rev/sec with a conservative rim speed of 2.7 mm/sec and an almost negligible drag power of $\sim 10^{-16}$ watts against the fluid, sorting small molecules at a rate of 10^6 molecules/sec with laminar flow. From Eqn. 18, the energy cost of small-molecule sortation at 310 K ranges from ~10 zJ/molecule at low pressures ($c_2/c_1 = 10$) up to ~40 zJ/molecule when pumping against the highest head pressures ($c_2/c_1 = 10^4$, ~30,000 atm for natural bloodstream concentrations of salt with osmotic $p_p \sim 3$ atm), consuming 0.01-0.04 pW per device in continuous operation. Rotors are fully reversible, so they can be used to load or unload target molecules depending on the direction of rotor rotation. Cylindrical rotors with many receptor rows are somewhat more energy-efficient, and rotor lifetimes should be $> 10^6$ sec.

Typical molecular concentrations in the blood for target molecules of nanomedical interest are $\sim 10^{-11}$ - 10^{-3} molecules/nm^3, which should be sufficient to ensure ~99% occupancy of rotor binding sites. Rotors targeting serum hormones and other low-concentration species at the parts-per-billion level must slow to <1 rev/sec to ensure complete receptor occupancy and to avoid exceeding diffusion limits.

Sorting of ionically charged species can use binding sites that display opposite charge, effectively increasing the affinity of the binding site for the charged species. Many ionic species dissolved in water are actually more complex than their symbolic representation would suggest. As an example, a naked proton (H^+) in water is always highly hydrated, usually as $H_5O_2^+$ or $H_7O_3^+$, or even as $H_9O_4^+$ in strong acid solutions. Small ions from Li^+ to I^- are also found in solvent cages, bound with energies $\geq$330 zJ relative to vacuum; for instance, Li^+ and I^- are coordinated to 46 water molecules. As a result, the design of an appropriate binding site or filtration process for such ions can likewise become more complex. The existence of biological channels that show remarkable selectivity for specific ions (e.g., Na^+ channels that largely exclude K^+ ions, and vice versa) provides one design approach for dealing with such species.

Drexler has also proposed cascades of sorting rotors to achieve high fidelity purification and a contaminant fraction of $<10^{-15}$. However, since only $\sim 10^{10}$ small molecules can be stored in 1 micron^3 volume (typical for nanomedical devices), 100% process purity requires a contaminant fraction of only $<10^{-10}$ which can be ensured in micron-scale nanosystems using at most 5 stages, starting from a dilute input substrate containing only 1 part per billion of the target molecule with each stage providing a concentration factor of $\sim 10^4$. For statistically pure extractions of more common molecules in blood and cytoplasm, a 3- or 4-stage cascade will usually suffice. Note that the optimal receptor structure may differ at different stages in a cascade, and that each 12-arm outbound rotor can contain binding sites for 12 different impurity molecules. The first-stage receptor will likely pass only a relatively small number of different contaminant species, so the number of outbound rotors in the entire system can probably be reduced to a small.

Internal Transport Streams

After pump mechanisms described above have reliably sorted externally-encountered target molecules into reservoirs filled with species of a single type, a medical nanodevice may require these molecules to be transported to specific internal locations for further processing. Bulk fluid flow or fluidized (solvated or suspended) transport through nanopipes may suffice for some purposes. However, in many cases it will be necessary to present reagent molecules to other subsystems as a well-ordered stream of precisely positioned moieties transported *in vacuo*, especially for mechanochemical operations.

For this purpose, Drexler proposes molecular mills—eutactic systems of nanoscale belts moving over rollers, with reagent-binding devices mounted on the belt surface. This class of device can be assembled into complex molecular transportation networks using conditional switching, crossed-axis belting, and transit speed/frequency multipliers, and may also be employed to drive mechanosynthetic chemical reactions. The benchmark mechanism uses 10-nm diameter rollers to carry closely packed reagent devices measuring 4 nm × 4 nm × 2 nm, or 32 nm^3. A 20-roller mill mechanism 1 micron long has a ~2 micron long belt with 500 reagent devices and delivers 10^6 molecules/sec at a belt speed of 4 mm/sec. Total power dissipation is ~1.4×10^{-18} watts, a rate of ~0.001 zJ per moiety (or per reagent device) delivered or ~10^{-6} zJ/nm traveled per reagent device. Total mill mechanism mass is ~6×10^{-20} kg.

An alternative to roller/belt mill mechanisms is a nonconnected stream of pallets pushed along tracks, also *in vacuo*. Such tracks may include merging junctions, distribution junctions, multiplane crossings and switching stations, as well as straight and curved sections. Assuming each pallet is a 32 nm^3 reagent device held to the track by pins in grooves resembling cam followers, energy dissipation by phonon scattering is given approximately by:

$$P_{drag} = \frac{4}{3} \frac{\varepsilon_p \, \omega_{therm} v^2}{v_{sound}} \qquad ...(19)$$

where ε_p = 2×10^8 joules/m^3 (phonon energy density), s_{therm} (a thermally-weighted scattering cross section) ~10^{-20} m^2 for reagent devices of mass m = 10^{-22} kg assuming a sliding contact of stiffness ~30 N/m in a moderately stiff medium, v = 4 mm/sec sliding speed, and v_{sound} = 10^4 m/sec (~speed of sound in diamond), giving P_{drag} ~4×10^{-21} watts per reagent device, or P_{drag} / v ~ 10^{-6} zJ/nm traveled per reagent device (pallet). Note that volume containerization of pallet-transported molecules is least efficient at the smallest scales, where surface area per unit enclosed volume is highest, since energy usage is proportional to the surface area of the carrier. Containerization of n >> 1 molecules for large-pallet transport is more efficient.

A less energy-efficient, but far more versatile, internal molecular transport device is the 100-nm telescoping manipulator arm has already been described. This flexible ~10^{-19} kg device employs a binding tip to pick and place small and large molecules alike, moving them at ~1 cm/sec with repeatable placement accuracy of 0.04 nm. Multiple devices can be used to establish an internal ciliary transport system; standardized volume containerization of molecules permits rapid stereotypical handoff motions and efficient parcel routing. Conveyance through a 100-nm arc takes 10^{-5} sec consuming 0.1 pW while the arm is in motion, or ~10 zJ/nm traveled per reagent molecule or per container transported (vs. ~1000 zJ per typical covalent bond).

In molecular cytobiology, vesicles and organelles are transported throughout the interior of a cell by riding on microtubular cables crisscrossing the cytosol. For example, neural vesicles show transport speeds up to 2-4 microns/sec.

Molecular Receptor Engineering

Molecular recognition requires a detailed surface complementarity between the target molecule and its receptor. The interplay of various molecular forces between ligand and receptor causes them to selectively bind together, typically engineered to occur in ~10^{-6} sec. It is useful first to briefly review and quantify the principal physical forces at work.

Physical Forces in Molecular Recognition

Covalent bonds, which occur when atoms share electrons, are the strongest bonds. Aside from metals and salts, most material objects are made of atoms held together by covalent bonds. The atoms comprising biological molecules like proteins, nucleic acids and lipids are strung together mostly by single, double, or triple covalent bonds, as are receptors and diamondoid nanomechanical structures. Interatomic bond strengths range from 181 zJ/bond for O-F up to 1785 zJ/bond for C-O. Covalent bond lengths range from 0.10-0.16 nm within CHON-atom molecules, giving a typical covalent bond rupture force of ~10 nN/bond.

But as Jean-Marie Lehn points out, "there is a chemistry beyond the molecule" —noncovalent supramolecular chemistry. The bonds employed in molecular recognition are weak noncovalent bonds. Noncovalent bonds are largely responsible for the secondary and higher order structure of macromolecules. On a per-bond basis, noncovalent bonds are 1-3 orders of magnitude weaker than covalent bonds. However, the possibility of combining within a limited area a great number of noncovalent bonds having complementary elements allows the formation of a large specific association whose affinity may be of the same order of magnitude as a covalent bond. The high combinatorial diversity provided by many complementary elements allows numerous orthogonal specific associations, enabling self-assembly of many components; by comparison, covalent chemistry offers a poor diversity of reactivities. An additional advantage is that the formation of noncovalent bonds often is not hindered by high energy barriers. At least five types of noncovalent bonds may be distinguished: electrostatic, hydrogen, van der Waals, π aromatic, and hydrophobic.

Electrostatic bond

The electrostatic bond between two charged particles (e.g., the "salt bridge" in proteins) is a dipole interaction whose energy E_e is given by Coulomb's law as:

$$E_e = \left(\frac{e^2}{4\pi\,\varepsilon_0}\right)\left(\frac{Z_1 Z_2}{\kappa_e r}\right)\exp(-K_{dh} r) \text{ (joules)} \qquad \text{...(20)}$$

where e = 1.60×10^{-19} coul (elementary charge), $\varepsilon_0 = 8.85 \times 10^{-12}$ farad/m (permittivity constant), Z_1 and Z_2 are the numbers of attractant charges, r is the distance between the charges, and κ_e is the dielectric constant (74.3 for pure water at 310 K, usually reduced to ~40 in a hydrophobic environment). The bond is strengthened if the charges are in a hydrophobic environment. Conversely, the presence of electrolytes weakens the bond energy due to a shielding effect, given by K_{dh}, the Debye-Huckel reciprocal length parameter, which has a value of 1.25 nm^{-1} for 0.15 M NaCl (~1% solution, ~human blood). Thus two unit charges separated by 0.3 nm produce an interaction energy of E_e = 19 zJ in a hydrophobic environment, 10 zJ in pure water, and 6.3 zJ in 1% salt water. Long-range electrostatic trapping has been observed in single-protein molecules at liquid-solid interfaces, raising the implication that the interaction of protein molecules with biological cell surfaces may be much more efficient than predicted by random diffusion.

Most isolated amino acids in neutral solution are zwitterionic—the molecule has no overall charge but carries both a negatively charged group (carboxyl, CO_2^-) and a positively charged group (amino, NH_3^+). In proteins the individual amino acids are polymerized, giving a peptide backbone which is electrically neutral except for the ends of the chain. Most of the standard amino acids found in proteins have uncharged side chains, although histidine, lysine and arginine each have a positive charge at neutral pH and both glutamic and aspartic acids normally carry a negative charge.

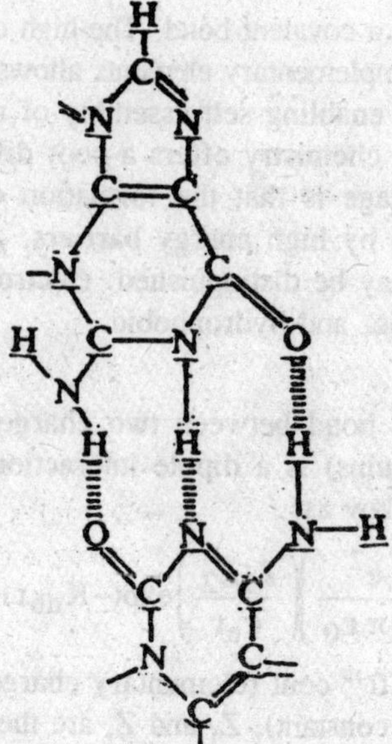

Fig. 6.7. Hydrogen bonds.

Hydrogen bond

A second important noncovalent interaction is the hydrogen bond, a dipole formed when a hydrogen atom covalently bonded to an electronegative atom is shared with a second electronegative atom

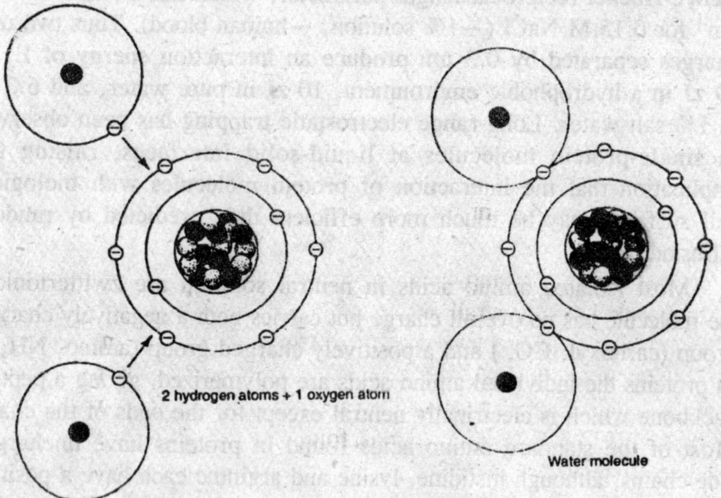

Fig. 6.8. Covalent bonds.

(typically an oxygen, nitrogen or fluorine atom), such that the proton may be approached very closely by an unshared pair of electrons. Hydrogen bonds are largely responsible for the unusual thermodynamic properties of water and ice, and the DNA double-helical and protein α-helical and β-structure conformations are extensively hydrogen bonded. Highest bonding energies occur when donor and acceptor atoms are 0.26-0.31 nm apart. Typical hydrogen bond strengths in proteins are 7-50 zJ.

Van der Waals Interaction

A third important noncovalent force is van der Waals interactions (London dispersion forces). There is an attractive component due to the induction of complementary partial charges or dipoles in the electron density of adjacent atoms when the electron orbitals of two atoms approach to a close distance. There is also a strongly repulsive component at shorter distances, when the electron orbitals of the adjacent atoms begin to overlap, commonly called steric hindrance. The van der Waals attractive bonding energy between two parallel plates of area A and separation z_{sep} is approximated by:

$$E_{vdW} = \frac{HA}{12\,\pi\,z_{sep}^2} \qquad ...(21)$$

where the Hamaker constant H = 37 zJ for water, 66 zJ for glycerol, 340 zJ for diamond. For A = 0.4 nm^2 (small molecule), z_{sep} = 0.3 nm, then E_{vdW} = 4 zJ (water) to 8 zJ (small organic molecules)—close to the mean energy of a thermally excited harmonic oscillator, kT ~ 4.3 zJ at 310 K. While van der Waals bonds are individually very weak, they are also very numerous since they involve all pairs of neighbouring atoms. For example, experimental analysis of an antigen molecule trapped in the anti-hen egg-white lysozyme monoclonal antibody Fv active binding site found 86 distinct interatomic contact points with antigen-antibody separations ranging from 0.25-0.46 nm, averaging 0.36 nm. Since intermolecular dispersion forces act on all molecules, there are probably no ligands with MW > 400 daltons which cannot be receptored. That is, van der Waals interactions ensure that virtually all molecules of nanomedical interest are theoretically bindable noncovalently.

Aromatic π bonds

A fourth type of interaction (π electron to π electron), called "aromatic" or "pi" bonding, occurs when two aromatic rings (conjugated π systems) approach each other with the plane of their aromatic rings overlapping, with successive π-bonded systems stacked like layers in a cake. This results in a noncovalent attractive force with a bond strength of ~40-50 zJ. (That is, the planes of conjugated π systems attract each other when superimposed.) π bond stacking forces contribute to nucleic acid stability at least as much as the hydrogen bonds between bases.

Hydrophobic forces

Finally, there are the strong hydrophobic forces, due entirely to solvent entropy changes. When two nonpolar residues approach each other, the surface area exposed to solvent is reduced, increasing the entropy of all the water present and decreasing the entropy of the residues, adding to the binding energy a hydrophobic free energy of ~17 zJ/nm^2 of contact surface area that was formerly exposed to water.

In designing an artificial binding site, the above forces may be combined to achieve the desired level of affinity and specificity for a given ligand. All forces are not equally useful in this regard, however. For example, hydrophobicity is the major factor in stabilizing protein-protein associations. But hydrophobicity is almost entirely nonspecific, hence contributes little to ligand discrimination. By contrast, the proper formation of hydrogen bonds and van der Waals contacts require complementarity of the surfaces involved. Such surfaces must be able to pack closely together, creating many contact points, and charged atoms must be properly positioned to make electrostatic bonds. Thus van der Waals and polar interactions may contribute little to the dynamic stability of the ligand-receptor complex, but they do determine which molecular structures may recognize each other. Other design elements of binding sites, such as directed channeling of substrates into the receptor, may also prove useful.

In analyzing molecular forces, note that at the nanoscale level, surface/surface, molecule/surface, and molecule/molecule interactions may feature very complicated behaviours. Nanodevices performing work may generate both thermodynamic and mechanical local nonequilibrium conditions, so calculations based on the general forms of interactions and on macroscopic expressions valid at equilibrium conditions should be taken only as basic estimates.

Ligand-Receptor Affinity

For a ligand binding to a receptor in a solvent, there will be a characteristic frequency with which existing ligand-receptor complexes dissociate as a result of thermal excitation, and a characteristic frequency with which empty receptors bind ligands as a result of Brownian encounters, forming new complexes, with the frequency of binding proportional to the concentration of the ligand in solution, c_{ligand} (molecules/nm^3). For simple processes, the equilibrium constant K_d, taken in the direction of dissociation, is:

$$K_d = k_d / k_a \text{ (molecules / nm}^3) \qquad ...(22)$$

where k_d is the dissociation rate constant (sec^{-1}) and k_a is the association rate constant (nm^3/molecule-sec). The k_a rate constant reflects mainly the molecular weight of the ligand, and thus varies little among antibody, enzyme, or other receptor systems. For example, Delaage notes that changing the solution pH for growth hormone from 7.5 to 4.0 increases K_d by a factor of 3000, due to k_d increasing by a factor of 1600 but k_a decreasing only by a factor of 1.7. Hence it is the rate constant of dissociation, k_d, which accounts for the vast bulk of affinity in receptor systems.

Thus receptor affinity is usually taken as the inverse of the dissociation rate constant, which may be placed in the context of the half-life of the ligand-receptor complex approximated by:

$$t_{1/2} \sim \ln(2) / kd \text{ (sec)} \qquad ...(23)$$

Observed half-lives range from <0.1 microsec ($k_d \sim 10^7$ sec^{-1}) for the enzyme catalase to a few months ($k_d \sim 10^{-7}$ sec^{-1}) for enzyme inhibitors such as the Kunitz inhibitor of trypsin and for avidin-biotin binding. The smaller the k_d (or the K_d), the greater the affinity and so the more firmly the receptor grasps the ligand.

The probability $P_{occupied}$ that a receptor will be occupied is given by:

$$P_{occupied} = \left(\frac{c_{ligand}}{K_d}\right) P_{unoccupied} \qquad ...(24)$$

where $P_{unoccupied} = 1 - P_{occupied}$. To ensure $P_{occupied} = 99\%$ receptor occupancy, K_d must $\sim c_{ligand} / 100$. For target molecules present at the 10^{-3} - 10^{-11} gm/cm^3 concentrations typically found in human blood, $c_{ligand} = 3 \times 10^{-3}$ molecules/nm^3 for glucose to $c_{ligand} \sim 10^{-11}$ molecules/nm^3 for female serum testosterone, giving a range of $K_d \sim 10^{-4}$ - 10^{-13} molecules/nm^3 to achieve 99% occupancy.

How much binding energy per receptor will this require? The free energy of dissociation ΔG_d of a ligand-receptor complex is related to its equilibrium dissociation constant K_d by:

$$\Delta G_d = -kT \ln(K_d / K_0) \qquad ...(25)$$

which refers to a standard reference state where all chemical species are 1 M (i.e., $K_0 \sim 0.6$ molecules/nm^3) and attributes a free energy of zero to a complex with a dissociation constant of 1 M.

For T = 310 K, the range of required K_d gives a range for ΔG_d of 39.4 zJ for glucose (at typical serum concentrations) to 128 zJ for female serum testosterone.

However, when ligand and receptor associate there is a loss of three degrees of freedom in each of translational and rotational entropy, which may be estimated using the classical Sackur-Tetrode equations, giving an entropic free energy range (for translation and rotation combined) of ΔG_s = 80 zJ for very small molecules (MW ~10 daltons), to 120 zJ (MW ~10^2 daltons), 200 zJ (MW ~10^4 daltons), and 280 zJ for large molecules (MW ~10^6 daltons).

Thus to form a ligand-receptor complex with a dissociation constant K_d, the receptor design must provide a free energy of binding of at least:

$$\Delta G_{total} = \Delta G_d + \Delta G_s \qquad ...(26)$$

or 120-410 zJ/molecule for designed receptors achieving 99% occupancy operating over the likely range of physiological concentrations and temperatures. This is consistent with Drexler's estimate of 161 zJ binding energy required to ensure reliable receptor occupancy for small plentiful molecules.

Ligand-Receptor Specificity

While affinity measures the strength of the binding of a ligand to a receptor, specificity defines the degree to which a receptor can distinguish between similar ligands. That is, the affinity of the target molecule for the receptor must be greater than the affinity of any other ligand in the environment that is competing for that same receptor, by some threshold multiple.

How much greater is enough? In natural dynamic cellular systems, the threshold multiple appears to be a factor of ~10^2-10^3. For example, the carrier which expels Ca^{++} from erythrocytes presents a variation in K_d of 10^{-6} to 10^{-3} in going from the interior to the exterior of the cell. The active transport of amino acids by hepatocytes normally involves $K_d \sim 10^{-1}$, but under conditions of deprivation a high affinity carrier with $K_d \sim 10^{-3}$ comes into play.

However, the key to assessing specificity in the nanomedical context would be to tally all the competing molecules in the *in vivo* environment, determine which are the nearest competitors, and then design to avoid them by imposing appropriate energy barriers. By 1998, only a very few competitive ligand-receptor binding analyses had been performed. Until a complete molecular inventory of the human body becomes available, the following crude estimate of nearest-neighbour differences must suffice.

The human body contains a minimum of $N_{prot} \sim 10^5$ distinguishable proteins. The maximum number of distinguishable proteins in the biosphere was given by Kauffman who estimates a useful biological catalytic task space of $N_{prot} \sim 10^8$ distinct protein forms, a tiny subset of the $\sim 20^{500}$ possible 500-residue protein sequences. If the average protein is constructed from $N_{residue} \sim 500$ amino acids (MW $\sim$50,000 daltons), then the average protein may differ from its most similar neighbour by $n_{var} \sim \log (N_{prot})/\log (N_{residue}) = 1.9\text{-}3.0$ residues. (The precise magnitudes of N_{prot} and $N_{residue}$ are not crucial to our conclusion.) Most proteins are confined to cells containing $N_{prot} \sim 5000$ different protein types each; given that evolution has probably optimized local specificity to ensure that closely competing crucial ligands rarely appear in the same cell, it seems reasonable to assume that the average closest neighbour ligand may differ from the average target ligand by at least $n_{var} \sim 1$ residue.

How much is receptor affinity reduced when binding molecules differing by $n_{var} \sim 1$ residue from the target ligand on receptor-accessible surfaces—the minimum threshold required to ensure specificity within a cell? In one experiment the relative affinity of an antibody constructed for succinylglycinamide-linked histamine (histamine-SGA), which was the target molecule, and the same molecule but with one methyl group or one carboxylic group removed and replaced with a hydrogen, was 1.45×10^4 or 2.5×10^5, respectively, due to steric hindrance. Similar investigations with antibodies for SGA-linked serotonin produced relative affinities of 500-1000, and with antibodies for single alanine substitutions in Human Growth Hormone (HGH), $\sim$1000. The relative affinity of a particular RNA oligomer for theophylline and caffeine, two ligands which differ by only a single methyl group, was measured experimentally as 10,900. Computational receptor experiments in which CH replaces N suggest a decline in relative affinity of $\sim 5 \times 10^4$. Indeed, the change of a single hydrogen atom on a ligand is usually sufficient to destroy its specificity for, or activity within, a particular enzyme. The single-residue affinity reduction at a receptor-accessible surface appears to be of order $\sim 10^3$ - 10^5.

Each increase of $\sim$10 zJ in bonding energy causes the reaction equilibrium constant to decline (hence receptor affinity to rise) by a factor of $\sim$10. If a difference in affinities of $\sim 10^3\text{-}10^5$ between the target molecule and its nearest competitor likely to be present in the environment provides sufficient receptor specificity for nanomedical purposes, this requirement corresponds to a binding energy differential affinity of ~30-50 zJ at 310 K between target and closest neighbour ligands.

Ligand-Receptor Dynamics

Diamondoid structures can exhibit a stiffness and rigidity one or two orders of magnitude greater than that available in protein structures. In general, stiffer structures permit greater specificity because they enhance exclusion of non-target ligands based on van der Waals overlap forces (called steric hindrance) and allow narrower tolerances in distinguishing acceptable ligands.

In less-stiff protein-based receptors, each of the atoms is engaged in relatively large, rapid jiggling movements. Experimental and theoretical work has been done on the atomic fluctuations within the basic pancreatic trypsin molecule, a small enzyme with 58 amino acids and 454 heavy (non-hydrogen) atoms. This work established that fluctuations increase with distance from the center of the molecule, with the magnitude of RMS fluctuations ranging from ~0.04 nm for backbone atoms to ~0.15 nm for the ends of long side chains (roughly one atomic diameter), and an average of 0.069-0.076 nm per atom over the entire molecule. A similar experimental analysis of reduced cytochrome c, a common metabolic enzyme, shows that RMS fluctuations of each of the 103 amino acid residues in the molecule averages ~0.11 nm with lattice disorder (~0.05 nm) included, and fluctuations range from 0.09-0.16 nm. Antibody core domain movements display RMS fluctuations of 0.04-0.19 nm. Hence it appears that the average atom within the typical protein receptor oscillates ~0.1 nm every ~10^{-12} sec, although frequently residues with long side chains (e.g., arg, lys) have much higher RMS deviations than average.

By contrast, in stiff diamondoid-based receptors each of the atoms is locked in a rigid crystalline structure and thus is subject to thermal displacements approximately 10 times smaller. The RMS displacement for a quantum mechanical harmonic oscillator is given by:

$$\Delta X = \left[\left(\frac{\hbar\omega}{k_1}\right)\left(\frac{1}{2}+\frac{1}{e^{\hbar m/kT}-1}\right)\right]^{1/2} \qquad ...(27)$$

where $\hbar$ = 1.055 x 10^{-34} joule-sec, kT = 4.28 zJ at T = 310 K, and angular frequency w = $(k_s/m_{red})^{1/2}$ rad/sec where k_s is mechanical stiffness and m_{red} is the reduced mass = $m_1m_2/(m_1 + m_2)$. For C-C atoms (e.g., in the receptor body), k_s = 440 N/m, m_1 = m_2 = 2 × 10^{-26} kg, thus w = 2.1 × 10^{14} rad/sec, so the RMS displacement of each atom is only ~0.005 nm every ~3 × 10^{-14} sec. For C-H atoms (e.g., on the hydrogen passivated receptor surface), k_s = 460 N/m, m_1 = 2 × 10^{-26} kg (C), and m_2 = 1.673 x 10^{-27} kg (H), thus w = 5.5 × 10^{14} rad/sec, so the RMS displacement of each atom is ~0.008 nm every ~1 × 10^{-14} sec. Similarly, at 310 K the RMS thermal displacement of a 1-nm wide, 10-nm long diamondoid rod is ~0.01 nm, including elastic and entropic contributions.

The ratio of RMS displacements for protein/diamondoid receptors is ~10:1, so the minimum addressable volume (hence inverse maximum specificity) of a diamondoid receptor should be ~10^3 smaller than for protein receptors, a ~30 zJ binding energy advantage for diamondoid receptors.

Diamondoid Receptor Design

Natural enzymes and antibodies are proteins folded into highly organized, preformed shapes that present a ready-made "keyhole" into which a target ligand will fit. The enzyme is folded in such a way as to create a region that has the correct molecular dimensions, the appropriate topology, and the optimal alignment of counterionic groups and hydrophobic regions to bind a specific target molecule. Tolerances in the active sites can be narrow enough to exclude one isomer of a diastereomeric pair. For example, D-amino acid oxidase will bind only D-amino acids, not L-amino acids.

These "keyholes" are extremely floppy, yet still achieve fair specificity. This is a consequence of "induced fit" in protein binding sites. That is, the interaction of the target molecule with an enzyme induces a conformational change in the enzyme, resulting in the formation of a strongly binding site and the repositioning of the appropriate amino acids to form the active site. The receptor flexes, balloons, hinges, or contracts by 0.05-1.0 nm in just the right places to maximize specificity as the selected ligand enters the site. In some cases such as O_2 and CO binding by myoglobin, ligands enter the receptor through a series of temporary voids that appear and disappear in the receptor as ~10 picosec dynamic structural fluctuations. Induced fit can reduce receptivity to undesired proteins that exploit relative geometry by bonding enough to bring portions of their surfaces into alignment with the same receptor sites that bind desired proteins. Folding transitions appear to be the most prevalent and to possess the most possibilities for adaptability or induced fit.

For smaller molecules, it is likely that recognition processes will be relatively inefficient, time-consuming, and more difficult to engineer if they involve a good deal of rearrangement of the receptor's shape.[382] Thus there is considerable interest among chemists in designing artificial receptors that have their cavities already formed into the shape appropriate for the intended substrate. For instance, rigid-cavity "spherand" receptors are exceptionally efficient at binding metal ions. Bowl-shaped molecules such as cryptaspherands, calixarenes, and carcerands can be lined with chemical groups along their walls and with charged groups along their rims to achieve high binding specificity. Self-assembling capsules or "container molecules" made of hydrogen-bonded subunits, capable of limited molecular recognition, have been synthesized. A designed receptor for creatinine was demonstrated in 1995, and in 1998, K. Suslick and colleagues designed metalloporphyrin-dendrimeric artificial receptors that can bind straight, skinny molecules but block out bent or fat molecules. Container molecules with 0.2-0.4 nm portals control entry to their interiors using "French door" and "sliding door" gates and hinges. Active binding sites for small simple molecules such as NO, CO, and C_2H_4 are well-known.

Ultimately, receptors will be designed to nanoscale precision and may be constructed using diamondoid materials. Electrostatic, hydrophobic, and hydrogen-bond forces will add immensely to artificial receptor specificity and are essential for binding small molecules. For example, using a 0.2-nm range in a saline environment, 10-40 charge contacts would be required to bind molecules of various sizes and concentrations using electrostatic force alone, which is ~0.5 charge/nm^2 over the entire surface of a 60,000 dalton globular protein (vs. ~1 charge/nm^2 for the surface of an isolated zwitterionic amino acid). Or, a 7 nm^2 hydrophobic cavity having the exact folded shape of the target ligand generates ~120 zJ binding energy as the molecule stuffs itself into the cavity to exclude its surface from solvent water.

But consider a theoretical receptor that employs van der Waals dispersion forces alone. Atoms comprising the typical protein or CHON target molecule in the human body have an average atomic mass of ~6 amu/atom and an average density of 1500 kg/m^3, giving a mean molecular volume of $\sim 6.7 \times 10^{-30}$ m^3 per atom in the target molecule. Assume for simplicity a spherical receptor surface that forms a negative image of the surface of the target molecule. The receptor surface lies ~1.5 × (minimum van der Waals contact distance) ~0.3 nm from the perimeter atoms of the target molecule, and completely encloses the target molecule (thus requiring at least one moving part). (Proteins are actually ellipsoidal with a much larger surface area $A_p = 0.111$ $MW^{2/3}$ nm^2, than if they were spherical, so figures are conservative for protein binding; proteins typically have a ~60-85% interior packing density.) Dispersion forces alone can provide the minimum required binding energy of ~120 zJ with A_p ~ 6 nm^2 of contact surface (MW > 400 daltons) at 0.3 nm mean range, and dispersion-force receptors can offer exceptionally high affinities for molecules >1000 atoms.

What about specificity? Given the ability to design diamondoid binding sites to at least localized <0.01 nm tolerances, chirality is readily detected and (purely as a design exercise) it may even be possible to distinguish diatomic nitrogen and oxygen on the basis of size alone. The molecular lengths (major axis) of N_2 and O_2 are 0.250 nm and 0.253 nm, but the molecular widths (minor axis) are 0.140 nm and 0.132 nm, respectively. (Diatomic molecules get longer and narrower at higher molecular weight.) Thus N_2 is distinguishable from O_2 on the basis of width (~0.01 nm). With a van der Waals energy well depth of 1.1-1.4 zJ for N_2 and O_2, in a tight receptor these gases are bound at an energy density corresponding to 3000-4000 atm of pressure.

Minimum Feature Size and Positioning Accuracy

Maximum displacement measurement accuracy in nanoscale devices is ~0.01 nm, and RMS thermal displacement in diamondoid bonds is ~0.01 nm at 310 K. Thermal displacements in 10-nm long diamondoid rods are ~0.01 nm at a 1-nm rod width, ~0.02 nm at 0.5-nm width, and ~0.10 nm at 0.3-nm width. However, it is possible to construct components to even narrower tolerances.

For instance, a single C-O bond inserted into a diamondoid rod in a collinear carbon chain extends rod length by 0.1402 nm, the C-O bond length. An adjacent rod into which an N-N bond is similarly inserted is extended by 0.1381 nm, the N-N bond length. By bonding these rods (aligned at one end) it is possible to build diamondoid structures having 0.002-nm features (at the other end), which at 310 K will nonetheless suffer thermal displacements of ~0.01 nm or more. Similarly tiny displacements can be induced in binding cavity surfaces by inserting a foreign atom deep inside the bulk diamondoid structure, causing dislocation strains that decline in magnitude at greater distances from the compositional disturbance.

It is also possible to translate diamondoid components through a picometer step size, much smaller than the unavoidable RMS thermal displacements, using any of several methods; for example:

Levers

Consider a 10-nm lever joined to a fixed bar by a pivot at one end, and driven axially by a ratchet interposed between lever and bar at the other end. Ratchet movements translate to smaller displacements at positions along the lever distant from the ratchet. Thus a follower rod attached to the lever 1 nm from the pivot and driven by a ratchet with 0.01-nm steps moves ~0.001 nm per ratchet step.

Screws

Consider a 3-nm diameter cylindrical screw with a 1-nm pitch. Rotating the screw through a 0.01-nm circumferential displacement causes the screw to move laterally by ~0.001 nm, which may be transmitted elsewhere in the machine by an attached follower rod. Of course, nanoscale screws or gears are sensitive to the precise cancellation of the potentials and hence cannot be perfectly smooth and circular, producing some unavoidable "knobbiness" under load.

Gear trains

Consider a 32-nm diameter worm gear with 1-nm teeth. One rotation of the gear requires 100 rotations of the worm; hence a 0.1 nm displacement applied to the worm produces a 0.001 nm displacement in the gear. More efficient (and coaxial) compound planetary gear trains commonly employed in transmissions achieve displacement ratios up to 10,000:1, a hundred times better than the above example.

Hydraulics

Consider a sealed, fluid-filled, tapered pipe. A piston 1000 nm^2 in area is mounted at one end; another piston 10 nm^2 in area lies at the other end. A displacement of 0.1 nm applied to the smaller piston produces a 0.001 nm displacement in the larger piston. (Here again the finite size of molecules may produce "knobby" performance as fluid particles slip from one stable configuration to the next.)

Compression

Consider a rod upon which a compressive force of 100 nN/nm^2 (near the maximum diamondoid strength) has been imposed. Affixed to the rod are two crossbars spaced 4 nm apart. If the force on the rod is increased to 101 nN/nm^2, the gap between the crossbars compresses by ~0.001 nm.

Receptor Configurations

Many different useful receptor configurations may be readily envisioned, of which the following brief descriptions are but a small sample. Note that most large target molecules of nanomedical interest are proteins which are probably floppy enough to permit entry into reasonably open multiply-concave rigid receptor structures; if not, hinges are easily added to the receptor design.

Imprint model

Molecular imprinting is an existing technique in which a cocktail of functionalized monomers interacts reversibly with a target molecule using only noncovalent forces. The complex is then crosslinked and polymerized in a casting procedure, leaving behind a polymer with recognition sites complementary to the target molecule in both shape and functionality. Each such site constitutes an induced molecular "memory," capable of selectively binding the target species. In one experiment involving an amino acid derivative target, one artificial binding site per $(3.8\ nm)^3$ polymer block was created, only slightly larger than the $(2.7\ nm)^3$ sorting rotor receptors described by Drexler. Chiral separations, enzymatic transition state activity, and high receptor affinities up to $K_d \sim 10^{-7}$ have been demonstrated, with specificity against closely competing ligands up to $\Delta K_d \sim 10^{-2}$ (~20 zJ).

Several difficulties with this approach from a diamondoid engineering perspective include:

1. A sample of the target molecule is required to make each mold.
2. It is currently unknown how to prepare diamondoid castings.
3. Once the imprint has been taken, the site cannot easily be further modified.

Solid mosaic model

In the solid mosaic receptor model, the precise shape and charge distribution of the target molecule is already known. Working from this information, a set of diamondoid components could be fabricated which, when fitted together like a Chinese puzzle box, create a solid object having a cavity in the precise shape of the optimum negative image of the target molecule. The mosaic may contain point charges, voids, stressed surfaces, or dislocations to achieve fine positional control. Mosaic components may be as small as individual atoms, so this model is conceptually similar to 3-D printing or raster-scan techniques in which the desired cavity formation is constructed atom by atom inside a nanofactory. This model, like the imprint model, cannot easily be reconfigured once it has been constructed because each of the many unique parts may contribute to the entire structure. The construction of receptors from parts of fixed size and shape is crudely analogous to members of the heterodimer receptor class (a two-component receptor) such as $GABA_B$.

M. Reza Ghadiri has designed a protein mosaic model using cyclic peptides that assemble spontaneously into nanotubes of predefined diameter; incorporation of hydrophobic amino acid side chains on the outside of these tubes leads to spontaneous insertion into bilayers, allowing the tubes to function as transmembrane ion channels. Other examples of mosaic model receptors are mesoporous silica filters with functionalized organic monolayers forming 36 nm sievelike pores, and zeolites and zeolitelike molecular sieves. Zeolites are artificial crystal structures with precise and uniform 0.4-1.5 nm internal void arrays which can also be used as shape-selective catalysts able to favour one product over another that differs in size by as little as 0.03 nm, such as pxylene and oxylene. By 1998, rational de novo computational design of artificial zeolite templates and crystal engineering had begun.

Tomographic model

In the tomographic receptor model, the receptor engineer again starts with a known target molecule topography and designs a series of thin planar sections which, when stacked together in the correct order (using positionally-coded docking pins) and bonded, create a solid object containing the desired optimum binding cavity. As in the mosaic model, point charges or dislocations in each planar segment can be used to manipulate cavity features and dimensions to precise tolerances. Unlike the mosaic model, a tomographic receptor can be reconfigured by partial disassembly and replacement of specific planar segments, each of which contributes only locally to the total receptor structure. Hybrid or modular artificial enzymes and two-dimensional sheetlike hydrogen-bonded networks are crude analogs in current research.

Pin cushion model

The pin cushion receptor, suggested independently by K.E. Drexler, is a hemispheroidal or hemiellipsoidal shell through which a number of rods protrude, each of which may be moved radially. When inserted through the shell to varying depths, the endpoints of the rods define a negative image surface which may be made to mirror the topography and charge distribution of a known *target molecule*. Rods may be tipped with positive, negative or no charge, or they may terminate in any number of functionalized surface segments designed to optimally match parts of the target molecule shape. Other configurations such as a rectangular box, hinged plates with protruding rods, counterrotating rollers, or time-varying rod positioners are readily conceivable. Pin cushion receptors are easily reconfigured to bind different target molecules, hence may be regarded as fully programmable "universal" binding sites. The principal difficulty with the pin cushion receptor is its excessive size (compared to other receptor models) and its greater complexity (since each rod may be controllable individually).

Pin cushion receptors can also be used to discover the shapes of unknown molecules: A target molecule is placed in the central cavity with all rods fully retracted, and the rods are slowly slid forward using nanopistons with force reflection feedback, until all pistons register zero force, indicating balance between attractive and repulsive van der Waals interactions, at which point all rod positions are recorded. Rods of differing end tip charge may then be tested for additional attractive potential. The final result is a precise mapping of the target molecule, which data may be stored or transmitted elsewhere for future use.

Construction costs

The active binding site of a receptor consisting of $(2.7 \text{ nm})^3 \sim 19 \text{ nm}^3$ of structural atoms and constructed with 0.001-nm feature sizes in theory requires information from 2×10^{10} voxels (volume pixels) for complete description. However, there are only $N_{atom} \sim 1000$ atoms involved in the structure and their locations cannot be arbitrarily chosen. Atomic scale ($\sim$0.1 nm) resolution would require $\sim$19,000 voxels; each voxel minimally requires an index number ($\sim \log_2(19{,}000) \sim 14$ bits), an atomic identifier ($\sim \log_2(92) \sim 7$ bits), and a charge identifier ($\sim \log_2(3) \sim 2$ bits), for a total of 23 bits/voxel which gives 4×10^5 bits/receptor at atomic scale resolution. Drexler estimates the number of configurational options per atom $N_{opt} \sim 150$; hence a description of the receptor could require as few as $N_{atom} \log_2(N_{opt}) \sim 7 \times 10^3$ bits. Assuming energy dissipation of $\sim$3 zJ/bit for rod logic register reading or $\sim kT \ln(2)$, then each time the receptor description is retrieved, stored or processed may require a minimum energy dissipation of $\sim 10^4$-10^6 zJ. Reversible computing may reduce this energy requirement by a factor of 10-100 or more. Construction of a receptor containing $\sim$1000 atoms using the nanomanipulator arm requires $\sim 10^{-2}$ sec and consumes $\sim$0.001 picojoule ($\sim 10^6$ zJ) of mechanical energy.

Receptor durability

Receptor durability is difficult to estimate ab initio. Molecularly imprinted polymer sites can be stored at least several years without loss of performance, but these receptors have only been tested to $\geq$100 cycles of use without any detectable loss of memory. The lifetime *in vivo* for metabolic enzymes often exceeds ~10^5 sec (~1 day), and taste cell receptors survive ~10^6 sec (~2 weeks), which suggests operational lifetimes for natural receptors on the order of 10^7-10^{12} cycles despite the relative fragility of protein structures. Diamondoid structures should be even more durable because of their superior physical strength, affirmative forcible ejection of bound ligands each cycle, and high resistance to chemical degradation thus reducing susceptibility to poisoning.

Ligand-Receptor Mapping

To achieve a fully general-purpose receptor capability in nanomedical systems, two classes of analytic function are essential.

First, presented with an arbitrary molecule, the system or user must be able to infer from the molecule's structure the shape and electronic configuration of an optimal receptor geometry that will efficiently bind it, with a particular affinity and specificity (as a design specification). This discovery procedure may involve a process akin to molecular imprinting, fluorescent dye affinity matching on testing chips, Structure Activity Relationships (SAR) by NMR techniques, or pin cushion receptor mapping.

Second, presented with an arbitrary protein-built binding site embedded in living tissue, the system or user must be able to infer the molecule(s) which the given receptor could bind, and to compute the affinity and specificity of that activity. This capability may require a rather diverse steric toolkit. Biological receptors are constructed in one of ~500-1000 distinct shapes or "domains," such as the well-known Y-shaped antibody immunoglobulin domain (~100 residues) and other assorted clefts, folds, kringles and coils. Several hundred distinct fold families are known, though it is believed there are only ~20 major domain types, and Chothia believes that "the large majority of proteins come from no more than one thousand families."

One mapping technique would employ a series of rodlike probes inserted into the receptor cavity like a pick gun or other lockpicking tool. After physically securing the receptor, the first crude probes quickly map the cavity to nanometer scale. Based on this preliminary information, subsequent probes having ever-finer discrimination chart smaller features as well as charge distributions by inserting a predetermined set of test rods with functionalized tips in a standard sequence, to rapidly prune the huge configurational space down to a single unique electrophysical shape using the minimum possible number of tests. (A 1 nm^3 volume of multi-element diamondoid receptor structure has $\sim 10^{148}$ possible distinct configurations, by one conservative estimate, requiring at least 492 binary tests to eliminate all but one configuration. Antibody domains contain $\sim 10^{50}$ possible configurations, requiring 166 binary tests.) Cavity Stuffer, a software package comprising an experimental design tool to investigate automated cram-packing of predefined cavities using randomly branched polymers, is a preliminary effort in this general direction, although the algorithmic task of discovering an unknown receptor contour may be considerably more challenging.

Another approach to receptor mapping would be the reversible chemical or mechanical denaturation of the receptor protein followed by precise nondestructive amino acid sequencing, from which tertiary structure and activity could then be computationally inferred. Algorithms to perform such computations are the subject of intense current research interest. Even imperfect tertiary structure predictions should greatly reduce the search space, so that only a partial residue sequencing may be necessary for unambiguous identification from a library of possible proteins. Once the target ligand structure has been inferred, the subsequent design and manufacture of receptor-specific agonists and antagonists, including catalysts and cofactors, activators and inhibitors, promoters and repressors, should be comparatively easy.

Large Molecule Binding, Sorting, and Transport

Is there any size limit for target molecules to be transported? Natural receptors have already been found for large molecules including low-density lipoproteins (LDLs) > 1,000,000 daltons and high-density lipoproteins (HDLs).

The methods described in earlier Sections can be adapted for binding large molecules (>1000 atoms), including molecules far wider than the binding device itself (e.g., ~200-nm diameter virus particles and larger). Making a binding site for a large molecule should be physically easier (albeit computationally more challenging) than making a binding site for a small molecule because of the greatly increased area of interaction. For example, a binding energy of 400 zJ may be realized by creating a dispersionforce binding area covering only ~25% of the surface of a 10,000-atom target molecule or a mere ~0.02% of a 200-nm virus particle.

This makes possible the concept of binding pads—small surfaces with dimples (concave or convex), each dimple consisting of precisely-placed nanometer-scale features that are complementary to specific patches (e.g., epitopes) on the surface of the large target molecule.

(Specificity is lost for portions of the molecule outside the particular patches.) Each dimple could effectively grasp the side of the large molecule without having to fully enclose it—a capability useful in nanorobot foot pads during cell walking and anchoring, in handles for nanociliary or nanomanipulator transport functions, and for chemotactic sensing. (For example, macrophage receptors for LDLs employ a "pad" consisting of three globular cysteine-rich domains.) The large compliance of target protein molecule subunits should prove curative for any misalignment problems caused by cumulative small errors in bond lengths across large diamondoid receptor structures.

Rather than using sorting rotors, which become unwieldy when large molecule binding pockets must be used, the shuttle pump may provide reasonably efficient large molecule sortation and transport. The shuttle pump consists of a diamondoid tube within which a receptor ring moves between iris diaphragms at either end. The receptor ring is constructed as two or more binding pad segments. For molecule pickup, the ring is pressed together, forming an annular binding region for the target large molecule, which binds and is shuttled to the other side. The receptor ring is then fragmented, destroying binding affinity and unlocking the target molecule, which escapes via diffusion. The shuttle returns to the pickup side, the receptor is pressed together again, and the cycle repeats. A biocompatible solvent environment is maintained during large-protein manipulation tasks.

Assuming a roughly spherical large molecule and laminar fluid flow at 1 atm forcing pressure, a 10-nm diameter molecule moves through a 20-nm long pump ($\sim 10^{-20}$ kg, $\sim 10^{6}$ atoms) in $\sim 10^{-6}$ sec at $\sim$0.02 m/sec, consuming $\sim$0.02 pW during transfer. A 200-nm virus-size target molecule moves through a 400-nm long pump ($\sim 10^{-17}$ kg, $\sim 10^{9}$ atoms) in $\sim 10^{-2}$ sec at $\sim$60 microns/sec, consuming $\sim 10^{-16}$ watts during transfer; at $\sim$0.0002 atm, release time is diffusion limited. The transfer force exerted on a 10-nm molecule is $\sim$1 pN, $\sim$600 pN on a 200-nm virion; a binding energy of 400 zJ at a 0.2-nm contact distance gives a binding force of $\sim$2300 pN, sufficient to hold a particle of either size firmly during transport and release.

J. Soreff points out that as protein size increases, so does the energy available for local minima in the binding. Desired proteins may become stuck in incorrect positions, or undesired proteins may become partially adhered to a receptor. Besides designing to minimize these possibilities, using a multireceptor cascade with different combinations of binding patches at each stage should allow complete exclusion of undesired large-molecule species.

7

Nanosensor and Nanoscale Scanning

Medical nanorobots need to acquire information from their environment to properly execute their assigned tasks. Such acquisition is achieved using onboard nanoscale sensors, or nanosensors, of various types. Nanosensors allow for medical nanodevices to monitor environmental states at three different operational levels:

1. Internal nanorobot states,
2. Local and global somatic states (inside the human body), and
3. Extrasomatic states (sensory data originating outside the human body).

The general physical limits to sensory perception are reviewed by Block and Bialek.

The specific nanosensor technologies required include sensors to detect chemical substances, displacement and motion, force and mass, and acoustic, thermal, and electromagnetic stimuli. Typical sensor device mass, volume, and sensitivity limits are summarized in each Section.

Chemical and Molecular Nanosensors

The area of chemical microsensors is well developed, and there is increasing research interest in nanoscale chemical sensors, "*chemosensors*" and "*biosensors*". The most common nanomedical application of chemical sensors will be to measure the concentration of specific molecules and biopolymers in aqueous solvent—whether in blood serum, interstitial fluid, or cytosol. This may be accomplished using ordered arrays of receptors or by using sorting rotors to directly count molecular populations in known sample volumes. Spatial and temporal concentration gradient sensing is also essential in navigation, communications, and in mediating rapid response to environmental stimuli. Chemotactic sensors may be used to sample the chemical composition of surfaces.

Broadband Receptor Arrays

One simple broadband concentration sensor consists of a graduated series of receptors having uniformly high specificity but engineered with progressively greater affinities (successively smaller equilibrium dissociation constants K_d (molecules/nm^3) for the target molecule, which is present in the test sample at concentration c_{ligand}. Exposure of the receptor array to the test sample for a time t_{EQ} necessary to reach diffusion-driven equilibrium gives a probability of receptor occupancy $P_{occ} \sim 0.91$ in a receptor with $K_d \sim 0.1c_{ligand}$, $P_{occ} \sim 0.50$ in a receptor with $K_d \sim c_{ligand}$, and $P_{occ} \sim 0.09$ in a receptor with $K_d \sim 10c_{ligand}$.

During each measurement cycle, all steric probes are simultaneously extended into their associated receptor volumes at a time t_{EQ} after presentation of sample to the array. Probes which reach full extension register an empty receptor; those which cannot fully extend register an occupied receptor. After registration, the probes are retracted and ejection rods are thrust into all receptors (typically requiring ~1 nanonewton (nN) of force for occupied receptors) to empty them while the test chamber is flushed clear in preparation for the next cycle to begin. Ideally, the probe rod pushing into the binding site should push the bound molecule further into the binding site to avoid complications that may arise due to competing kinetic barriers, and ejection rods should push molecules far enough to ensure that the next resample is independent.

Consider a series of N steric probe units as proposed by Drexler. Each probe measures 8 nm × (2.5 nm)^2 ~50 nm^3 and has mass ~2 × 10^{-22} kg. The ratio of the dissociation constants of adjacent units is $k = K_{di}/K_{di+1} > 1$, with $K_{d1} \sim 10^{-4}$ molecules/nm^3 and $K_{dN} \sim 10^{-13}$ molecules/nm^3 typically in the human body, and so $N = 1 + \{\log_{10}(K_{d1}/K_{dN})/\log_{10}(k)\}$.

Minimum measurement error occurs when c_{ligand} exactly matches the K_d of a probe unit, that is, when $c_{ligand} = K_{di}$ (i.e., $P_{occupiedi} = 0.5$). Maximum measurement error occurs when c_{ligand} lies exactly midway between two probes such that $K_{di} > c_{ligand} > K_{di+1}$, or, more specifically, at the geometric midpoint $c_{ligand} = (K_{di}\ K_{di+1})^{1/2} = K_{di}\ \kappa^{-1/2}$. Now the equation becomes

$$P_{occupied_i} = \left(\frac{c_{ligand}}{K_{d_i} \kappa^{-1/2}} \right) P_{unoccupied_i} \qquad \text{...(1)}$$

where $P_{unoccupied_i} = 1 - P_{occupied_i}$. Additionally, sampling error is $\sim N_m^{-1/2}$ when N_m independent measurements are taken (i.e., N_m measurement cycles employing all N probe units during each cycle), for $N_m >> 1$, establishing an error bound on the probability of receptor occupancy of $P_{occupied_i} \pm N_m^{-1/2}$. If two concentrations $c_1 = P_{occupied_i} / (1 - P_{occupied_i})$ and $c_2 = (P_{occupied_i} + N_m^{-1/2}) / (1 - (P_{occupied_i} + N_m^{-1/2}))$ may be distinguished, where $P_{occupied_i}$ is given by Eqn. 1, then the minimum detectable concentration differential $\Delta c / c$ is

$$\Delta c / c = [c_2 / c_1] - 1 = \frac{\left(1 + \kappa^{-1/2}\right) N_m^{-1/2}}{1 - \left(1 + \kappa^{-1/2}\right) - N_m^{-1/2}} - 1 \qquad \text{...(2)}$$

For $k >> 1$, Eqn. 2 reduces to: $(\Delta c/c)_{large} \sim (1 + k^{1/2})/N_m^{1/2}$. In the limit as k approaches 1 and $N_m >> 1$, Eqn. 2 reduces to: $(\Delta c/c)_{small} \sim k^{-1/2} (\Delta c/c)_{large}$.

For $\kappa = 10$ (an $N = 10$-probe sensor, sufficient to span the entire $K_d = 10^{-4}$ to 10^{-13} molecules/nm³ range using one probe per decade), $\Delta c/c = 0.94$ (94%) at $N_m = 100$, 0.20 (20%) at $N_m = 1000$, (0.06) 6% at $N_m = 10{,}000$, and 0.006 (0.6%) at $N_m = 10^6$ measurement cycles. Little additional discrimination is obtained by using more than $N \sim 10$ probes in the sensor. For $N_m = 1000$, $\Delta c/c = 0.89$ (89%) at $\kappa = 178$ (N = 5 probes), 0.20 (20%) at $\kappa = 10$ (N = 10 probes), 0.136 (13.6%) at $\kappa = 1.23$ (N = 100 probes), and 0.135 (13.5%) at $\kappa = 1.0021$ (N = 10,000 probes). Note that $\Delta c/c = 0.20$ (20%) is equivalent to detecting a pH variation of ~0.08 (e.g., $10^{(0.08)} - 1 \sim 0.20$).

Sensor cycle time $\sim t_{EQ}$ may be crudely approximated by the number of random ligand-receptor encounters ($N_{encounters} \sim 100$; Drexler) necessary to ensure binding divided by the number of ligands striking the receptor surface per second, or (sec/cycle)

$$t_{EQ} \sim \left(\frac{N_{encounters}}{A c_{ligand}} \right) \left(\frac{\pi MW_{log}}{2kT\, N_A} \right)^{1/2} \text{(sec/ cycle)} \qquad \text{...(3)}$$

where A = active receptor cross-sectional area ~0.1 nm² for small molecule (MW ~100 gm/mole) receptors to ~10 nm² for large molecule (MW ~ 10^5 gm/mole) receptors; $c_{ligand} = 10^{-2}$ nm⁻³ for the most common molecules to 10^{-12} nm^{-3} for rare molecules in the human body; MW_{kg} = target ligand molecular weight in kg/mole, k = 0.01381 zJ/K (Boltzmann constant), T=310 K (human body temperature), and N_A = 6.023 × 10^{23} molecules/mole (Avogadro's number).

For common small molecules, $t_{EQ} \sim 10^{-6}$ sec; for common large molecules, $t_{EQ} \sim 0.2 \times 10^{-6}$ sec, but a sample chamber ~(25 nm)^3 in size is diffusion-limited to a minimum $t_{EQ} \sim 10^{-5}$ sec for $c_{ligand} > \sim 10^{-4}$ nm^{-3}. For the rarest small molecules in the human body, $t_{EQ} \sim 800$ sec, ~200 sec for rare large molecules, and a larger chamber may also be required. Preconcentration of the test sample by dehydration using sorting rotors to rapidly remove water from the sample chamber reduces t_{EQ} to 10-40 sec for very rare molecules.

Steric probes using a driving force of ~100 pN, over 2 nm of motion per cycle require 200 zJ/measurement. If 1000 measurement cycles of a 10-probe sensor suffice to determine concentration to ~2 significant figures, this information costs ~0.002 pJ to acquire, a power consumption of from ~10 pW for large common molecules to ~10^{-9} pW (~0.6 kT/sec) for small rare molecules while the sensor is working. Power dissipation may run up to ~10 times higher, depending upon the number of occupied receptors that must be cleared via ejection rods during each cycle.

Narrowband Receptor Arrays

If the concentration of the target ligand varies only slightly within known limits, similar results may be obtained using a narrowband sensor that employs a repeating array of a single receptor type tuned to the maximum expected concentration c_{max} of the target ligand. If the sensor consists of N_r receptors, choosing $K_d > \sim c_{max} N_r^{-1}/(1-N_r^{-1})$ ensures that all receptors are occupied at $c=c_{max}$, thus avoiding array saturation at concentrations below c_{max}. At minimum sensitivity, $K_d \sim c_{min} (1 - N_r^{-1})/N_r^{-1}$, and so $(1 + (\Delta c/c)) = (c_{max}/c_{min})^{(Nr-1)}$ for the entire array. Thus the minimum detectable concentration differential $\Delta c/c$ for the array is

$$\Delta c / c = (N_r - 1)^{(2N_r)} - 1 \qquad ...(4)$$

If A = receptor unit active area (~50 nm^2), L = receptor unit length (~10 nm), and receptor arrays are packed in parallel sheets with separation Δx (~10 nm, for small-molecule diffusion in ~10^{-6} sec) to allow sample access, sensor volume $V_s = AL\, N_r (1 + (\Delta x / L))$ and sensor scale ~ $V_s^{1/3}$. Thus a sensor with N_r = 100 receptor units achieves Δc / c = 0.10 (10%) with a (46 nm)^3 sensor; N_r =

1400 receptor units achieves $\Delta c / c = 0.01$ (1%) with a (112 nm)^3 sensor. A dedicated micron-scale nanodevice employing a (1.44 micron)^3 sensor array having 3 million receptor cells could at best distinguish 0.001% concentration differentials in ~3 sec (from Eqn. 6) for common molecules like serum glucose.

The ion channel switch (ICS) biosensor (aka. the "Australian sensor") can detect small-molecule concentrations as low as ~10^{-14} nm^{-3} in a ~600 sec measurement time.

Counting Rotors

Concentration may be determined by varying exposure time to the sample. One energy-efficient design uses an input sorting rotor running at varying speeds (according to target concentration) synchronized with a counting rotor (linked by rods and ratchets to a data storage register device) to assay the number of molecules of the desired type that are present in a known volume of fluid. The counting rotor uses a steric probe, moved up and down in synchrony with the arrival of a binding site under the test probe, to count the number of molecules transferred by the sorting rotor. The fluid sample is drawn from the environment into a 4000 nm^3 reservoir ($L_{reservoir}$ ~16 nm) with the equivalent of 10^4 refills/sec on a flow-through basis, using paddlewheel pumps or reciprocating pistons operated slowly enough to avoid damaging molecules in the sample volume. This sensor, which measures 45 nm × 45 nm × 10 nm comprising ~500,000 atoms (~10^{-20} kg), should count 'N ~ 10^4 - 10^5 molecules/sec of small common molecules present at typical human serum concentrations of 10^{-3} - 10^{-2} nm^{-3}, such as glucose, with the rotor spinning at ~1% of normal speed.

For rare molecules with c ~ 10^{-11} nm^{-3}, the count rate for a single sensing rotor falls to 'N ~ 10^{-4} molecules/sec. However, a (340 nm)^3 reservoir (a ~10^4 larger chamber volume) serviced by a bank of 10^4 rotors (4 × 10^{-18} kg additional mass) improves the count rate for rare molecules to 'N ~ 1 molecule/sec; that is,

$$N \sim 2500\, c_{ligand} L_{reservoir} \text{ (molecules / sec)} \qquad \ldots(5)$$

Sample preconcentration may allow further improvement by a factor of up to 10-100. Energy requirements are comparable to those for sorting rotors except for the need to retrieve ~all target molecules from a known sample volume, which slows the sorting process, requiring ~10 zJ/count for common molecules and ~100 zJ/count for the rarest molecules in the human body. Large-molecule concentrations may be measured using a shuttle pump and a counting device.

Chemical Assay

It may also be useful to design a sensor that is capable of detecting as many different chemical species as possible, in contrast to detecting the precise concentration of a single species, for use in chemomessaging, chemonavigation, cellular diagnosis, and other *in vivo* assay work. Counting rotors may be used for this purpose. Each 12-receptor sorting rotor/counting rotor pair has a minimum 400 nm^3 volume. Allowing an additional ~800 nm^3 per pair to account for power, control, mechanical attachments and housings, which includes ~300 nm^3 per pair for sample chamber volume and sample access, requires ~100 nm^3 per receptor, roughly the same as for the steric probe units already described. A dedicated 1 $micron^3$ chemical assay nanorobot, which includes ~0.25 $micron^3$ sample volume, with ~10^7 pairs (receptors) could thus continuously scan for as many as ~10^7 different chemical species, counting 'N ~ 1 large molecule/sec at concentrations of $c_{ligand} \sim 10^{-10}$ molecules/nm^3.

Chemical Nanosensor Theoretical Limits

Berg and Purcell estimated that an ideal concentration sensor, limited only by diffusion constraints and drawing through a spherical boundary surface of radius r_s, provides a minimum detectable concentration differential $\Delta c/c$ of

$$\Delta c / c = (1.61\, \Delta t\, D c r_S)^{-1/2} \quad \text{...(6)}$$

where Δt = time (sec), D is diffusion coefficient ~10^{-9} m^2/sec for small molecules, and c = concentration (molecules/m^3). Using a sample chamber with r_s = 10 nm, a concentration of c = 3 × 10^{-3} nm^{-3} for serum glucose could be measured with 1% uncertainty in Δt ~ 260 microsec, or with 0.01% uncertainty in 2.6 sec. A dedicated micron-scale nanodevice using a 10-sec sampling time could at best distinguish a 0.0005% concentration differential for small common molecules like serum glucose but only a 27% concentration differential for large rare serum molecules like somatotropin.

Berg and Purcell also found that the capture of small molecules at nanorobot surfaces can be surprisingly efficient. Specifically, for a spherical nanodevice of radius R, across whose surface are uniformly distributed N_r receptor spots of radius r_r, the maximum diffusion current absorbed by the receptor array, J_{array}, is

$$J_{array} = J\left(\frac{N_r r_r}{N_r r_r + \pi R}\right) \quad \text{...(7)}$$

where J = maximum diffusive intake current. For r_r = 1 nm and R = 1 micron, the J_{array} is fully 50% of J using only N_r = 3100 surface receptors, which occupy a mere 0.1% of the total device surface area.

For a purely sensory spherical nanodevice, a uniform distribution of a relatively small number of chemoreceptors confers optimal sensitivity. However, in applications where sensors (which attract and then release) are in competition with pumps (which strongly attract and then remove) for the same target molecules at the nanodevice surface, greater sensing efficiency may be achieved by clustering the sensors, as in a "nose" organ.

Proteins may be modified as they age within biological cells. For instance, under steady state conditions, ~10% of protein molecules may exhibit carbonyl (oxidation) modifications, glycosylation (non-enzymatic), and dysfunctional proteins may be ubiquitinized and phosphorylated under enzyme control. Heat shock also modifies proteins. Chemosensor receptors must be carefully designed to accommodate changes in target ligand structures or sites that are likely to be subject to such alterations.

Spatial Concentration Gradients

Medical nanodevices can detect both extremely steep and extremely shallow spatial concentration gradients. For example, a receptor array sensor consisting of 10 steric probe units has a total volume of ~1000 nm^3 including its prorated share of drive systems, housings, etc. In just a few measurement cycles, adjacent sensors can detect a maximum concentration change from 10^{-2} nm^{-3} to 10^{-11} nm^{-3} across a mean nanosensor separation of 10 nm on a nanodevice surface, a steep chemical spatial gradient of ~10^{10}% per nanometer. J. Soreff points out that maintaining a steep concentration gradient of a freely diffusing species implies a power dissipation of ~kT per molecule per time to diffuse a concentration-halving distance. For a common, small, tethered molecule with c_{ligand} ~ 3 × 10^{-3} molecules/nm^3 that diffuses ~1 nm in t_{diff} ~ 1 nanosec, implied power density D_{mol} ~ kT c_{ligand}/t_{diff} ~ 10^{13} watts/m^3, a power density that might possibly be encountered in the vicinity of chemoelectric transducers. (Such high power densities never occur on macroscale dimensions, and heat rapidly diffuses at nanoscale dimensions.)

Similarly, counting rotor sensors for small common serum molecules like glucose deployed at opposite ends of a 1-micron nanodevice can detect a concentration differential of ~10^{-4} over a ~1 sec measurement period, a shallow chemical spatial gradient of only

$\sim 10^{-5}$ % per nanometer. Even lower gradients may be detected by communication between physically separated cooperating nanodevices.

Temporal Concentration Gradients

Koshland and Macnab have shown that bacteria possess a mechanism for sensing small temporal concentration gradients. A bacterium detects a spatial gradient of attractant not by comparing the concentration at its head and tail, but rather by traveling through space and comparing its observations over time. *E. coli* can measure a ~0.01% gradient during the time it travels one body length (~2 microns) at top speed (20-40 microns/sec), a minimum temporal gradient of 0.1% per second. This appears very efficient, since the counting rotor sensor is not much better, achieving ~0.01% per second. As for peak gradients, the 10-probe receptor array sensor can detect a maximum concentration change (e.g., from an already-known rare concentration) in $\sim 10^{-4}$ sec, a peak temporal gradient of $\sim 10^{15}$ % per second.

Chemotactic Sensor Pads

Besides concentration sensing, chemical sensors may be used to determine the chemical characteristics of surfaces. A binding pad coated with an array of reversible, possibly reconfigurable, artificial receptors may be pressed against test surfaces such as cell membranes, or the perimeters of very large molecules, in order to detect the presence of specific chemical ligands. If pad receptor specificity is high for the target ligand, then the degree of pad "stickiness" to the test surface is a good measure of the population of target ligands on that surface. Probing two-dimensional surfaces with receptors may be constrained by orientation effects, since binding may require a specific relative orientation of ligand and receptor. Compliant pad receptors can reduce this problem but may impair lateral resolution—there is a three-way tradeoff among angular tolerance, lateral resolution, and multioriented multicopy receptors. A physical contact sensor prevents confusion of surface-bound and free-solvated ligands of the same species. The selective and reversible stickiness of such pads can also be exploited for locomotion.

Molecular adhesion experiments in chemical force microscopy and interfacial force microscopy using AFM probes with functionalized tips typically detect maximum adhesive noncovalent force differentials of 40-160 piconewton (pN)/ligand for competing ligands at a peak-force distance of ~0.2 nm. Forces >10 pN are readily measured by nanosensors, so binding pad sensors should give an exact count of the number of target ligands present per unit area of test surface if (receptor areal number density) > (ligand areal number density) > (1 ligand/pad area), and if contact time is adequate. Receptors for many different ligands can be mounted on moving rotors or belts, or employed as interchangeable tool tips, thus allowing a single sensor pad to reconfigure its specificity or sensitivity under external control.

Receptor Sensors

Ligand mimics may also be used to allow a medical nanodevice to detect a particular receptor in its environment. Each such "receptor sensor" is a protruding physical structure that mimics all or part of the electrosteric structure of the ligand to which the target receptor has maximum specificity. For example, probes consisting of artificial structures that simulate antigen peptides (8-15 amino acids long) bound in an MHC-like structure could be used to sense the presence of specific T-cell receptors.

Sensing pads displaying ligand mimic structures are brought into contact with the test surface. Any target receptors present will bind to a mimic structure, thus may be counted as the sensor sweeps slowly across the surface. Immediately after recognition has occurred, the mimic structure may be reversibly fragmented, destroying receptor/mimic affinity and allowing the receptor to release the mimic. Mimics are mounted on the sensor pad with an appropriately elastic compliance to prevent structure detachment during a measurement scan.

Receptor sensing is most useful in detecting receptors for large molecules. In the case of receptors for small molecules, there may sometimes be insufficient space in the target molecular volume to match the surface electrosteric structure, to accommodate ligand fragmentation control rods or cables, or to build internal boundaries sufficient to permit ligand fragmentation without creating undesirable reactive species. Some of these problems with small-molecule receptors may be partially overcome by employing an annular repulsive ring through which a mimic ligand may be forcibly retracted without fragmentation, but receptor orientation effects will present additional constraints.

Displacement and Motion Sensors

The ability to precisely measure position and displacement, velocity and acceleration, rotation and orientation are essential requirements for nanoscale locomotion, physical manipulation, and navigation. A comprehensive analysis is beyond the scope of this book, but the following approximations should prove useful in nanomedical design.

Displacement Sensors

Diamondoid parts may contain features, or be positioned, in picometer (1 pm = 10^{-12} m) steps, much smaller than the unavoidable RMS thermal vibrations. The nanomanipulator robot arm may be moved in picometer increments, and the RMS thermal longitudinal displacements at the tip of a 20-nm long, 10-nm wide diamondoid lever is ~1 pm. Displacement sensitivity of 10 pm is routinely achieved in STMs. It is claimed this can be improved down to the ~1 pm level; STM resolution of 2 pm has been demonstrated experimentally. A C_{60} molecule used as the active element in an electromechanical amplifier transmits ~100 times more electrical current when physically compressed by 100 pm, suggesting a minimum deformation detection limit of a few picometers. (A strain gauge and a vibration sensor using a (17,0) carbon nanotube (~2000 atoms) has been proposed.)

Thermally-induced positional uncertainty in the displacement of nanoscale components is approximated by the classical value

$$\Delta x \sim (kT / k_S)^{1/2} \qquad \text{...(8)}$$

where T is temperature (K) and k_s is the Hooke's law spring constant, or restoring force stiffness (N/m). At 310 K, this classical approximation is accurate to within < ~10% of the quantum mechanical treatment for harmonic oscillators with RMS displacements Dx > ~ 10 pm. Spring stiffness k_s is ~0.1 N/m for nonbonded (noncovalent) interatomic interactions, ~30 N/m for covalent bond angle bending, ~400 N/m for covalent bond stretching, and ~1000 N/m for solid 1 nm^3 diamondoid blocks. At 310 K, springs of such stiffness produce minimum displacement uncertainties of 200 pm, 10 pm, 3 pm and 2 pm, respectively.

The log ratio of detection energy to noise energy, the signal/noise ratio (SNR), for a displacement sensor is derived from the harmonic potential (1/2) k_s x^2 as

$$\text{SNR} \sim \ln\left(\frac{\text{signal energy}}{\text{noise energy}}\right) \sim \ln\left(\frac{k_S\, \Delta x^2}{2kT}\right) \qquad \text{...(9)}$$

Assuming a very stiff k_s = 600 N/m, a minimal SNR = 1 gives a minimum detectable displacement of 6 pm, or 10 pm at a more reasonable SNR = 2 (20 dB). The conservative conclusion is that Δx_{min} ~ 10 pm displacements should be reliably detectable by medical nanosensors *in vivo*, in a measurement time t_{meas} ~ 10^{-9} sec. This compares favourably with the stereocilia of the inner ear, which can only detect 100 pm displacements in 10^{-5} sec.

Velocity and Flow Rate Sensors

Typical thermal velocities of nanoscale diamondoid components at 310 K, are ~60 m/sec for 1 nm^3 objects, ~2 m/sec for $(10\ nm)^3$ objects, 0.06 m/sec for $(100\ nm)^3$ objects, and 0.02 m/sec for 1 $micron^3$ objects. These thermal motions are in degrees of freedom whose range of motion is largely confined, by design, to sub-nanometer distances within nanoscale measuring devices. However, objects of nanomedical interest may have velocities ranging from ~10^{-9} m/sec to ~10^3 m/sec (e.g., the speed of sound in water), and may include biological elements in the environment or internal machine components. Thus it is useful to explore the fundamental limitations involved in measuring the full range of this physical variable.

There are at least two simple ways to determine a velocity. First, velocity may be measured by transferring the kinetic energy of a moving object or fluid volume to a sensor element. The movement of the sensor element is then clocked as it traverses a known distance. Second, and more efficiently, the passage of the moving object may sequentially trip two latches located a known distance apart, again revealing the moving object's velocity; alternatively, detecting the presence or absence of a protruding feature on a passing body using probe rods can determine velocity without any requirement for large kinetic energy transfers. The need for physical gating movements by sensor components and computational register-shift operations using diamondoid rods moving at ~GHz clocking speeds suggests a conservative minimum stable sensor cycle time of ~10^{-9} sec.

The two principal restrictions on the maximum detectable velocity v_{max} are the maximum linear dimension of the sensor element, or L_{sensor}, and the minimum event time that may be accurately discriminated, or Δt_{min}. For a sensor element with thermal noise energy $\Delta E \sim 10\ kT = 43$ zJ at 310 K, the quantum mechanical limit is $\Delta t_{min} > \sim \hbar / (2\ \Delta E) \sim 10^{-15}$ sec. Δt_{min} is also limited by the thermal variation in length of the individual latches. If the minimum detectable latch displacement is $\Delta x \sim 10$ pm, and the maximum speed of latch displacement is approximately the speed of sound in the latch material (e.g., $v_{sound} \sim 17{,}300$ m/sec for diamond), then $\Delta t \sim \Delta x / v_{sound} \sim 10^{-15}$ sec, near the quantum mechanical limit. Δt is further restricted by the minimum detectable phase variance between latches located at either end of the sensor. An acoustically-transmitted clock pulse emitted from a centrally-located source that is Δx closer to one of the two latches can discriminate a phase variance of $> \sim \Delta x / L_{sensor} \sim 10^{-5}$ taking

$L_{sensor} \sim 1$ micron, so the minimum discriminable time is again $\Delta t_{min} \sim 10^{-5}\ L_{sensor}/v_{sound} \sim 10^{-15}$ sec; hence $v_{max} \sim L_{sensor}/\Delta t_{min} \sim 10^{9}$ m/sec $\sim c$ (= 3×10^{8} m/sec, the speed of light).

However, unlike photons, physical objects moving in an aqueous medium are normally limited to the speed of sound in that medium, e.g., $v_{sound} \sim 1500$ m/sec in water at 310 K. If v_{max} is taken as $\sim 10^{3}$ m/sec, then for a sensor of size $L_{sensor} \sim 1$ micron the required time discrimination is only $\Delta t_{min} \sim L_{sensor}/v_{max} \sim 10^{-9}$ sec.

The minimum detectable velocity v_{min} is limited by the minimum measurable linear displacement $\Delta x \sim 10$ pm and the limits of our patience in making the measurement. For a maximum measurement time of $t_{meas} \sim 1$ sec, $v_{min} \sim \Delta x/t_{meas} \sim 0.01$ nm/sec *in vacuo*. In a nanomedically-relevant aqueous medium, the systematic motion of a free-floating (untethered) object of radius R is masked by added Brownian noise, giving the limiting $v_{brownian} = \Delta X/\tau >\sim 260$ nm/sec for R = 10 micron and $v_{brownian} >\sim 2.6$ micron/sec for R = 100 nm, where $\tau = t_{meas} = 1$ sec. The thermal motions of tethered objects moving along surfaces (e.g., vesicles transported along microtubules by kinesin motors) are more restricted, hence Brownian velocity masking is reduced.

Modest accuracy may be achieved using acoustic reflection in a Doppler velocimeter. Employing acoustic waves of frequency ν, the minimum measurable velocity change $\Delta v = v_{sound}/(\nu\ t_{meas})$ = 150 microns/sec in an aqueous medium at 310 K with v_{sound} = 1500 m/sec, $\nu \sim 10$ MHz, and t_{meas} = 1 sec, corresponding to the shifted frequency $\nu\ (1 + (\Delta v/v_{sound}))$ having enough time to make one extra cycle during t_{meas}. Even at such long wavelengths (~ 150 microns), significant backscattering of ultrasound from ~ 7.82-micron red cells is observed experimentally.

Velocity sensors may also be used to estimate volumetric flow rates in fluidic channels and blood vessels by measuring the speed at which fluid passes a nanodevice resting on the vessel wall. The parabolic velocity profile of v is a function of radial distance r from the center of a cylindrical vessel of radius R for Hagen-Poiseuille flow:

$$v = \frac{\left(R^2 - r^2\right)}{4\eta L_v} \qquad \text{...(10)}$$

where absolute viscosity η = 1.1 centipoise (1.1×10^{-3} kg/m-sec) for plasma at 310 K and ΔP is the pressure change along the length of a vessel of length L_v. A 1-micron nanodevice obtains measurements of v at various distances from the vessel wall, then fits these data to Eqn.

10 to find the choice of R that minimizes the variation in estimated $\Delta P/L_v$, which should be roughly constant under nonaccelerative conditions. Once R is determined, $\Delta P/L_v$ is known as well as the cross-sectional area (πR^2) of the vessel lumen, and a simple integration of the radial velocity distribution provides the flow rate in m^3/sec. Unfortunately, the usefulness of this technique is limited by boundary layer effects.

Acceleration Sensors

Accelerations inside nanoscale devices are often measured in the trillions of g's (g = 9.81 m/sec²). For example, an L = 1 nm diamondoid block ($m = 3.5 \times 10^{-24}$ kg) with thermal energy ~kT experiences random thermal accelerations of $a_{therm} \sim kT / g\ m\ L = 10^{11}$g's; a 10-nm logic rod with vibrational energy ~kT experiences ~10^{12} g oscillations; a 200-nm wide hoop spinning axially with 1000 m/sec rim speed has rotational acceleration of a ~ 10^{12} g's.

Box-spring accelerometers

As one member of a class of devices, a simple nanoscale omnidirectional accelerometer is conceptualized as a reaction mass suspended by three pairs of reaction springs (one pair per orthogonal directional axis) with each spring connected to the walls of an evacuated box. Acceleration in any direction extends some springs and compresses others, which is measured by force or displacement sensors attached to each spring. The signal/noise ratio for this sensor is

$$SNR = \ln\left(\frac{E_{spring}}{E_{noise}}\right) = \ln\left(\frac{m^2 a^2}{2kT\ k_s}\right) \quad \text{...(11)}$$

For a sensor mass $m = 10^{-17}$ kg (~78 nm-wide platinum or ~140 nm-wide diamond cube) suspended with k_s = 1 N/m floppy springs at T = 310 K, the minimum detectable acceleration $a_{min} = 1.6 \times 10^6$ g's for SNR = 1 and 2.6×10^6 g's for SNR = 2 (undamped resonant acceleration ~ $L\ k_s/m$ ~ 10^9 g's for L = 100 nm). Undamped resonance frequency $w_{res} = (k_s/m)^{1/2} = 3 \times 10^8$ radians/sec, which is measurable.

Sensor mass exchanges reciprocally with acceleration sensitivity in Eqn. 11, so the lower limit even for a dedicated micron-scale platinum-core box-spring accelerometer is a_{min} ~1000 g's at SNR ~ 2 (undamped resonant acceleration ~ 10^6 g's for L = 2 microns). The maximum detectable acceleration requires m ~ 10^{-23} kg (~1.4 nm diamond cube) and k_s ~ 1000 N/m, giving a_{max} ~5×10^{13} g's at SNR = 1.

If the accelerometer is installed near the center of mass of a nanorobot, the sensor may avoid confusing centrifugal and translation forces. Otherwise, information from multiple sensors must be combined to obtain an unambiguous measurement. Multiple measurements can improve the SNR for constant accelerations of long duration.

In 1998, ~100-nm vibrating-beam silicon accelerometers were under development by Analog Devices and Daimler-Benz. Tunneling accelerometers had been fabricated with ~10^{-6} g resolution at a 350 Hz bandwidth or ~10^{-8} g/$Hz^{1/2}$. Silicon-based capacitive accelerometers (5-1000 g full scale) manufactured by Silicon Designs (Issaquah, WA) were widely used in automotive airbag safety systems; in 1998, Motorola shipped its ten-millionth accelerometer, and the global market for automotive silicon accelerometers was expected to reach $463 million by 2002.

Displacement accelerometers

Formed elements in the human bloodstream (e.g., red cells, white cells, platelets) typically are buffeted by much smaller accelerations. For example at 310 K, instantaneous random thermal accelerations of $a_{thermal}$ ~ 500 g's are experienced by a 0.2-micron virus particle, ~0.05 g's by a 2-micron platelet, ~10^{-4} g's by a 10-micron neutrophil — and ~0.4 g's by a 1 $micron^3$ spherical nanorobot. Force-based sensors cannot easily detect such low accelerations, since the smaller the sensor, the larger the a_{min} for a given SNR (Eqn. 11). An alternative approach uses displacement sensors to determine an object's velocity twice in rapid sequence, allowing acceleration to be computed from the difference.

Consider a constantly accelerating object that triggers three clocking latches at positions x_1, x_2, and x_3, in sequence. Time and position of the object are measured as (x_1, t_1), (x_2, t_2), and (x_3, t_3). The measured constant acceleration is then given by

$$a = \frac{v_{32} - v_{21}}{t_{32} - t_{21}} (m / sec^2) \quad ...(12)$$

where $v_{32} = (x_3 - x_2) / (t_3 - t_2)$, $v_{21} = (x_2 - x_1) / (t_2 - t_1)$, $t_{32} = (t_3 + t_2) / 2$ and $t_{21} = (t_2 + t_1) / 2$. If adjacent latches are an equal distance x_{latch} apart so that $x_{latch} = (x_3 - x_2) = (x_2 - x_1)$, and taking $t_1 = 0$.

Assume that a slowly accelerating external object passes three clocking latches that are fixed to the exterior of a nanorobot of radius R. Measurements of the object's velocity incur an unavoidable error due to nanorobot thermal motion. The object is observed to pass between latch pairs during a measurement time $\sim t_{meas}$. During this same interval, the nanorobot translates a distance $\Delta X = (kT\, t_{meas} / 3\, \pi\, \eta\, R)^{1/2}$, incurring a maximum velocity measurement error of $\sim \Delta X / t_{meas}$, and thus from Eqn. 12 a maximum acceleration measurement error of $\sim 2\, \Delta X / t_{meas}{}^2$ if $(t_3 - t_2) \sim (t_2 - t_1) \sim t_{meas}$. Thus the minimum detectable acceleration is

$$a_{min} \sim \left(\frac{4kT}{3\pi\eta R\, t_{meas}^3} \right)^{1/2} \quad (m / sec^2) \qquad ...(13)$$

Taking T = 310 K, $\eta = 1.1 \times 10^{-3}$ kg/m-sec and R = 1 micron, then $a_{min} \sim 10^{-7}$ g's for t_{meas} = 1 sec; $a_{min} \sim 0.004$ g's for a more reasonable $t_{meas} = 10^{-3}$ sec measurement time in the *in vivo* environment.

Fluid acceleration sensors

Biological fluid accelerations may also be inferred from measurements of spatial pressure gradients. Ignoring frictional and gravitational forces and assuming laminar flow, fluid acceleration a_{fluid} is given by

$$a_{fluid} = \frac{\Delta P}{\rho \Delta x} (m / sec^2) \qquad ...(14)$$

where ρ = fluid density (e.g., blood plasma is an almost incompressible Newtonian fluid, with $\rho \sim 1025$ kg/m^3 at 310 K) and ΔP is the measured change in pressure over a distance Δx. The mitral and aortic valves in the heart ($\Delta P / \Delta x \sim 10^5$ N/m^3, $a_{fluid} \sim 10$ g's, $\Delta x \sim$ a few cm), as well as valves in the veins and lymphatics, operate as bistable analog decelerometers employing this principle. Micron-scale medical nanorobots traveling across blood vessel surfaces could use pressure sensors capable of 10^{-6} atm resolution to infer local blood flow accelerations to an accuracy of $\pm$ 0.01 g from sequential measurements taken 1 mm apart.

Pivoted gyroscopic accelerometers

Consider a cylindrical gyroscope spinning at an angular velocity w (radians/sec), with radius r and thickness h, whose center of gravity lies a distance L_p from a frictionless pivot point along the rotational axis. The pivot point defines the center of a spherical sensor grid of radius R $(r^2 + 4h^2)^{1/2}$ having distinguishable grid elements located a distance $\Delta x \sim 1$ nm apart. Under uniform acceleration a, the gyroscope precesses around its spin axis at an angular velocity $\omega_p = 2\, a\, L_p / \omega$

$r^2 = \Delta x / R\ t_{meas}$, where t_{meas} is measurement duration. Hence the minimum detectable acceleration is

$$a_{min} = \frac{\Delta x\, r^2\, \omega}{2\, L_p R t_{meas}} (m / sec^2) \quad ...(15)$$

by comparing precessional motion against the grid.

In choosing an ω with which to measure a_{min}, there is an upper and a lower limit. The upper bound is the bursting strength condition

$$\omega = \omega_{max} < \left(\frac{2\sigma_w}{\rho r^2}\right) (rad / sec^2) \quad ...(16)$$

where σ_w is a safe working stress = 10^{10} N/m² (~0.2 tensile strength) for diamond and ρ is the density of the gyroscope material (~3510 kg/m³ for diamond). The lower bound is the minimum spin angular velocity below which the pivoted gyroscope cannot spin stably about the vertical axis and begins to wobble, thus ruining the measurement. The lower bound is given by

$$\omega = \omega_{min} = \left(\frac{4 a_{min} L_p \left(r^2 + \frac{1}{3} h^2\right)}{\pi\, h\, \rho\, r^2}\right)^{1/2} \quad ...(17)$$

If $\omega_{max} / \omega_{min} > 1$, then nonwobbling spin velocities are available below the bursting speed of the gyroscope cylinder. Since this ratio scales with R, there is a minimum sensor size R_{min} below which no useful spin velocities are available. Combining Eqns. 15 and 17 and solving the quadratic in ω gives

$$R_{min} = \frac{2\, \Delta x \left(r^2 + \frac{1}{3} h^2\right)}{\pi\, h\, t_{meas} r^3 (2\sigma_w\, \rho)^{1/2}} \quad ...(18)$$

For $t_{meas} = 10^{-3}$ sec, $R_{min} >\sim 69$ microns (optimum r = 48 microns, h = 25 microns); for t_{meas} = 1 sec, $R_{min} >\sim 6.9$ microns (optimum r = 4.8 microns, h = 2.5 microns). Thus it appears that pivoted gyroscopic accelerometers likely will be employed only in devices >> 10 microns in size. Microscale gyroscopes implemented in silicon have been discussed by Greiff and others.

Accelerative onset

Aside from pure acceleration, the rate of accelerative onset—sometimes called jolt, surge, or jerk—can produce additional damaging effects on macroscale mechanical and biological structures. For example, the effects of accelerative onset ('a) on the human body were studied by Col. John Paul Stapp in a series of aggressive rocket sled experiments conducted during the 1950s. For human subjects, cardiovascular shock was entirely absent for 'a $> \sim$ 500-600 g/sec but began to appear at 'a $> \sim$ 1100-1400 g/sec, at 1-40 g accelerations. Human tolerance to linear decelerative force is determined both by rate of onset and peak acceleration; the endurance limit for well-restrained healthy young males has been given as $< \sim 500$ g/sec and $< \sim 50$ g for $> \sim 0.2$ sec durations.

By 1998, the effects of accelerative onset on nanomechanical structures had not been extensively studied, but similar limits might apply to nanomachines. For example, a diamond rod of length L_{rod} = 1 micron, cross-sectional area S_{rod} = 0.01 micron^2, mass m_{rod} = 3.5 $\times 10^{-17}$ kg, and working strength $W_{rod} \sim 5 \times 10^{10}$ N/m^2 fails when subjected to a static acceleration $a_{fail} > \sim W_{rod} S_{rod} / m_{rod} \sim 10^{12}$ g. But accelerative shocks can produce dynamic amplification in the response spectrum of elastic systems. Dynamic stress rises to a peak of roughly twice the corresponding static value when '$a_{peak} > \sim 4\ n_{res}\ a_{max}$ for a trapezoidal pulse shape, where n_{res} is the frequency of natural vibration of the system, because the stresses from the free oscillations of the system, induced by the changes in the acceleration of the system, can add to the static stresses from the acceleration, with damage done by the total stress. Taking $k_s \sim 1$ N/m for the rod described above, then $n_{res} \sim 30$ MHz and '$a_{peak} \sim 10^{20}$ g/sec at $a_{max} \sim 10^{12}$ g. For comparison, the teragravity nanocentrifuge has mass m $\sim 10^{-17}$ kg and is spun up to $a_{max} = 5 \times 10^{11}$ g's in $> \sim 10^{-6}$ sec, giving 'a $< \sim 5 \times 10^{17}$ g/sec $<<$ 'a_{peak}.

Angular Displacement

There are circumstances in which it is necessary for a medical nanodevice to measure, establish or maintain a preferred orientation, as for example nanorobots with a clearly defined "top" and "bottom" or nanodevices performing lengthy intracellular operations requiring precision stationkeeping. Measurement of rotation rates may also be required.

Gimballed nanogyroscopes

Pivoted gyroscopes of the type already described are impractical for orientational sensing in all but the largest medical nanodevices. Setting w_{max} (Eqn. 16) equal to w_{min} (Eqn. 18), taking a_{min} = g = 9.81 m/sec^2 (gravitational acceleration), and optimizing for minimum sensor size gives a critical nonwobbling nonbursting spin velocity $\omega_{crit} \sim 6.8 \times 10^7$ radians/sec at a minimum device radius of ~35 microns. If a pivoted gyroscope is smaller than this minimum radius, then spinning it slower than ω_{crit} causes the gyro to tumble helplessly, while spinning it faster than ω_{crit} tears the gyro apart.

However, a triaxially gimballed nanogyroscope may be regarded as having its pivot point near its center of mass, so L_p becomes very small and w_{min} becomes very slow, making nanogyroscopes feasible in medical nanorobots. Taking h = 1 micron, r = 0.5 micron, ρ = 3510 kg/m^3 and a_{min} = g in Eqn. 17, then if the gimbals are aligned such that L_p = 1 nm between the pivot point and the effective center of gravity along the rotation axis, then $\omega_{min} \sim$ 400 rad/sec. If gimbal tolerances are improved to L_p = 0.1 nm, then $\omega_{min} \sim$ 125 rad/sec.

How stable is a nanogyroscopic orientation standard? There are two principal considerations.

First, Brownian thermal rotation in all three angular degrees of freedom (θ, φ and ψ) gives rise to a small nutation around the nanogyroscopic precession axis of magnitude $\Delta\theta_{nutate} \sim p_\varphi / p_\psi$, where p_φ and p_ψ are angular momenta of the gyro. The minimum value for p_φ is approximated by $(1/2)\, kT \sim (1/2)\, p_\varphi^2 / I_1$, where $I_1 = (1/4)\, m\, r^2 + (1/12)\, m\, h^2$, the moment of inertia around φ, and $m = \pi\, r^2\, h\, \rho$. The maximum value for p_φ is given by the burst-strength condition $\omega < \omega_{max}$ (Eqn. 16) as $p_\psi \sim I_3\, w$, where $I_3 = (1/2)\, m\, r^2$, the moment of inertia around ψ. Hence:

$$\Delta\theta_{nutate} \sim \left(\frac{kT\left(r^2 + \frac{1}{3}h^2\right)}{2\,\pi\,\sigma_w\, h\, r^4} \right) \text{(radians)} \qquad \text{...(19)}$$

For T = 310 K, r = 0.5 micron, h = 1 micron, $\sigma_\omega \sim 10^{10}$ N/m^2, $\Delta\theta_{nutate} \sim$ 0.8 microradian.

Second, the gimbal bearings have small frictional losses, exerting small torques on the gyro and causing a small precession away from the original orientation. The angular velocity of precession $\omega_{precess} \sim T_{gimbal} / (I_3\, \omega)$ (rad/sec) where T_{gimbal} is the frictional torque caused by an imperfect gimbal bearing. Assume that the gimbal is driven by external forces to oscillate at some frequency ν_{gimbal}, and that a gimbal bearing dissipates power at the rate of P_{gimbal} (watts), so that each oscillatory motion of the gimbal dissipates $E_{gimbal} = P_{gimbal}/\nu_{gimbal}$ (joules). If the angular amplitude of the gimbal motion is α_{gimbal} (radians) and the gimbal radius is r_{gimbal}, then the gimbal bearing travels $X_{gimbal} = \alpha_{gimbal}\ r_{gimbal}$ (meters) per gimbal oscillation and the force applied in dissipating an energy E_{gimbal} is $F_{gimbal} = E_{gimbal}/X_{gimbal}$. Thus, the torque applied by each gimbal oscillation event is $T_{gimbal} = F_{gimbal}\ r_{gimbal} = P_{gimbal}\ /\ (\nu_{gimbal}\ \alpha_{gimbal})$; $P_{gimbal} = k_p\ v_{gimbal}^{\ 2}$, where $v_{gimbal} = \alpha_{gimbal}\ \nu_{gimbal}\ r_{gimbal}$ is the sliding speed of the gimbal bearing surfaces during movement and k_p is a constant that depends solely on the geometry of the bearing.

The change in orientation angle caused by each gimbal oscillation event is $\Delta\theta_{osc} = \omega_{precess}\ /\ \nu_{gimbal}$. The number of gimbal oscillations N $= t\ \nu_{gimbal}$, where t is elapsed time since the gyro was last calibrated. If gimbal oscillations are independent and randomly distributed, then the total change in orientation angle is very roughly $\Delta\theta_{precess} \sim \Delta\theta_{osc}\ N^{1/2}$; assuming $r_{gimbal} \sim r$, then:

$$\Delta\theta_{precess} \sim \frac{2\ k_p\ \alpha_{gimbal}\left(t\ \nu_{gimbal}\right)^{1/2}}{\pi\ r^2\ h\ \rho\ \omega} \text{(radians)} \quad ...(20)$$

From Drexler, $k_p = 2.7 \times 10^{-14}$ watt-sec^2/m^2 as a conservative value for a small (~2 nm) stiff bearing. Random hydrodynamic flows induced by thermal fluctuations inside biological cells have a characteristic duration $t_{fluct} \sim 10$ millisec, which suggests $\alpha_{gimbal} \sim 0.06$ radians (~3°) for an R ~ 1 micron in cyto nanorobot. Taking r = 0.5 micron, h = 1 micron, $\rho = 3510$ kg/m^3 for diamond (giving m ~ 3×10^{-15} kg), $\omega = 1 \times 10^9$ rad/sec (just below the $\omega_{max} \sim 4.8 \times 10^9$ rad/sec bursting speed), and assuming $\nu_{gimbal} \sim t_{fluct}^{\ -1} = 100$ Hz (to which the results are not terribly sensitive), then $\Delta\theta_{precess} \sim 10^{-8}\ t^{1/2}$ (radians). Thus a $\Delta\theta_{precess} \sim 1$ microradian orientation shift takes ~1 hour; a $\Delta\theta_{precess} \sim 4$ microradian shift takes ~1 day; and a $\Delta\theta_{precess} \sim 70$ microradian (~10 arcsec) shift takes ~1 year. (Total angular error $\Delta\theta_{total} = \Delta\theta_{precess} + \Delta\theta_{nutate}$.) Thus a gimballed nanogyroscope may be sufficiently stable to serve as a readily storable, easily transportable, and highly accurate onboard orientation standard.

Nanopendulum orientation sensing

A nanopendulum may also be employed for orientational sensing. As one example of a large class of related devices, consider a simple rigid spherical nanopendulum of radius r and mass m, with a large hemispherical bob free to swing through the entire solid angle about a central pivot point, with a gimballed housing to allow active avoidance of the pivot support beam. The nanopendulum bob moves along a spherical surface located far enough from a concentric spherical sensor grid of radius $R = r + z_{sep}$ to minimize van der Waals interactions, with bob and sensor surfaces electrically neutral or uniformly mutually repulsive. Much like the pin cushion receptor model, the sensor grid consists of a spherical array of diamondoid sensor rods $\Delta x \sim 1$ nm wide that are rapidly and simultaneously extended into the evacuated sensor cavity to determine bob position during a measurement cycle of duration t_{meas}. Radial sensor rod velocity should exceed typical tangential bob velocity by at least 1-2 orders of magnitude. For the example given below, a sensor stud on a 500-nm bob that is turning at $v_{turn} \sim v_{thermal} = 3.6$ mm/sec passes a ~1 nm sensor rod in ~280 nanosec; a sensor rod displacing ~1 nm in the radial direction in ~1 nanosec travels at $v_{rod} \sim 1$ m/sec. This nanopendulum may serve both as an orientation sensor and as a rotation rate sensor, using principles similar to vestibular (semicircular canal) mechanics.

If a simple nanopendulum spherical sensor housing with bob initially at rest in a uniform vertical gravity field g is rotated through the smallest detectable angle θ_{min} (radians) around the center pivot, then ignoring friction, the bob's position over the sensor grid moves a distance Δx, hence

$$\Delta\theta_{min} = \Delta x / r \qquad ...(21)$$

For $\Delta x = 1$ nm and $r = 500$ nm, $\Delta\theta_{min} \sim 2$ milliradians (~0.1 deg). For a practical sliding interface (e.g., between bob and sensor surface) of contact area S and turning velocity v_{turn}, Drexler estimates frictional dissipation, primarily driven by shear-reflection drag and band-stiffness scattering, as $P_{drag} \sim k_1 v_{turn}^2 S$, where $k_1 \sim 400$ kg/m^2-sec, hence the energy dissipated between the bob and the nearest sensor rod (the one interacting with the bob) is $E_{drag} \sim k_1 v_{turn} S \Delta x$. Bob gravitational energy $E_{bob} = m g r \sin \Delta\theta_{min} \sim m g r \Delta\theta_{min}$ for small displacements. The bob swings freely at the minimum displacement if $E_{bob}/E_{drag} > 1$. If $m = 10^{-15}$ kg, $r = 500$ nm, and $S = 1$ nm^2, then $v_{turn} < 2.5$ m/sec or $\omega < 49$ megaradians/sec for the bob to swing freely (vs. $v_{thermal} = 3.6$ mm/sec and $\omega_{res} \sim 9900$ rad/sec for this pendulum).

The bob-up/bob-down energy differential is ~2 m g r = 9.8 zJ ~2.3 kT for $m = 10^{-15}$ kg and $r = 500$ nm. To distinguish opposite orientations ($\Delta\theta \sim \pi$) requires $N_{meas} \sim (kT / 2 m g r)^2 \sim 1$ measurement for this sensor. To distinguish $\Delta\theta_{min} \sim 2$ milliradians requires averaging $N_{meas} \sim (kT / E_{bob})^2 \sim 190{,}000$ independent measurements (of ~0.002 kT/measurement) or $t_{meas} \sim 190$ microsec using a sensor rod clock cycle of ~1 nanosec. Thus, after subtracting out the effects of translation displacements as determined by linear accelerometers, a nanodevice can measure angular displacements in its orientation to within $\pm \Delta\theta_{min}$ in a measurement time $t_{meas} \sim 10^{-4}$ sec. With sensor rods moving at ~1 m/sec, energy dissipation for each rod sliding in its housing is ~0.01 zJ/cycle. If only ~100 test rods need be triggered to continuously track the moving stud, then a sensor rod clock cycle of ~1 nanosec implies a continuous nanopendulum power dissipation of ~1 pW; additional drag power loss between stud and contacting rod is only $P_{drag} \sim 0.0004$ pW.

Many other configurations of this basic device are readily conceived and may be more volume-efficient, including nested spheres or ovoids, two-dimensional rocking disk sections for in-plane rotational measurements (three devices required for monitoring all angular degrees of freedom in a rigid rotator), and multiaxial circumferential tracks or tubes containing rolling balls or sliding plugs. J. Logajan suggests that a pair of linear three-dimensional translation accelerometers should be able to unambiguously extract the 3 translation and 3 -otation components of all possible acceleration vectors. Rotation about any axis will result in distinctly different mass displacements while accelerated translation along any axis will result in identical displacements.

Nanopendulum tachometry

The smallest angular velocity detectable using a simple nanopendulum is given approximately by $\omega_{min} \sim \Delta x / r\, t_{meas} = 2$ milliradian/sec (~0.1 deg/sec) for $\Delta x = 1$ nm, $r = 500$ nm, and $t_{meas} = 1$ sec. The largest detectable angular velocity $\omega_{max} \sim 2\pi / t_{meas}$ if t_{meas} is the minimum time required to count one complete rotation of the sensor; the smallest $t_{meas} \sim 2\pi / \omega_{max} \sim 100$ nanosec for $\omega_{max} \sim 49$ megaradians/secor ~5% of the bursting speed for an $r = 1$ micron spherical diamondoid nanodevice.

What rotational displacements will normally be encountered by an *in vivo* medical nanorobot? A 1-micron device experiences instantaneous Brownian rotations of $\omega_B = (kT / 4\pi\eta R^3 \tau)^{1/2}$ radians/sec, where $\tau = (MW_{kg} / 32\pi kT N_A c_{H2O} R^4)^{1/2}$, the mean time between molecular collisions with the device (mostly by water molecules), from Eqn. 3, where $c_{H2O} \sim 3.3 \times 10^{28}$ molecules/m^3. For $R = 0.5$ microns and $T = 310$ K, $\tau \sim 6 \times 10^{-20}$ sec/collision and $\omega_B \sim 7 \times 10^9$ rad/sec, with each displacement of $\sim \tau\omega_B = 4 \times 10^{-10}$ radian, none of which are detectable. However, the net of all Brownian rotational displacements is one rotation every 16 sec, ~0.4 radian/sec.

The largest measurable angular displacements *in vivo* are caused by tumbling due to differential shear forces in the bloodstream. Normal vessel wall shear rates in physiological bloodflow range from 100-1400 radians/sec in the larger arteries to 500-4000 radians/sec in the smaller arteries and capillaries. The measured rolling velocity of white cells during vascular "tank tread" locomotion on venule walls is ~40 micron/sec, or ~8 radians/sec, which may be encountered by nanodevices employing cytocarriage. The human vestibular mechanism has a frequency response range of 0.048-260 radians/sec.

Thus, the maximum range of physiologically relevant rotation rates is 10^{-2} to 10^4 radians/sec. Since the natural harmonic frequency of a simple nanopendulum is $\omega_{res} = (g/r)^{1/2} = 4400$ radians/sec for r = 500 nm, the possibility of forced oscillations and resonance need be faced only when measuring the highest likely physiological rotation rates, in which case the addition of a damping force $F_{damp} \sim m\ v_{turn}\ \omega_{res}$ during resonance events should eliminate the problem. F_{damp} for critical damping peaks at ~mg ~ 0.01 pN for the m = 10^{-15} kg,

Force Nanosensors

Piconewton forces may be applied and must be detected by medical nanodevices. Pendulum- and cantilever-based sensors may be used to probe nanoscale forces, to monitor the ambient gravitational field, to measure molecular-scale masses to single-proton accuracy, and to discriminate among molecules having different isotopic composition.

Minimum Detectable Force

Extremely small forces have been measured using macroscale instrumentation, particularly the atomic force microscope (AFM). In 1998, experimental forces of 1 nN in the contact regime and 1 pN in the noncontact regime were routinely measured. In 1994, D. Rugar achieved a force sensitivity of 5×10^{-16} N using an AFM with a 90-nm thick cantilever at room temperature, and in 1998 Rugar reported a force resolution of 7×10^{-18} N using a magnetic resonance force microscope (MRFM) with a 230-micron long, 60-nm thick, silicon cantilever. The theoretical limit on the force sensitivity of the AFM has been estimated as 10^{-18} - 10^{-19} N for temperatures near absolute zero using fiber optic interferometry or other techniques. However, many of these methods are impractical or impossible to employ in medical nanodevices.

What is the minimum force F_{min} likely to be detectable by nanorobot sensors? Several lines of reasoning produce consistent answers. For example, force may be measured by the displacement Δx of a beam against the known resistive pressure provided by a spring of minimum stiffness k_s; that is, $F_{min} \sim k_s \Delta x$. However, the energy stored in the spring must exceed thermal noise energy, with SNR ~ ln ($k_s \Delta x^2$ /2 kT) (Eqn. 9), giving F_{min} = 2 kT e^{SNR} / Δx. For T = 310 K and SNR = 2, F_{min} = 100 pN at Δx = 0.6 nm, 10 pN at 6 nm, and 1 pN at 60 nm. Indeed, the natural molecular motors (kinesin, dynein, and actin-myosin) have isometric force outputs on the order of 1-14 pN per protein molecule, which are routinely observed using microscale cantilever spring sensors measuring total forces of 10-100 pN. Breaking antigen-antibody or receptor-ligand interactions generally requires rupture forces in the 50-300 pN range. A force of ~160 pN separates biotin that is seated in its avidin receptor; ~85 pN pulls iminobiotin out of the avidin receptor. A force of ~300 pN ruptures a hydrogen bond between two isolated water molecules, but >3000 pN are likely to be needed to rupture a single C-C covalent bond. The rupture strength of single covalent bonds has been directly measured experimentally by AFM as, for instance, 2000 ± 300 pN for Si-C (528 zJ/molecule dissociation enthalpy, 0.185 nm bond length) and 1400 ± 300 pN for Au-S.

Force sensing may also require using measurement devices that involve sliding interfaces of two mutually inert periodic surfaces. This motion may be characterized by the potential energy of interaction, ~0.1-1 zJ as determined by Monte Carlo simulations of typical diamondoid material at 0.2-0.5 nm separations. Assuming a minimum 0.5 zJ energy barrier, for small sliding contacts the typical force amplitude $F_{min} \sim 1.7 \times 10^{10} \Delta V_{barrier} \sim 8.5$ pN.

A force of 10 piconewtons across a small-molecule receptor (0.3 nm)^3 of surface area 0.09 nm^2 represents an energy density of ~10^8 J/m^3 × 0.027 nm^3 = 2.7 zJ ~ 0.6 kT. And the nanorobotic manipulator arm is capable of ~picometer step sizes, representing a minimum applicable force of F_{min} ~ 10 pN assuming 0.04 m/N compliance for the entire mechanism. Thus as a conservative estimate, the smallest force that is reliably detectable (and applicable) by nanodevices is probably ~10 pN.

Finally, a nanosensor of dimension L able to distinguish two forces differing by ΔF newtons has a probability of erroneous measurement of

$$P_{err} \sim \exp (L \Delta F / 4 kT) \qquad ...(22)$$

For ΔF = 10 pN and T = 310 K, P_{err} = 1% for L = 8 nm, or 0.01% for L = 16 nm, so reliable nanoscale picoforce sensors may be ~10-20 nm in size.

Nanogravimeters

Research in space medicine has discovered that micron-scale elements of the immune system have extraordinary sensitivity to gravity. One study found that bone marrow-derived (B6MP102) macrophage single cells respond after 8 seconds to exposure to a 10^{-2} g hypogravity environment as indicated by increased cell spreading, a well-known marker of cell activation. Individual immune system cells placed in microgravity exhibit enhanced growth but depressed differentiation, decreased activation by concanavalin A, a 20% reduction in glucose metabolism, increased resistance to antibiotics, increased number of pseudopodia on monocytes, and changes in cytoplasmic streaming velocity and frequency.

As of 1998, the exact mechanism by which the leukocyte "gravity sensor" operates remained unknown. It has been speculated that differences in densities of several organelles, such as the nucleolus, the ribosomes, and the centrioles, could generate detectable pressures on the structure of the cytoskeleton under gravitational loading mediated through mechanosensitive stretch-activated ion channels or the extracellular matrix. While indirect gravitational effects that modify the cellular environment (such as reduced sedimentation or decreased thermal convection due to hypogravity) could also be responsible, a direct interaction of the gravitational force with cellular structures appears to be the most likely sensor mechanism. For instance, if a change in gravitational loading equivalent to the force generated by a natural molecular motor ($F \sim 1$ pN) is detectable by the cytoskeleton, and this force is applied to the entire mass of a (10 micron)^3 cell ($m \sim 10^{-12}$ kg), then the minimum detectable gravitational acceleration is $g \sim F/m \sim 1$ m/sec^2 = 0.1 g, which should be adequate to detect the onset of hypogravity.

A nanomechanical gravity sensor of similar size to a macrophage cell may be at least 100,000 times more sensitive. For example, the energy of a tethered mass swinging freely in a uniform gravity field is approximately $E_g \sim m\, g\, \Delta h$, where Δh is the maximum vertical amplitude of the motion. To be detectable, oscillator energy must exceed thermal noise energy, so SNR $\sim \ln (m\, g\, \Delta h\, N_{meas}^{1/2} / kT)$, giving $g_{min} = kT\, e^{SNR} / m\, \Delta h\, N_{meas}^{1/2} \sim kT\, e^{SNR} / \rho\, L^4\, N_{meas}^{1/2}$ for a bob mass of density ρ and size L^3, N_{meas} independent measurements, and assuming $\Delta h \sim L$. If T = 310 K, SNR = 2, N_{meas} = 1, and ρ = 21,450 kg/m^3 (Pt), then g_{min} = 0.1 g for $L \sim 1$ micron, 0.01 g for $L \sim 2$ microns, 10^{-5} g for $L \sim 11$ microns, and 10^{-6} g for L = 20 microns, the size of the average human tissue cell.

Regarding nanogravimeter design, for a pendulum of length L in a gravity field g ~ 9.81 m/sec^2 (1 g), the resonant period of the motion $T_{res} = 2\pi (L/g)^{1/2} = 2 \times 10^{-3}$ sec for L = 1 micron, 9×10^{-3} sec for L = 20 microns. If Δt_{min} = 1 nanosec is the minimum detectable decrease in resonant period caused by an increase in gravity from g to g + Δg, then the minimum detectable change in gravitation force is

$$\Delta g / g = 1\left(\frac{L}{g k_g^2}\right) \sim \frac{10^{-9}}{L^{1/2}} \qquad \ldots(23)$$

for a medical nanorobot, where $k_g = (\Delta t_{min} / 2\pi) + (L/g)^{1/2}$. For L = 1 micron, $\Delta g / g \sim 10^{-6}$; for L = 20 microns, $\Delta g/g \sim 2 \times 10^{-7}$.

Single-Proton Massometer

A nanopendulum may also be used to measure molecular-scale masses. Torsion, helical spring, and vibrating rod mechanisms all will work, but one of the most efficient designs for weighing the smallest masses may be the coiled suspension-spring cantilever massometer, which is analogous in principle to the traditional quartz crystal microbalance. In this design, a beam of length L_{beam} and circular cross-sectional radius R_{beam} is hinged at one end and carries a sample holder of solid volume V_{sample} at the other end. The beam is supported (a variable distance L_{spring} from the hinge) by a spring consisting of a movable suspension rod of total length l_{rod}, circular cross-sectional radius r_{rod}, and spring constant $k_s \sim \pi^2 r_{rod}^4 E G / (G l_{rod}^3 + E \pi r_{rod}^2 l_{rod})$ using spring (rod) material of density ρ, Young's modulus E, and shear modulus G. The natural resonance period for this unloaded system is $T_0 = (2\pi L_{beam}/L_{spring}) (m_0/k_s)^{1/2} = Z m_0^{1/2}$, where $Z = (2 L_{beam} / L_{spring} r_{rod}^2) ((G l_{rod}^3 + E \pi r_{rod}^2 l_{rod}) / E G)^{1/2}$ and $m_0 = \rho (V_{sample} + \pi L_{beam} R_{beam}^2 + \pi l_{rod} r_{rod}^2)$. Gravitational loading is negligible for suspension system mass $m << (k_s L_{spring}^2 / g L_{beam}^2) \sim 10^{-15}$ kg; $m \sim 10^{-20}$ kg in the exemplar system.

After adding a test mass Δm to the sample holder, the resonance period increases to $T_m = Z (m_0 + \Delta m)^{1/2}$, which may be detected if $T_m - T_0 \geq \Delta t_{min} \sim 1$ nanosec, the minimum convenient time interval. Hence the minimum detectable molecular weight ΔMW, expressed in proton masses ($m_p = 1.67 \times 10^{-27}$ kg), is

$$\Delta MW = \Delta m / m_p = m_p^{-1}\left[m_0^{1/2} + \Delta t_{min} / Z\right] \text{(protons)} \qquad \ldots(24)$$

L_{beam} and l_{rod} are chosen to minimize total coil area $A_{coil} = 2 R_{beam} L_{beam} + 2 r_{rod} l_{rod}$. Using aluminum for the rod and beam (E = 6.9 ×

10^{10} N/m², G = 2.6 × 10^{10} N/m², ρ = 2700 kg/m³), L_{beam} = 1200 nm, l_{rod} = 1920 nm, r_{rod} = 0.5 nm, V_{sample} = 1 nm³, and Δt_{min} = 10^{-9} sec, then ΔMW ~ 1 proton for L_{spring} = 1 nm >> Dx_{min} = 0.01 nm, the minimum measurable displacement. (For L_{spring} > 1 nm, physiological ~1 g accelerative loads on a ~1 micron nanorobot in which the sensor is embedded produce undetectable accelerative beam displacements < Δx_{min}.)

Moving the suspension rod closer to the sample holder reduces device sensitivity. For L_{spring} = L_{beam}, minimum ΔMW ~ 1200 protons, permitting moderately large molecules to be weighed in minimum time (~10^{-6} sec). Using a very thin sensor configuration (h_{sensor} ~20 nm) and coils occupying ~50% of sensor cavity space, sensor dimensions are (79 nm)² × 20 nm giving a massometer volume ~10^5 nm³ or ~0.01% of the total volume of a 1 micron³ nanorobot. Sensor mass is ~ 3 × 10^{-19} kg, including diamondoid housing and prorated share of support mechanisms. If one measurement cycle including sample loading/unloading operations can be completed every ~10 microsec, then the device processes its own mass of glucose molecules in ~10 sec. Note that biological molecules weighed in air or in vacuo may lose water of hydration, whereas weighing in water presents other complications that merit further investigation.

J. Soreff points out that for a molecule on a spring, the precision of the measurement is linked to the sharpness of the resonance, which is optimized to the degree the vibrational frequency of the spring/molecular mass system is kept well below the vibrational resonance frequencies of the molecule itself. The following discussion reports the results of a simple lumped-element analysis of the effects of internal vibration on the massometer completed for the author by J. Soreff, which confirms the previously described performance estimates for the massometer.

Suppose that it is desired to weigh some mass M to a precision ΔM by attaching it to a weak measuring spring (spring constant k_{weak}) and measuring the shift in resonant frequency. The shift in frequency due to coupling to an internal mode of the sample must be less than (1/2) (ΔM / M) (1 / 2 π) $(k_{weak}/M)^{1/2}$. If the sample has some lowest internal mode frequency n_{int}, then approximating the sample as two M/2 masses separated by x_{strong} connected by a strong spring (k_{strong}) gives ν_{int} = (1/π) $(k_{strong}/M)^{1/2}$. If this sample is attached to a weak measuring spring of length x_{weak}, which is in turn attached to a rigid support, then the equations of motion are:

$$(M/2)\ddot{x}_{strong} = -k_{weak}\, x_{strong}\left(x_{strong} - x_{weak}\right) \quad ...(25)$$

$$(M/2)\ddot{x}_{strong} = k_{strong}\left(x_{weak} - x_{strong}\right) \quad ...(26)$$

Setting $x_{weak} = e^{st}$ and $x_{strong} = R\, e^{st}$ where t = time, s = time constant, and $R = -(1 + (2\, k_{weak} / M\, s^2))$, and since the angular frequencies of the two springs are $\omega_{strong} = 2\,\pi\,\nu_{int} = 2\,(k_{strong}/M)^{1/2}$ and $\omega_{weak} = (k_{weak}/M)^{1/2}$, then the lower frequency solution for $\omega_{weak} << \omega_{strong}$ may be approximated by $s = i\,\omega_{weak}\,(1 - \omega_{weak}^{\,2} / 2\,\omega_{strong}^{\,2})$ and so the (undesired) fractional change in frequency due to the flexibility of the object to be weighed is

$$\omega_{weak} / \omega_{strong} < (\Delta M / M)^{1/2} \quad ...(27)$$

For a molecule sample size of L_{sample}, $\nu_{int} \sim \nu_{sound} / L_{sample} \sim 500$ GHz for $\nu_{sound} \sim 1500$ m/sec and $L_{sample} \sim 3$ nm; $M = \rho\, L_{sample}^{\,3} = 2.7 \times 10^{-23}$ kg for $\rho = 1000$ kg/m^3. To detect the mass difference of a single proton in a sample molecule, sensor $Q = M / m_p = 16{,}000$; from Eqn. 27, the sensor may operate at a factor of $Q^{1/2}$ below ν_{int}, or $\nu_{spring} \sim \nu_{int} / Q^{1/2} \sim 4$ GHz. A period of 0.25 nanosec is slightly too short to allow a phase shift of $\Delta t_{min} \sim 1$ nanosec, since the period is only one-quarter of this. However, a phase shift of π radians should accumulate in $t_{meas} = Q / 2\,\nu_{spring} \sim 2$ microsec and should be detectable by looking at magnitudes of forces. Interestingly, an SPM microprobe of circular cross-section and exponential profile with a 0.1 nm^2 tip has been proposed that could also reach ~proton mass detection sensitivity at room temperature and 1 atm pressure with an eigenmode frequency of 10 GHz; an electrically-driven carbon nanotube-based resonant-beam balance demonstrated ~10^{-17} kg (~10^{10} proton or ~0.01 micron^3 mass) sensitivity in 1999.

Existing surface acoustic wave (SAW) devices are chemical sensors that can measure areal mass densities as small as 80 picogram/cm^2, or 0.01 N_2 molecule per nm^2 of surface, though current devices are limited to ~micron^2 areas. A proposed quartz microresonator microbalance could detect a 10^{-6} molecular monolayer, the equivalent of 5×10^{-9} molecule per nm^2. Detection of ~400 million protons using an MRFM vibrating rod ~60 nm thick was demonstrated by Rugar in 1998.

Isotope Discrimination

It will often be useful in nanomedicine to distinguish molecules containing different isotopes of the same chemical elements, for example to count the number of organic molecules in a sample that contain radioactive C^{14} atoms in place of stable C^{12} atoms or to make direct assays of heavy radiometal-contaminated tissues. Nanodevices may also sort isotopically heteronuclear molecules into homonuclear fractions, for example to separate deuterochemicals from hydrochemicals or to acquire isotopically pure materials with which to construct:

1. Small, high-speed rotors with precisely balanced mass distributions;
2. Components with extremely fine size differences;
3. Nanostructures with maximum thermal conductivity, or mixed isotopes to reduce thermal conductivity; or
4. Long-lived isotopically pure structures that generate no internal radioactivity, thus promoting longer device lifetimes.

The most reliable sensor for discriminating among isotopes within target molecules is the single-proton massometer. Isotopes of any chemical element differ in mass by at least ~1 neutron (~protonic) increments. Thus a massometer with single-proton resolution can resolve all isotopes starting from chemically pure samples. Isobars (atoms having the same mass number) and isotones (atoms having the same number of neutrons) may be finally resolved cnemically; nuclear isomers differing in mass by ~0.001 amu cannot be resolved. Teragravity centrifugation and time-of-flight particle-beam deflection techniques may also prove useful.

Isotopes vary in many subtle ways due to their mass differences, although it is not yet clear how a potential isotope nanosensor might exploit some or all of these distinctions:

Line shifts

The electronic energy states of atoms or molecules depend on the reduced nuclear mass, causing a spectral line shift of ~1.0005 between hydrogen and deuterium, less among heavier isotopes, detectable spectroscopically.

Hyperfine structure

Isotopes differ in electrical quadrupole moment, nuclear spin and magnetic moment, giving rise to hyperfine structure in the optical spectra, especially in the heavier elements. For example, the magnetic moments of hydrogen and deuterium (2.79268 and 0.857387 nuclear magnetons, respectively) are distinguishable via NMR. The NMR frequency in a 10 kilogauss magnetic field is 42.5759 MHz for H, 6.53566 MHz for D. Of course, not all isotopes are NMR sensitive.

Vibrations and rotations

Molecular spectra are especially sensitive to isotopic changes, since the quanta of vibrational and rotational energies are directly dependent upon the masses of the atoms involved, and there are also some subtle effects from differences in nuclear spin. Separations are commonly achieved by laser excitation.

Cross-sections

Isotopes have different thermal neutron capture cross sections; for instance, the cross-section of C^{12} is 3.7×10^3 barns (1 barn = 10^{-24} cm^2) compared to 9×10^{-4} barns for C^{13} and $<10^{-6}$ barns for C^{14}. (Electron scattering might be better, since the electron is less likely to cause the nucleus to fission and it is generally difficult to use neutrons to probe samples on the molecular scale.) Similarly, the neutron cross-section for H_2O is 0.6 barns compared to 9.2×10^{-4} barns for D_2O.

Other properties

Isotopes differ slightly in various bulk properties, many of which could be measured in submicron-scale aliquots by nanosensors. The ability of both large mammals and micron-size photosynthetic plant cells to preferentially concentrate C^{12} or C^{13} in organic matter, and the predicted ability of carbon nanotubes to selectively extract tritium from low-concentration H_2-T_2 mixtures, constitute proofs of principle that at least some of the bulk properties may become useful in achieving nanoscale isotopic separation or purification.

Pressure Sensing

This Section describes the ability of medical nanodevices to detect and measure fluid pressure variations while functioning inside the human body. Generation of acoustic signals for power transmission, communication, and navigation are considered elsewhere.

Minimum Detectable Pressure

A change in the ambient pressure in a fluid environment (ΔP) may be determined by measuring the change in volume (ΔV) of a fixed quantity of gas at constant temperature, relative to a known sensor volume at a known reference pressure. (Thermal sensors elsewhere in the nanorobot provide readings by which corrections can be made for temperature variations.) If the sensor volume consists of a cylinder of fixed cross-sectional area, the change in volume is converted into a linear displacement of a piston located at the interface between gas and fluid environment. The change in Gibbs free energy is $\Delta G = \Delta P \, \Delta V$, while thermal noise $\sim kT$. Pressure sensor signal/noise ratio SNR $\sim \ln (\Delta P \, \Delta V \, / \, kT)$, so at the minimum ΔG the minimum detectable pressure ΔP_{min} is measured at the maximum possible volume change, or V_{sensor} (the sensor volume), given approximately by

$$\Delta P_{min} \sim kT e^{SNR} / V_{sensor} \qquad ...(28)$$

using the classical ideal gas formulation. At SNR = 2, a $(22\ nm)^3 = 10^4\ nm^3$ sensor ($\sim 5 \times 10^5$ atoms or $\sim 10^{-20}$ kg for a cube with 1 nm thick walls) can detect a minimum pressure change $\Delta P_{min} \sim 0.03$ atm; a $(680\ nm)^3 = 0.3\ micron^3$ sensor ($\sim 3 \times 10^9$ atoms, $\sim 6 \times 10^{-17}$ kg) detects $\Delta P_{min} \sim 10^{-6}$ atm, probably near the practical limit for mobile in vivo medical nanodevices. Note that ΔP_{min} scales approximately as the inverse cube of mean sensor dimension, and immunity to thermal noise scales exponentially with sensor volume. Current silicon micromachined pressure sensors are $\sim 1\ mm^3$ with 10^{-3} atm sensitivity, a factor of $\sim 10^{12}$ poorer sensitivity than the theoretical minimum for their size. (A field-effect chemical sensor with an active sensing volume of 0.1 mm^3 that measures pressure indirectly has detected a 10^{-11} atm step in hydrogen gas with $t_{meas} \sim 3$ sec, only a factor of 1000 above the theoretical minimum for a sensor of that size.) The human ear can detect pressure waves as small as 2×10^{-10} atm.

A less sensitive approach relies on the observation that a change in pressure P alters both density and bulk modulus of a compressed fluid, variations that may be detected by measuring the change in the speed of sound v_s in the working fluid, given by

$$v_s = (B / \rho)^{1/2} \qquad ...(29)$$

where B is bulk modulus of the fluid and ρ is fluid density. (dB/dP)/B may be up to several times larger than (dρ/dP)/ρ for molecular fluids, but there is some $(dv_s/dP)/v_s$ for each fluid. Differentiating Eqn. 29 with respect to P and using dρ/dP = ρ/B, then $dv_s/dP = (dB/dP - 1) / 2\ \rho^{1/2} B^{1/2}$. For water at 1 atm, dB/dP ~ 8.4 atm/atm, ρ ~ 1000 kg/m^3 and B = 22,000 atm, giving $dv_s/dP \sim 0.25$ m/sec-atm. If Δv_s is the measured change in acoustic speed, then:

$$\Delta P_{min} = \Delta v_s / (dv_s / dP) \qquad ...(30)$$

Velocity can be measured to ~0.1 mm/sec accuracy using a micron-scale velocity sensor and $t_{meas} \sim 1$ sec, so a minimum $\Delta v_s \sim 0.1$ mm/sec implies $\Delta P_{min} \sim 0.0004$ atm.

Spatial Pressure Gradients

Two 0.3-micron³ pressure sensors located at either end of a 1 micron nanorobot could detect pressure changes as small as 10^{-6} atm per micron, a ~1 atm/m spatial gradient. The maximum detectable gradient would involve adjacent measurements of minimum (10^{-6} atm) and maximum ($\sim 10^4$ atm) absolute pressure, a theoretical peak spatial gradient of $\sim 10^{16}$ atm/m, though a practical device could not survive exposure to such extreme shear force.

Temporal Pressure Gradients

Driving the sensor at a frequency near ν_{res} may cause large amplitude excursions unless the motion is heavily damped. The lowest possible undamped resonance frequency ν_{res} for piston-type acoustic sensors of size $\sim$ L with minimum spring constant $k_s \sim 2\ kT / L^2$ for a maximally sensitive spring having an energy of $\sim$kT when stretched by L, and piston mass m, is

$$\nu_{res} = \left(\frac{1}{2\pi}\right)\left(\frac{k_s}{m}\right)^{1/2} \qquad ...(31)$$

where $m = \rho L^2 h$ for a piston of area L^2, height h, and density ρ. Assuming h/L = 0.1 and $\rho \sim 1000$ kg/m³, ν_{res} = 21 MHz for an L = 22 nm sensor ($k_s = 1.8 \times 10^{-5}$ N/m) and ν_{res} = 4 KHz for an L = 680 nm sensor ($k_s = 1.9 \times 10^{-8}$ N/m). Resonance frequency may be adjusted by selecting appropriate k_s and m. In practical devices resonance will occur at a slightly lower frequency than ν_{res} and with some broadening of the peak, because m should also include the masses of the shank of the piston plus a time-varying portion of the fluid mass occupying the piston's chamber (depending upon the time-varying position of the piston).

Consider the surface of a piston of height h, area L^2 and mass m to which a pressure spike of amplitude P (N/m²) and frequency ν_{driven} is applied, producing a stroke length X_{stroke}. The solution for the equation of motion of an undamped forced harmonic oscillator has a maximum amplitude of $X_{stroke} = F_{spike} / \text{abs}\ (m\ (\omega_{driven}{}^2\ \omega_{res}{}^2))$ where $\omega = 2\pi\nu$, $F_{spike} = P L^2$ and m is defined above. The driving frequency that produces stroke length X_{stroke} = L is given by:

$$\nu_{driven} = \left(\frac{1}{2\pi}\right)\left[\left(\frac{P}{\rho L h}\right) - 4\pi^2 \nu_{res}^2\right]^{1/2} \qquad ...(32)$$

For an L = 22 nm sensor driven to full throw by a minimum detectable ΔP_{min} = 0.03 atm pressure spike (again taking h/L = 0.1 and $\rho \sim 1000$ kg/m³), ν_{driven} = 34 MHz; ν_{driven} = 45 GHz when the sensor is driven to full throw by the maximum possible pressure spike P_{max} = 39,000 atm consistent with water remaining in the liquid state.

The piston may be driven at these frequencies, or slower, using triangular-wave pulse trains of pressure spikes of these magnitudes.

For an L = 680 nm sensor driven to full throw by a minimum detectable $\Delta P_{min} = 10^{-6}$ atm pressure spike, $\nu_{driven} = 6$ KHz; $\nu_{driven} =$ 1.5 GHz when the sensor is driven to full throw by the maximum pressure spike $P_{max} = 39{,}000$ atm.

What is the maximum tolerable pressure spike in aqueous media? The requirement of subsonic piston motion in water gives the highest maximum: $P_{max} < (1/2)\ \rho\ v_{sound}^2$ (N/m^2), if $\rho = 3510$ kg/m^3 for a solid diamondoid piston and the speed of sound $v_s \sim 1450$ m/sec in water for bulk modulus $B_{water} = 2.2 \times 10^9$ N/m^2, $\rho_{water} = 1000$ kg/m^3, giving $P_{max} < 39{,}000$ atm. However, rapid dissipative pressure spikes $P_{max} > 26{,}000$ atm add sufficient energy to boil 310 K water, and isothermal static compression of pure water at 310 K causes crystallization into Ice VI near ~11,500 atm.

Ullage Sensors

It will often be necessary to determine the degree to which a fluid storage tank is full or empty. In most circumstances, an ullage sensor may consist of a simple static pressure sensor embedded in the tank wall, with readout calibrated according to the van der Waals equation for gases and liquids. A fractal network of chemical sensors permeating the internal volume of the tank could provide continuous readouts of local concentration which could be integrated to estimate total storage in the tank. The force required to rotate a paddle inside the tank would increase as a tank was filled, due to the increase in net viscosity of the contents. Transmission time of thermal or acoustic pulses might also give some measure of fluid concentration and physical phase. C. Phoenix suggests storing liquids in a piston, allowing direct and precise molecular volume readout, in designs where this is convenient; in theory, a piston that slides the minimum measurable distance of ~10 pm within the 2-nm-diameter throat of a water-filled cylinder could detect the removal of a single water molecule from the column.

Thermal Nanosensors

The ability of medical nanodevices to measure both absolute temperature and changes in temperature is crucial for monitoring *in vivo* physiological thermoregulatory mechanisms and intracellular energy transactions. Precision thermal sensing is also important within nanoscale devices to provide corrective input for pressure, chemical, and displacement sensors, and to improve the stability of onboard clocks.

Minimum Detectable Temperature Change

A sensor system consisting of N atoms has (3N-6) internal coordinates and thus ~(3N-6) oscillators, each with an average energy of ~kT assuming no modes are high enough to be frozen out. However, the standard deviation of the energy of the oscillators around the mean energy is ~kT, and these energy fluctuations are uncorrelated (ignoring coupling between the oscillators), so (3N-6) of them will have an average fluctuation of kT $(3N-6)^{1/2}$. (By comparison, the energy in one 10-micron infrared photon is ~5 kT.) As a result, an instantaneous measurement of the total energy of all N atoms gives a minimum detectable temperature change ΔT_{min} of

$$\Delta T_{min} / T = (3N-6)^{-1/2} - (3N_{meas}\, n_d\, V_{sensor})^{-1/2} \quad ...(33)$$

for a set of N_{meas} independent temperature measurements using a sensor of volume V_{sensor} constructed with material of atomic number density n_d, which might be maximized using diamond (n_d = 1.76 × 10^{29} carbon atoms/m^3).

Thermal equilibration time t_{EQ} ~ $L^2 C_V/K_t$ for a sensor of size L, of heat capacity C_V (1.8 × 10^6 joules/m^3-K for diamond) and of thermal conductivity K_t (2000 watts/m-K for diamond). Thus, t_{EQ} ~ 10^{-13} sec for a sensor of size L = 10 nm, 10^{-11} sec for L = 100 nm, and 10^{-9} sec for L = 1 micron, so thermal sensors up to 1 micron in size should easily achieve thermal equilibration within a typical measurement cycle time Δt_{min} ~ 10^{-9} sec. Sensor measurement time $t_{meas} = N_{meas}\, \Delta t_{min}$.

A (57 nm)^3 sensor can detect ΔT_{min} / T = 10^{-4} (~31 millikelvins at 310 K) in a single measurement (N_{meas} = 1), with measurement time t_{meas} ~ 1 nanosec. A sensitivity of ΔT_{min} / T = 10^{-6} (~310 microkelvins at 310 K) may be achieved using either a ~1 micron^3 sensor with a single measurement (N_{meas} = 1, t_{meas} = 1 nanosec) or a (124 nm)^3 sensor with N_{meas} = 1000 independent measurement cycles and t_{meas} = 1 microsec. Finally, a 1 micron^3 sensor can detect ΔT_{min} / T = 3 × 10^{-9} (~1 microkelvin at 310 K) with N_{meas} = 100,000 independent sensor cycles, giving a measurement time t_{meas} ~ 100 microsec.

These figures are confirmed by experimental estimates of detector noise temperature, such as ΔT_{min} / T = $(k / L^3 C_V)^{1/2}$ ~ 10^{-6} for a silicon nitride detector with C_V = 5.2 × 10^6 joules/m^3-K and L ~1 micron. It has been calculated that sensitivity of ~1 microkelvin is theoretically possible using quartz electronic microresonators as precision thermometers. Thermocouple probes with 100 nm tips have already shown ~100 microkelvin sensitivity, tunneling thermometers capable of measuring thermoelectric potential localized to atomic-scale dimensions have been proposed, and experiments leading toward "yoctocalorimetry" were being pursued in 1998.

For comparison, heat sensors in human skin have $\Delta T_{min} / T \sim 3 \times 10^{-4}$ (~90 millikelvins), and the infrared sensor pit of the rattlesnake is sensitive to an energy intensity of ~0.8 pJ/micron^2 in a measurement time t_{meas} ~35 millisec and achieves $\Delta T_{min} / T \sim 3 \times 10^{-6}$; mosquitos register $\Delta T_{min} / T \sim 6 \times 10^{-6}$ at a distance of ~1 cm.

Piston-Based Temperature Sensors

Consider a coiled cylinder of cross-sectional area A filled with gas at pressure P, with a piston at one end whose movement is resisted by a constant-force spring. Increasing gas temperature from the coldest temperature at which the sensor will operate (piston at maximum ingress), T_0, to some warmer temperature T_1 causes the piston to move from position x_0 to position x_1 while the spring holds pressure constant at P inside the cylinder during the measurement. If Δx is the smallest measurable piston displacement and $\Delta T = T_1 - T_0$, then the poorest accuracy is

$$\Delta T / T_0 = \frac{x_1 - x_0}{x_0} = \frac{\Delta x}{x_0} = N_{meas}^{-1/2} (N+1)^{-1/2} \qquad \text{...(34)}$$

because Δx has a thermal noise component as well, and the number of gas molecules $N = n_d V_{sensor}$ where $n_d \sim 10^{28}$ gas molecules/m^3 at P = 1000 atm. Sensitivity is maximized at the largest feasible x_0; for $\Delta T / T_0 = 10^{-6}$ (~300 microkelvins at 310 K) and $\Delta x = \Delta x_{min} = 1$ nm, then $x_0 = 1000$ microns, $N = 10^9$ gas molecules at 1000 atm (using the van der Waals equation of state) and 310 K taking $N_{meas} = 1000$, giving $V_{sensor} = 10^{-19}$ m^3 and thus a cylinder cross-sectional area of $A = V_{sensor} / x_0 = (10 \text{ nm})^2$. There are ~1000 gas molecules per nanometer of cylinder length, each traveling with mean thermal velocity $v_t \sim 500$ m/sec. Tightly coiled into a cubical volume, the folded sensor size is $L_{sensor} \sim V_{sensor}^{1/3} \sim (464 \text{ nm})^3$ and measurement time $t_{meas} = N_{meas} \Delta t_{min} \sim 1$ microsec. Sensor mass is ~10^{-16} kg.

The coiled tube design, though not strictly necessary, is nonetheless convenient because it allows a continuous sensor element to be arbitrarily distributed through the nanorobot volume, to be concentrated in multiple specific interior regions, or to be placed inside nanorobot components that will be required to flex such as metamorphic protuberances. If instead the working gas is placed in a cubical box of side L_{box} with a sensor piston of area A_p, any change in temperature causes the piston to move a distance $\Delta x = (L_{box}^3/A_p)(\Delta T / T_0) \sim (k T_0 / k_s)^{1/2}$, where $k_s = (A_p/L_{box}^2)(N kT_0 / L_{box}^2)$, from Drexler; hence

$$\Delta T / T_0 - \left(A_p / NL_{box}^2\right)^{1/2} \qquad ...(35)$$

which argues for a small piston, a large box, and a high operating pressure to achieve maximum sensitivity. For $A_p = 100\ nm^2$, $N = 10^9$ molecules, and $L_{box} = 464$ nm, $\Delta T / T \sim 10^{-6}$.

Thermal-Expansion Temperature Sensors

Consider a thin rod of length x and width w, constructed as a sandwich of two dissimilar materials each of thickness h, having coefficients of linear expansion α_1 and α_2, respectively. When heated, the two materials expand differentially to lengths of $x(1 + \alpha_i \Delta T)$ (neglecting the quadratic and higher-order terms in a_i), producing a cantilever deflection of $\theta_t = (x/h)(\alpha_1 - \alpha_2)$ (radians) from simple geometry. The minimum detectable temperature change $\Delta T_{min} = 2\pi \Delta x_{min} / x \theta_t$ or

$$\Delta T_{min} = \frac{8 \pi h^3 w^2 \Delta x_{min}}{(\alpha_1 - \alpha_2) V_{rod}^2} \qquad ...(36)$$

where volume of the sandwich rod $V_{rod} = 2$ h w x. Taking $\Delta x_{min} = 10$ pm, $\alpha_1 = 2.38 \times 10^{-5}$ /K for aluminum, $\alpha_2 = 8 \times 10^{-7}$ /K for diamond at 310 K, h = w = 5 nm, and assuming sensor volume $V_{sensor} \sim 2 V_{rod}$ in a tightly packed three-dimensional coil configuration, $\Delta T_{min} = 1$ microkelvin for an $L^3 \sim (227\ nm)^3$ sensor ~ 1% of the volume of a 1-micron nanorobot. Use of a suitable negative-coefficient material for α_2, such as the ceramics zirconium tungstate and cordierite, can further reduce L by 5%-10%.

Measuring the volume of the sensor is essentially like taking an inventory of total stored thermal energy, which has a kT $N^{1/2}$ fluctuation term from adding the N uncorrelated kT fluctuations; hence

$$\Delta T / T_0 \sim \left(N_{meas}\, n_d\, V_{sensor}\right)^{-1/2} \qquad ...(37)$$

Hence, to achieve $\Delta T \sim 1$ microkelvin at $T_0 = 310$ K with this sensor may require $N_{meas} \sim 47$ million independent measurements of $\Delta t_{min} = 1$ nanosec each, or $t_{meas} \sim 47$ millisec.

Micromechanical silicon cantilever heat sensors (microcalorimeters) of ~20,000 micron³ volume have already achieved $\Delta T_{min} = 10$ microkelvins at room temperature ($\Delta T_{min} / T \sim 3 \times 10^{-8}$), responding to ~1 pJ of heat in a measurement time $t_{meas} \sim 1$ millisec. Thermal-expansion nanosensors might also exploit the temperature sensitivity of viscoelastic materials (e.g., modulus of relaxation).

Mechanochemical Temperature Sensors

Since binding rates are temperature sensitive, a chemical concentration sensor using steric probes immersed in a sample containing a precisely known concentration of target ligands can be used as a temperature sensor. Similarly, enzyme reaction rates increase with a change in temperature according to

$$\Delta T = \frac{10 \ln(k_2 / k_1)}{\ln(Q_{10})} \quad \text{...(38)}$$

where $\Delta T = T_2 - T_1$, k_1 and k_2 are the reaction rate constants at temperatures T_1 and T_2, and Q_{10} is the well-known temperature coefficient — the ratio of the rates of reaction measured at two temperatures 10°C apart. $Q_{10} \sim 2$ for most enzymes, but may range up to 4 for some enzymes that catalyze reactions with unusually high activation energies. Assuming $Q_{10} \sim 4$, the (112 nm)^3 sensor with $\Delta c / c = 0.01(1\%)$ can distinguish temperatures 0.07 K apart; the huge (1.44 micron)^3 sensor with $\Delta c / c = 10^{-5}$ (0.001%) resolves $\Delta T \sim 72$ microkelvins. Because of their relative volumetric inefficiency, such sensors are best used in large ΔT applications. Similar considerations apply to thermal sensors based on the temperature sensitivity of protein folding (as in heat shock protein mediators of cellular responses to a 5 K change), protein thermal contraction, and thermal conformation rate constants for protein a helices that typically coil in $\sim 10^{-6}$ sec. The quantum tunneling transmission coefficient for ion channels is also exquisitely temperature-sensitive and thus could serve as the basis of a nanoscale thermal sensor.

Spatial Thermal Gradients

Two (200-400 nm)^3 thermal sensors located at either end of a 1-micron nanorobot can detect temperature changes as small as 10^6 K per micron, a ~ 1 kelvin/m spatial gradient. The maximum detectable gradient in nanomedical systems would involve adjacent measurements differing by ~ 100°C, a peak spatial gradient of $\sim 10^8$ kelvins/m. The detection event must be completed before the two sensors can thermally equilibrate, e.g., $t_{meas} < t_{EQ} = L^2 C_V / K_t$. If the material between the sensors is mostly diamondoid, then $t_{EQ} \sim 10^{-9}$ sec; if mostly water, then $t_{EQ} \sim 10^{-5}$ sec, for L = 1 micron.

Temporal Thermal Gradients

On the timescale of nanomachine operations, typically 10^{-9}-10^{-6} sec, the human body provides an essentially isothermal operating regime. However, medical nanorobots circulating in the bloodstream may encounter ΔT of ~3 K during one circulation time ~60 sec under normal circumstances, a temporal thermal gradient of ~0.1 kelvin/sec, or >10 kelvin/sec at 1 mm tissue depth for a fingertip placed on a 600 K stove. Thermal nanosensors can take independent measurements differing by as much as ~100°C in ~10^{-9} sec, giving a maximum detectable thermal gradient of 10^{11} kelvins/sec, far in excess of physiological rates. Detection of the minimum ~1 microkelvin change between two measurements taken 1 sec apart probably represents the minimum practical temperature gradient, ~1 microkelvin/sec. For comparison, the record-holder in the insect world is the eyeless cave beetle *Speophyes lucidulus*, whose antennae can detect thermal change rates as small as ~3000 microkelvins/sec.

The extraordinary thermal conductivity of diamond ensures that endogenous nanorobot waste heat is rapidly conducted to the surrounding aqueous medium. For example, a 1 micron^3 diamondoid nanorobot generating 10 picowatts of onboard power at its core experiences a temperature differential of ~10^{-8} K between core and perimeter (using thermal conductivity K_t = 2000 watts/m-K for diamond), well below minimum detectable limits.

Electric and Magnetic Sensing

Electric Fields

The electric field E (volts/m) caused by a single point charge q (e.g., an electron or singly-charged ion) is given by Coulomb's law as

$$E = \frac{q_e}{4 \pi \varepsilon_0 \kappa_e r} \qquad \text{...(39)}$$

where q_e = 1.60 × 10^{-19} coul (one charge), ε_0 = 8.85 × 10^{-12} farad/m (permittivity constant), κ_e = dielectric constant of the matter traversed by the electric field (κ_e = 74.31 for pure water at 310 K, 5.7 for diamond, 1 for vacuum or air), and r = distance from the charge, in meters. The field at a distance of 1 nm from a point charge is ~10^9 volts/m in air, 2 × 10^7 volts/m in water. The force between two unit charges $F = q_e E$ = 230 pN in air, 3 pN in water, at 1-nm separation.

What is the magnitude of electric fields likely to be encountered in nanomedical situations? Electric fields in the human body may be generated from internal or external sources. Internally, ions and individual molecules may carry static electric fields. For example, the surface of an isolated charged amino acid has ~1 charge/nm^2. Debye-Huckel shielding due to counterion flow in salty fluids (as are found in the human body) reduces these fields very rapidly with distance. The static field from a single charged amino acid floating in human plasma falls from ~6 × 10^6 volts/m at a range of 1 nm to 0.7 volts/m at r = 10 nm. In plant cells, photosynthetic "receivers" detect gradients in the electrochemical potential of the order of 1 eV across distances on the order of 1 nm, an instantaneous ~10^9 volts/m.

The most important internal electrical sources are the electrochemical gradients caused by gated channel and transporter molecular pump operations at the intracellular level; muscular, membrane, digestive, and neural activity at the intercellular and organ level; and piezoelectric fields generated by movement of collagenous tissues (e.g., tendon and elastin), bones, and many other biological materials. Typically these sources generate potentials of 10-100 millivolts over distances of 0.01-10 microns, producing local fields in the range of 10^3-10^7 volts/m. For example, the 50 millivolt transmembrane potential necessary to open a sodium channel corresponds to an electric field of 5 × 10^6 volts/m. Interestingly, implanted electrical devices such as pacemakers may produce up to 600 volts/m one millimeter from the casing.

External sources may also contribute to detectable nanomedical electric fields within the human body, though usually to a lesser degree. Placing a hand flat between two plates charged to E_p = 3 × 10^6 volts/m (near the breakdown voltage for air) induces an instantaneous field of E ~ E_p / k_{water} = 42,000 volts/m transversely through the limb. (A DC field induces ion current flows which accumulate surface charge and cancel out the internal field after ~one RC time.) Electric fields directly beneath high-tension power lines are 2,000-11,000 volts/m, or 50-1000 volts/m at a lateral displacement of 25 meters along the ground. Household appliances produce 200-300 volt/m fields; 20 amperes flowing through standard Romex copper wiring in the walls of a house produces 100-600 volts/m in the middle of a room, up to 24,000 volts/m at the wall (E = B / μ_0 ε_0 c; c = the speed of light, B from Eqn. 43); static sparks from carpet walking are ~10,000 volts/m, and ~10^5 volts/m makes hair stand on end. Quiescent atmospheric charge is ~100 volts/m, rising to 5000 volts/m during severe thunderstorms oscillating from positive to negative over a ~10,000 volt/m range. Lightning bolts (~20 coul 0.002-sec discharge) produce 60,000 volt/m spikes at a distance of 10 meters from the strike, 600 volts/m at a range of 1 kilometer. Electron-hole avalanches in semiconductors are initiated by fields >~5 × 10^7 volts/m. Penal electrocutions use ~1500 volts/m.

Direct force sensing of these electric fields is ineffective; from Eqn. 39, the minimum detectable field $E_{min} = F_{min} / q = 6 \times 10^7$ volts/m using $F_{min} = 10$ pN and a single-charge sensor. Particle deflection seems more efficient. Consider a collimated stream of singly-charged particles, each of mass m and initial velocity V_0 entering an evacuated test chamber of length L and experiencing a uniform electric field E, causing a lateral displacement $\Delta x = q E L^2 / 2 m V_0^2$ measured by the sensor. For this measurement, SNR $\sim \ln (m V_0^2 / 2 kT)$, so the minimum detectable electric field is

$$E_{min} = \frac{4 kT \Delta x_{min} e^{SNR}}{qL^2} \text{ (volts / m)} \qquad ...(40)$$

For $\Delta x_{min} = 10$ pm, T = 310 K and SNR = 2, $E_{min} = 10{,}000$ volts/m for a sensor of size L = 28 nm, or 100 volts/m for L = 280 nm, allowing measurement of all nanomedically relevant fields that are homogeneous across the entire device. Using the larger 280-nm sensor, time of flight is $\sim 10^{-12}$ sec for electrons or $\sim 10^{-9}$ sec for U^{238} ions, permitting $t_{meas} \sim 10^{-9}$ sec and dissipating at least ~ 30 pW waste heat (due to kinetic impact) in continuous operation. Time-varying electric fields can be measured up to $t_{meas}^{-1} \sim$ GHz frequencies, and electric current may be determined using standard techniques because current density is scale-invariant. Care must be taken to eliminate measurement error due to exogenous magnetic fields. Piezoelectric or other electromechanical transducers can convert electricity into mechanical signals appropriate for all-mechanical computation and control systems.

Another class of electric field detector of similar size and sensitivity is the electrosensitive gel sensor. For example, the negatively charged heparan sulfate proteoglycan network found in the secretory granule matrix expands 50% in volume within milliseconds of exposure to a -2 volt/micron field, and the response appears linear with voltage. Thus the length of a 400-nm long gel-filled sensor of fixed cross-sectional area should expand a minimally-detectable 10 pm (= Δx_{min}) in a 100 volt/m field. Similar sensitivities to small periodic fields, analogous to shark electroreceptors (maximum sensitivity $\sim 10^{-7}$ volts/m), are found in ion-channel embedded artificial membranes that selectively amplify electric signals in noisy environments via stochastic resonance. Stochastic resonance in neurons permits detection of electric fields as weak as 10 volts/m.

The theoretical limits have been analyzed by Weaver and Astumian and by Block, who conclude that the minimum detectable static field for a hollow spherical sensor of radius r_{sen} and shell membrane thickness d_{mem} (capacitance $C_e = 4\pi\varepsilon_0 \kappa_e r_{sen}^2 / d_{mem}$) is

$$E_{min\,DC} = \left(\frac{2}{3r_{sen}^2}\right)\left(\frac{kTd_{mem}}{4\pi\varepsilon_0\kappa_e}\right)^{1/2} \quad (\text{volts/m}) \qquad ...(41)$$

Using $\kappa_e = 74.31$ for water at T = 310 K, r_{sen} = 0.5 micron and d_{mem} = 10 nm gives E_{minDC} = 200 volt/m. Increasing r_{sen} to 5 microns reduces E_{min} to ~2 volts/m. To detect a periodic field of frequency ν_e by sampling over ($\nu_e\, t_{meas}$) cycles for a measurement time t_{meas}, Block gives

$$E_{min\,AC} = \frac{E_{min\,DC}}{(\nu_e\, t_{meas})^{1/2}} (\text{volts/m}) \qquad ...(42)$$

Using ν_e = 1 MHz and t_{meas} = 1 sec gives E_{minAC} = 0.2 volt/m. Cylindrical sensors are 10-100 times more sensitive, and sensitivity may possibly be further enhanced by coupling the periodic electric potential to a Michaelis-Menten type enzyme or a similar nanomechanism embedded in the sensor shell to achieve tuning. Fields as small as ~10^{-4} volts/m may be detectable by ~20-micron mammalian living cells; such detection may trigger biochemical observables which may be monitored or eavesdropped by resident nanomedical devices. Submicron-scale electrometers capable of ~0.1 electrons/$Hz^{1/2}$ (charge per unit bandwidth) have been demonstrated experimentally, and also single-electron transistors (SETs) to 10^{-4} electron/$Hz^{1/2}$, with the potential to reach 10^{-6} electron/$Hz^{1/2}$. Galvanotropic bacteria have been reported in the literature.

Magnetic Fields

The most significant source of biomagnetic fields, generated by electric charges moving as currents through the body, is neural impulses. The magnetic field generated by these currents is given by Ampere's law as

$$B = \frac{\mu_0 \kappa_m i}{2\pi r} \qquad ...(43)$$

where B is magnetic flux density (teslas), μ_0 is the permeability constant (~$4\pi \times 10^{-7}$ henry/m), relative permeability κ_m = 1 for most biological materials and *in vacuo*, i is electric current (~0.2 microamperes for human nerve impulses), and r = distance from the current-carrying conductor. Hence B ~ 0.04 microtesla for r = 1 micron, adjacent to an axonal conductor. For comparison, magnetic fields directly beneath high-tension power lines are 10-40 microtesla, or 2-8 microtesla at a lateral displacement of 25 meters along the ground. Household appliances produce 1-1000 microtesla at a distance of 3 cm and 0.01-2 microtesla at a 1-meter range (electric shavers produce 0.4-60 microtesla at a range of 15 cm). 20-amp Romex wiring in the walls of a house can produce 2 microtesla in mid-room, up to 80 microtesla at the wall, though in 1998 the average strength in 98% of U.S. homes from artificial sources was 0.05-0.09 microtesla, typical exposures throughout the day were 0.1-1 microtesla, and the federal workplace standard was a maximum of 100 microtesla. Magnetotactic marsh bacteria, birds and insects can detect the geomagnetic field vector (identifying Earth's magnetic poles), which is 30-70 microtesla, depending on geographic position. Behavioural thresholds have been measured as 0.005-0.02 microtesla for pigeons and 0.001-0.01 microtesla for honeybees, and no more than 1000-2000 microteslas for human subjects. Some biochemical changes have been observed in human cells exposed to ~10-100 microtesla 60 Hz fields, although the data are controversial; it appears the fields pump iron atoms through cell membranes, potentially causing peroxidation or other damage to occur. In 1998, transcranial magnetic stimulation was used to expedite recovery of motor skills for stroke patients, and was being explored as a treatment for depression and a means of altering patient mood. Lightning bolts (i ~ 10,000 amp/discharge) produce 200 microtesla spikes at a distance of 10 meters from the strike, 2 microtesla at a range of 1 kilometer.

Computer simulations of a fully heterogeneous human body divided into 37,000 $(1.31 \text{ cm})^3$ voxels produced a maximum induced current of 8.84 amp/m^2 (average ~1 amp/m^2) when the body was exposed to a uniform 60 Hz 1-tesla magnetic field. Patients undergoing Echo-Planar MRI scanning report "tickling" and "pain" when exposed to field changes exceeding 60 tesla/sec; ventricular fibrillation may result from myocardial current densities of 4 amp/m^2, which is induced at 250 tesla/sec.

The simplest magnetosensor is a freely pivoting permanent magnet that experiences a restoring force when the magnet's axis is not aligned with the external field. This "compass effect" is also employed by magnetotactic bacteria, which have ~0.1 micron ferromagnets in magnetosomes in their cells that allow them to detect and track the geomagnetic field vector. Permanent magnets of cross-sectional area A_{magnet} and length l_{magnet} made of Alnico or Alcomax ($\rho \sim 8000$ kg/m^3)

can provide $B_{magnet} \sim 1.4$ tesla. (In 1998, the record flux density was 13.5 tesla for dipole magnets, ~27 tesla for resistive magnets.) The force on such a magnet immersed in an external field of flux B is $F = B\, B_{magnet}\, A_{magnet} / m_0$, so the minimum detectable flux B_{min} is

$$B_{min} = \frac{\mu_0\, \kappa_m\, F_{min}}{B_{magnet}\, A_{magnet}} \quad \text{...(44)}$$

If the minimum detectable force is $F_{min} = 10$ pN, $B_{min} = 50$ microtesla for $A_{magnet} = 0.4$ micron^2, about right for magnetotactic bacteria, and $B_{min} = 2$ microtesla for $A_{magnet} = 2$ micron^2.

The limitations of force sensing may be avoided by suspending the bar magnet from its center of mass and allowing it to oscillate in vacuo about its stable equilibrium position in the external field B, making a simple harmonic oscillator with period

$$t_{magnet} = 2\pi \left(\frac{\mu_0\, \kappa_m\, I_{mag}}{B\, B_{magnet}\, A_{magnet}\, l_{magnet}} \right)^{1/2} \quad \text{...(45)}$$

where the moment of inertia $I_{mag} = m_{magnet}\, ((A_{magnet}/4) + (l_{magnet}^2/12))$. Oscillator energy $E_m \sim B\, B_{magnet}\, A_{magnet}\, l_{magnet} / \mu_0\, \kappa_m$, so the minimum detectable external field is

$$B_{min} = \frac{\mu_0\, \kappa_m\, kT\, e^{SNR}}{B_{magnet}\, A_{magnet}\, l_{magnet}} \quad \text{...(46)}$$

For $B_{magnet} = 1.4$ tesla, $A_{magnet} = (660 \text{ nm})^2$, $l_{magnet} = 660$ nm, T = 310 K and SNR = 2, then $B_{min} = 0.1$ microtesla and $t_{magnet} = 6 \times 10^{-4}$ sec ($\sim t_{meas}$), allowing a ~KHz sampling frequency. Magnet mass $m_{magnet} \sim 2 \times 10^{-15}$ kg. Several classes of molecular magnets have been described in the literature.

Another class of magnetosensor detects the change in dimensions when a magnetic body is magnetized, which is called magnetostriction. For instance, iron-cobalt ferrite expands by $\Delta L / L \sim 7.5 \times 10^{-4}$ at saturation magnetostriction in a ~1 tesla field, permitting detection of ~10 millitesla field changes using a 1 micron sensor and $\Delta x_{min} \sim 10$ pm. Small magnetostrictive volume changes make possible a sensor of even greater sensitivity. A field of ~0.2 tesla causes a fractional volume increase of ~10^{-6} in iron and up to 40×10^{-6} in some iron-nickel (~30% Ni) alloys. A (370 nm)^3 magnetostriction sensor monitored by stretch sensors with displacement sensitivity $\Delta x_{min} = 0.1$ nm (~1 atomic radius) can in theory detect $B_{min} \sim 0.1$ microtesla.

Magnetostriction is nonlinear with applied field, and there are significant hysteresis effects which may confuse data interpretation. However, multisensor arrays combining positive- and negative-magnetostriction materials should enhance signal extraction. Since the change in volume is small, temperature compensation may be required.

Hall effect probe microscopes have demonstrated 10 microtesla sensitivity with 350 nm spatial resolution, and alternative nanoscale magnetic detectors such as direct-current SQUID (superconducting quantum interference device) magnetometers, stray field devices, and magnetic spin sensors are being developed. In 1999, microSQUID could measure magnetic moments as small as $\sim 10^4$ Bohr magnetons or $\sim 10^{-19}$ J/tesla, roughly 7×10^6 protons or a 14-nm iron cube. Magnetoencephalographs using SQUID magnetometers to measure external electromagnetic fields produced by neural traffic in the brain can register 10^{-6} - 10^{-3} microtesla signals. Other biomagnetic measurement techniques have helped to investigate electrophysiological disturbances in heart muscle and in the brain. By 1998, microcoils as small as 100 microns in diameter had been constructed using microcontact printing of 25 micron wide silver wires, producing magnetic flux densities >0.4 tesla with a 10 milliamp current, switchable in $<10^{-3}$ sec.

Optical Sensing

In biosystems and nanosystems alike, the detection of an optical photon is a high-energy event. A single photon of red light at $\lambda = 700$ nm carries an energy of 280 zJ, a blue photon at 400 nm conveys 500 zJ, and an ultraviolet (UV) photon at 200 nm transmits 1000 zJ to its receptor. These energies, ranging from 65-234 kT, exceed thermal noise by a wide margin, hence are all easily detectable reliably by suitable nanosensors. Note that optical photon detection does not require a nanosensor the size of one wavelength or larger—for example, 1-nm chlorophyll molecules have no difficulty absorbing 660-nm photons.

Biological systems provide numerous examples of single-photon transducers, including photopigment molecules such as chlorophyll and rhodopsin which mediate electron transfers in $< \sim 1$ picosecond by undergoing conformational changes, triggering a series of biochemical events. An estimated 100-500 cell membrane channels are affected by the detection of a single optical photon. The first documented case of a physiological role for chlorophyll in animals has been found in the fish *Malacosteus niger*, which uses a chlorophyll derivative to see far-infrared light. Mammalian cells also act as indirect UV sensors, as for instance when energy absorption induces perturbations of membrane structure or a conformational change in membrane proteins, activating the JNK cascade below 300 pJ/micron^2, and centrosomes may serve as cellular directional IR sensors in single fibroblast cells.

Artificial "bioelectronic" cyanine-quinone chromophore molecules ~1 nm in size can detect optical photons of wavelengths λ = 580-630 nm, and can trigger and reset in 1-3 picosec (governed by the tunneling times associated with reformation of the zwitterionic states of the inputs), in theory permitting ~100 GHz modulations to be received. Exciton-generating poly(p-phenylene vinylene) polymer switches have the potential to achieve ~THz switching speed; a birefringent Kerr medium also switches in a few picoseconds. Chalcogenide glass has viscosity $>10^{13}$ kg/m-sec in darkness, but $\sim 5 \times 10^{11}$ kg/m-sec under illumination. Proposals for specific implementations of photoergic transducers and optical cables are elsewhere.

Particulate and High-Energy Radiation

High energy photons including X-rays (λ ~ 2 nm, ~10^5 zJ or 0.0001 picojoule) and gamma rays (λ ~ 0.002 nm, ~0.1 picojoule), and ejecta from the decay of radioactive nuclei such as α-particles (helium nuclei, ~1 picojoule) and β-particles (free electrons, 0.01-0.1 picojoule) may be extremely destructive to nanomechanical systems and are difficult to reliably measure quantitatively. For example, a charged particle with 0.01 pJ of kinetic energy (K.E.) traversing a decelerative electrostatic field of 10^9 volts/m requires K.E./qE ~ 60 microns to slow to a halt. However, particle-induced visual sensations have been observed by dark-adapted astronauts in space, and Ruderfer has suggested that even solar neutrinos may be detectable by the human brain (though most physicists would find this implausible). The nucleoelectric transducer could be adapted to serve as a directional a particle sensor.

Cellular Bioscanning

The goal of cellular bioscanning is the noninvasive and non-destructive *in vivo* examination of interior biological structures. One of the most common nanomedical sensor tasks is the scanning of cellular and subcellular structures. Such tasks may include localization and examination of cytoplasmic and nuclear membranes, as well as the identification and diagnostic measurement of cellular contents including organelles and other natural molecular devices, cytoskeletal structures, biochemical composition and the kinetics of the cytoplasm.

The precision and speed of medical nanodevices is so great that they can provide a surfeit of detailed diagnostic information well beyond that which is normally needed in classical medicine for a complete analysis of somatic status. Except in the most subtle cases a malfunction in any one component of the cellular machinery normally provokes a cascade of pathological observables in many other subsystems. Detection of any one of these cascade observables, if sufficiently unique and well-defined, may provide adequate diagnostic information to plan a proper reparative procedure.

DNA is one of the few cellular components that is regularly inspected and repaired. Most homeostatic systems adopt a more simple philosophy of periodic replacement of components regardless of functionality. Nanomedicine allows the philosophy of inspection and repair to be extended to all cellular components. The following discussion briefly describes a few of the sensory techniques that may be useful in achieving this objective.

Cellular Topographics

Tactile topographic scanning provides the most direct means for examining cellular structures *in vivo*. For example, *ex vivo* live cell scanning in air by commercially available atomic force microscopy using a 20-40 nm radius tip allows nondestructive feature resolutions of ~50 nm across the top 10 nm of the cell. Investigators have used the scanning tip to punch a hole in the cell, then pull back and scan the breach, observing the membrane heal itself via self-assembly in real time. Up to ~KHz scanning frequencies are possible, with up to 1024 data points per scan line. The nanomechanical manipulator arm achieves comparable tip velocities and positional accuracies.

Assuming a scan rate of ~10^6 pixels/sec, a micron-scale nanodevice, once securely anchored to the surface of a (~20 micron)^3 human cell, could employ a tactile scanning probe to image the 0.1% of plasma membrane lying within its (1.4 micron)^2 vicinity in ~2 sec to ~1 nm^2 resolution (~1 mm/sec tip velocity), or ~50 sec to ~0.2 nm (e.g., atomic) resolution (~0.2 mm/sec tip velocity). Inside the cell, and again post-anchoring, an entire 6 micron^2 mitochondrial surface could be imaged to atomic resolution in ~100 sec; the surface of a 100-nm length of 25-nm diameter microtubule could be atomically resolved in 0.2 sec — though of course in a living cell these structures may be changing dynamically during the scanning process. Continuous power dissipation of a (0.1 micron)^2 scan head moving at 1 mm/sec through water is ~0.002 pW (~kT/pixel at ~10^6 pixels/sec), though of course the energy cost of sensing, recording, and processing each pixel must be at least ~10 kT/pixel (ignoring the possibility of pre-computational image compression), so the total scanner power draw could be as high as 0.02-0.1 pW in continuous operation. Special scanning tips and techniques should allow topographic, roughness, elastic, adhesive, chemical, electrostatic (charge density), conductance, capacitance, magnetic, or thermal surface properties to be measured.

For large (~0.1-1 micron) cellular components, identification and preliminary diagnosis of improper structure may be possible using measured surface characteristics which may be matched to entries in an extensive onboard library, perhaps combined with dynamic monitoring of anomalies in cross-membrane molecular traffic using chemical nanosensors. Most membranes are self-sealing, so it should also be possible to gently insert a telescoping member into the target organelle, which member then slowly reticulates and extrudes smaller probes with sensory tips in a fixed pattern and step size, allowing the acquisition of detailed internal structural and compositional information.

Small (~1-10 nm) protein-based components of the cell such as enzymes, MHC carriers, and ribosomes are self-assembling or require only modest assistance (e.g., molecular chaperones) to self-assemble. Reversible mechanical denaturation of these smaller proteins, using diamondoid probe structure to displace water molecules (thus reducing hydrophobic forces) and by using specialized handling tools akin to functionalized AFM tips and molecular clamps, may be followed by precise nondestructive amino acid sequencing to allow identification and diagnostic compositional analysis. Afterward, the protein molecule may be refolded back into its original minimum-energy conformation, possibly with the assistance of chaperone-like structures in some cases.

Transcellular Acoustic Microscopy

Acoustic microscopy is another noninvasive scanning technique that can provide nanomedically useful spatial resolutions. Frequencies are in the GHz range, far exceeding the relaxation time of the protoplasm. A cryogenic acoustic microscope operated at 8 GHz has demonstrated 20 nm lateral resolution in liquid helium. By 1998, the best resolution achieved with water as the coupling fluid has been 240 nm at a frequency of 4.4 GHz. Operated in water at 310 K, a nanomechanical 1.5 GHz acoustic microscope would achieve a far-field minimum lateral resolution $\lambda/2 \sim 500$ nm, $\sim 10^5$ voxels per human cell, sufficient to locate and count all major organelles and a few intermediate-scale structures. Scanning acoustic microscopes (SAM) operated at 2 GHz in reflection mode (which allows detection of interference effects) achieve 30-50 nm resolution in the direction of the acoustical axis, and the Subtraction SAM approach reveals topographical deviations of 7.5 nm at 1 GHz. It has been proposed that picosecond ultrasonics could be used to obtain an image of the cytoskeleton with detail comparable to that of conventional X-ray images of a human skeleton.

Power requirements are a significant constraint on acoustic reflection microscopy (echolocation). In the simplest case, consider an acoustic emitter of radius r_E which converts an input power of P_{in} into acoustic power of intensity $I_E \leq I_{max}$ = 1000 watts/m^2 with efficiency e%, conservatively taken here to be 0.50 (50%). The emitter produces a train of omnidirectional pressure pulses of amplitude $A_p = (2\ \rho\ v_{sound}\ I_E)^{1/2}$ (N/m^2) and frequency ν which travels a distance X_{path} to a target of radius r_T. As the signal reaches the target, its amplitude has been reduced by $(r_E\ /\ X_{path})$ due to the $1/r^2$ dependence of intensity in spherical waves and by $e^{-\alpha_{tiss}\ \nu\ X_{path}}$ by attenuation with $\alpha_{tiss} = 8.3 \times 10^{-6}$ sec/m for soft tissue.

Upon reaching the target, a fraction $f_{reflect}$ of the signal is reflected as a point source echo; if the target surface has an acoustic impedance similar to liver tissue and the medium is similar to water, then from Eqn 53, $f_{reflect} \sim 0.05$ (5%). The echo then travels a distance X_{path} back to a receiver (which may or may not be located near the emitter), the amplitude again losing $(r_T\ /\ X_{path})$ by geometry and $e^{-\alpha_{tiss}\ \nu\ X_{path}}$ by attenuation. The echo is finally detected by the receiver which can measure a pulse of minimum pressure amplitude $\Delta P_{min} \sim 10^{-6}$ atm.

Combining these relations gives the following result:

$$P_{in} \geq \frac{\pi \Delta P_{min}^2 X_{path}^4}{2\ \rho\ v_{sound}\ e\%\ f_{reflect}^2\ r_T^2\ e^{-\alpha_{tiss}\ \nu\ X_{path}}} \text{(watts)} \quad ...(47)$$

To scan the entire interior of a (20 micron)^3 cell, whether from within or without, requires $X_{path} > \sim 20$ microns. A frequency of ν = 1.5 GHz allows $v_{sound}/n \sim 1$ micron spatial resolution. The fastest possible pulse repetition time is $X_{path}/v_{sound} \sim 13$ nanosec. Assuming $r_E = r_T = 0.5$ micron, $P_{in} \sim 7$ pW or ~5 watts/m^2, well within the safe intensity range for acoustic radiation. At maximum safe transmitter intensity I_{max}, the longest scannable path length at 1.5 GHz is $X_{path} \sim 46$ microns.

Power constraints are somewhat less severe in the case of acoustic transmission microscopy (acoustic tomography), a technique which requires a minimum of two physically separated components (e.g., a transmitter and a receiver). Consider an acoustic emitter radiating a series of omnidirectional pressure pulses of frequency ν which travel across a tissue cell of width X_{path}, to be detected by a receiver on the other side after a signal transit time $t_{cell} \sim X_{path} / v_{cell}$, where v_{cell} is the speed of sound in the cytosol. In the simplest case, assume that the cytosol is clear except for one target of interest along the path (e.g., an organelle) in which the speed of sound is $v_{target} \# v_{cell}$. The minimum detectable difference in travel times between a signal that passes through the target and one that does not is Δt ($\sim 10^{-9}$ sec for diamondoid systems), so the minimum resolvable target size is $r_{min} \sim v_{target}\ v_{cell}\ \Delta t / 2\ \mathrm{abs}(v_{target} - v_{cell})$ when $r_{min} >\sim v_{cell} / \nu$. For soft tissue targets in water at 1.5 GHz, $r_{min} \sim 30$ microns; for tooth enamel in water, $r_{min} \sim 1$ micron. However, the use of sampling gates will permit the detection of phase shifts that are much less than the characteristic time constant of the system in incoming waves, hence the minimum thickness of objects in theory detectable by acoustic tomography can be far smaller than the values for r_{min} estimated above.

Computing acoustic tomographic power requirement P_{in} in a similar manner as for echolocation, except that the transmitted signal rather than the echo is detected and assuming the signals are of approximately equal strength regardless of the path taken, gives:

$$P_{in} \geq \frac{\pi\ \Delta P_{min}^2\ X_{path}^4}{2\rho\ v_{cell}\ e\%\ r_T^2\ e^{-2\alpha_{min}\nu X_{path}}} \text{(watts)} \qquad \ldots(48)$$

Using the same values as the previous example, $P_{in} = 1$ pW for $X_{path} \sim 45$ microns; for $X_{path} \sim 140$ microns, $P_{in} = 1000$ pW, near I_{max}.

In either mode of operation, data gathering is accelerated by positioning more than one receiver around the target. Sensitivity is boosted by increasing emitter size, taking multiple measurements, or by illuminating the target with some large external source located elsewhere in the tissue or even outside of the organ. However, cells and body tissues are mesoscopic "junkyards" — highly heterogeneous media which may produce large numbers of nontarget scattering events, thus increasing the difficulty of extracting signal from noise. Additional complications arise due to:

1. Scattering on rough surfaces;
2. Rapid prèssure variations with range in the Fresnel zone or near field of the transmitted signal;

3. Cytoplasmic viscosity inhomogeneities due to the asymmetric arrangement of cytoskeletal structure, granules, vacuoles, and the endomembrane system; and
4. Variation in cytoplasmic Young's modulus due to time-varying tensions in semirandomly-distributed cytosolic fibrillar elements (e.g., a relaxed or contracted state) which alter elasticity and thus the local speed of sound.

Magnetic Resonance Cytotomography

Electrostatic scanning is largely ineffective because of Debye-Huckel shielding. Magnetic stray field probes allow resolution of 10 nm physical features, but only for materials with substantial magnetic domains — which most biological substances lack. But nuclear magnetic resonance (NMR) imaging may allow cellular tomography by creating 3-D proton (hydrogen atom) density maps. Atomic density maps of other biologically important elements with nonzero nuclear magnetic moments (including D, Li^6, B^{10}, B^{11}, C^{13}, N^{14}, N^{15}, O^{17}, F^{19}, Na^{23}, Mg^{25}, P^{31}, Cl^{35}, K^{39}, Fe^{57}, Cu^{63} and Cu^{65}) may also be compiled. For example, sodium imaging is already used clinically to assess brain damage in patients with strokes, epilepsy, and tumours.

In a hypothetical NMR cytotomographic nanoinstrument, a large permanent magnet is positioned near the surface of the cell or organelle to be examined. This creates a large static background magnetic field that polarizes the protons. The spatial gradient of this field establishes a unique resonant frequency, called the Larmor frequency, within each isomagnetic surface throughout the test volume. A second time-varying magnetic driver field (e.g., vibrating permanent magnet or weak rf field) is then scanned through the full range of resonant frequencies, exciting into resonance the protons (nuclear spins) in each isomagnetic surface, in turn. Depending on the sensor implementation chosen, each resonance detected may cause absorption of driver field energy, an increase in measured impedance, or even a return echo of magnetic energy (if the driver field is operated in pulse echo mode) as the excited protons relax to equilibrium in ~ 1 sec. The large polarizing magnet is then rotated to a new orientation, moving the isomagnetic surfaces to new positions within the test volume, and the scan is repeated. A 3-D map of the spatial proton distribution may be computed after several scan cycles.

Only those regions within a linewidth of resonance ΔB will generate an appreciable signal, and the expected spatial resolution (say, along the z-axis) $\Delta z = \Delta B$ / (dB/dz), where dB/dz is the spatial flux gradient of B. For a polarizing field B = 1.4 tesla placed adjacent to a (20 micron)^3 cell, the cross-cell gradient dB/dz ~ 7 × 10^4 tesla/m; placed next to a (1 micron)^3 organelle, dB/dz ~ 1.4 × 10^6 tesla/m. Assuming an effective NMR linewidth ~ 0.002 tesla, minimum spatial resolutions are 29 nm and 1.4 nm, respectively.

Unfortunately, the smallest reliably detectable energy in the sensor element is ~ kT e^{SNR}. The energy required to flip a single proton is $E_{flip} = 4 \pi \nu_L L_{proton}$, where $L_{proton} = 5.28 \times 10^{35}$ joule-sec (the quantized spin angular momentum of the proton) and the Larmor resonance frequency $\nu_L = g_{proton}$ B; g_{proton} is the gyromagnetic ratio (4.26 × 10^7 Hz/tesla for protons) and B is the polarizing magnetic flux. However, the population difference between spin-up and spin-down nuclei in NMR is very small. For small energy differences, the Boltzmann distribution only allows a fraction E_{flip} / 2 kT to be flipped before the upper and lower state populations are made equal and the absorption disappears. In order to flip $N_{flip} \sim kT\, e^{SNR} / 4 \pi \nu_L L_{proton}$ protons, the sample volume musι include at least:

$$N_{min} - 2e^{SNR}\left(\frac{kT}{4 \pi \nu_L L_{proton}}\right)^2 \text{ (protons)} \qquad \ldots(49)$$

For T = 310 K, SNR = 2 and B = 1.4 tesla, ν_L = 60 MHz and $N_{min} \sim 1.7 \times 10^{11}$ protons.

When scanning a cell or organelle, the vast majority of the protons present are in water molecules. Water at 310 K has a proton density $n_{water} = 6.7 \times 10^{28}$ protons/m^3, compared to $n_{fat} \sim 6.4 \times 10^{28}$ protons/m^3 (palmitic acid). Distinguishing fat from water thus requires a minimum sample volume of $N_{min} / (n_{water} - n_{fat}) \sim 55$ micron^3.

Proteins range from 4.4 × 10^{28} protons/m^3 (aspartic acid minus one water) to 7.6 × 10^{28} protons/m^3 (leucine minus one water), average $n_{protein} \sim 5.6 \times 10^{28}$ protons/m^3; $n_{carbohydrate} \sim 5.8 \times 10^{28}$ protons/m^3 (glucose minus one water), allowing NMR cellular tomography to resolve minimum feature sizes of 2-4 microns (560 micron^3). This resolution may include intracellular structures such as the endoplasmic reticulum (rough ER ~1100 micron^3, smooth ER ~400 micron^3), the Golgi complex (~500 micron^3), the nucleus (~270 micron^3), possibly the nucleolus (~50 micron^3), and even large localized ferritin granule concentrations.

In 1998, the minimum sample size for microcoil (~470 micron diameter) NMR scanning was ~10,000 micron^3 or ~10^{16} protons for a 1-minute measurement cycle. Individual T-cell vacuoles ~1 micron in diameter labeled with dextran-coated iron oxide particles have been imaged by MRI.

Near-Field Optical Nanoimaging

Electromagnetic waves of optical wavelength l that interact with an object are diffracted into two components, called "far-field" and "near-field." The propagation of electromagnetic radiation over distances $z > \lambda$ acts as a spatial filter of finite bandwidth, resulting in the familiar diffraction-limited resolution $\sim\lambda/2$. Classical optics is concerned with this far-field regime with low spatial frequencies < 2/l, and conventional optical imaging will be difficult at the cellular level *in vivo*. (Short-wave X-rays will damage living biological cells.) Information about the high spatial frequency components of the diffracted waves is lost in the far-field regime, so information about sub-wavelength features of the object cannot be retrieved in classical microscopy.

However, for propagation over distances $z << \lambda$, far higher spatial frequencies can be detected because their amplitudes are then of the same order as the sample ($z = 0$). This second diffraction component is the "near-field" evanescent waves with high spatial frequencies $>2/\lambda$. Evanescent waves are confined to subwavelength distances from the object. Thus a localized optical probe, such as a subwavelength aperture in an opaque screen, can be scanned raster fashion in this regime to generate an image with a resolution on the order of the probe size.

The original Near-field Scanning Optical Microscope (NSOM) surpassed the classical diffraction limit by operating an optical probe at close proximity to the object. The NSOM probe uses an aluminum-coated "light funnel" scanned over the sample. Visible light emanates from the narrow end (~20 nm in diameter) of the light funnel and either reflects off the sample or travels through the sample into a detector, producing a visible light image of the surface with ~12 nm resolution at $\lambda = 514.5$ nm provided the distance between light source and sample is very short, about 5 nm, with signal intensities up to 10^{11} photons/sec (~50 nanowatts). This represents a resolution of $\sim\lambda/40$. The near-field acoustic equivalent is found in the medical stethoscope, which exhibits a resolution of $\sim\lambda/100$.

Applications include dynamical studies at video scanning rates, low-noise high-resolution spectroscopy, and differential absorption measurements. Optical imaging of individual dye molecules has already been demonstrated, with the ability to determine the orientation and depth of each target molecule located within ~30 nm of the scanned surface. Molecules with nanometer-scale packing densities have been resolved to ~0.4-nm diameters using STMs to create photon emission maps (an electric current generates photons in the sample), and laser interferometric NSOMs have produced clear optical images of dispersed oil drops on mica to ~1 nm resolution. Live specimen 250-nm optical sectioning for three-dimensional dynamic imaging has also been demonstrated.

Submicron laser emitters have been available since the late 1980s. It should be possible to use NSOM-like nanoprobes to optically scan the surfaces of cells or organelles to <~1 nm resolution, mapping their topography and spectroscopic characteristics to depths of tens of nanometers without penetrating the surface. However, a proper membrane-sealing invasive light funnel ~20 nm in diameter might not seriously disrupt some cellular or organelle membranes and thus could be inserted into the interiors of these bodies or through the cytoskeletal interstices to permit deeper volumetric scanning. The thermal conductivity of water at 310 K is 0.623 watts/m-K and the energy per 500-nm photon is 400 zJ, so for 1 micron^3 of watery tissue the maximum scan rate is ~35 micron^3/sec-K, if e^{SNR} photons are used to image each 1 nm^3 voxel with SNR = 2. Thus a 1-sec volumetric optical scan of an aqueous ~1-micron^3 sample volume to 1 nm^3 resolution requires a ~1 nanowatt scanner running at ~GHz bit rates, raising sample volume temperature by ~0.03 K.

NSOM permits the determination of five of the six degrees of freedom for each molecule, lacking only the optically inactive rotation around the dipole axis. In principle, it should be possible to produce <~1 nm resolution near-field optical scans of in situ protein molecules, since with atomically precise fabrication and single lines of atoms as conductors a minimum light guide (metal-dielectric-metal) is 3 atoms wide. Given a molecular laser and adequate collimation, photons passing through folded proteins will scatter according to the molecular structure. Detection of sufficient photons comprising these scattering patterns should allow the noninvasive determination of protein structure; polarized photons provide information on chirality. Absorption and fluorescent signals will be visible from phenyl rings, tryptophan, and bound cofactors such as ATP and adenine (which is fluorescent). Positions of monoclonal antibodies on virus surfaces are now identifiable experimentally using NSOM; it should be possible to map binding sites on virus and cell surfaces using fluorescently labelled antibodies. Single molecule detection has been proposed as a tool for rapid base-sequencing of DNA.

Other optically-based cellular imaging techniques must be distinguished. Optical Coherence Tomography (OCT) uses Michelson interferometry to achieve ~10 micron spatial resolutions over tissue depths of 2-3 mm in nontransparent tissue in the near infrared. However, OCT requires numerous physical components not easily implemented on a micron-size detector (e.g., beam splitter, lens-grating pair, galvanometer mirror, optical prism), femtosecond pulse shaping, and illumination power levels of ~10^7 watts/m^2 >> 100 watts/m^2 "safe" continuous limit in tissue. Bioluminescence techniques in which the light source is placed inside the tissue (e.g., a transgenic mouse with a luciferase gene in every cell of its body) has a spatial resolution limited to ~10% of the depth, or ~10 microns resolution at a ~100 micron depth. Coherent anti-Stokes Raman scattering (CARS) already permits organelle imaging in living cells and can in theory create a point-by-point chemical map of a cell using two intersecting lasers.

Three-dimensional observation of microscopic biological nonliving structures by means of X-ray holography requires a high degree of spatial coherence and good contrast between target and surroundings. Good contrast may be achieved in the wavelength range between the K absorption edges of carbon (l = 4.37 nm) and oxygen (l =2.33 nm), the "water window" where carbon-containing biological objects absorb radiation efficiently but water is relatively transparent. A 50-micron diameter emitter has been tested that uses near-IR 5-femtosecond laser pulses impinging upon a helium gas target to create a well-collimated (<1 milliradian) beam of coherent soft X-rays at a 1 KHz repetition rate producing a brightness of 5×10^8 photons/mm^2-milliradian^2-sec, a peak X-ray intensity of >10^{10} watts/m^2 on the propagation axis behind the He target.

Cell Volume Sensing

Many cellular parameters need not be measured directly in order to be detected by a medical nanodevice. Cell volume sensing is a case in point. An intracellular nanodevice can indirectly monitor changes in the volume of the cell in which it resides by one of two methods. First, measurements of the mechanical deformation of the cellular membrane, stretch-activated channels, or cytoskeletal strains and structural changes are quite sensitive to alterations in total cell volume. Second, concentration or dilution of the cytoplasmic environment through cell shrinkage or swelling leads to the activation of various volume-regulation responses which may be detected by the nanorobot. Changes in the concentration of soluble cytosolic proteins may nonspecifically affect enzyme activity via "macro- molecular crowding". Minor changes in cell volume can cause severalfold changes in ion transport. Cellular signalling entities that have been linked to the transduction and amplification of the primary volume signal include Ca^{++} transients, phosphoinositide turnover, eicosanoid metabolism, kinase/phosphatase systems such as JNK and p38, cAMP, and G-proteins. These signal amplification pathways may be monitored using chemical concentration sensors aboard the medical nanodevice, allowing the nanorobot to eavesdrop on the natural sensory channel traffic of the cell.

Noninvasive Neuroelectric Monitoring

In many ways the neuron is the most nanomedically important class of cell in the human body. The following is a brief summary of noninvasive (i.e., no axonal membrane penetration) nanotechnological methods for monitoring the electrical traffic of individual neurons. Noninvasive measurement of axonal traffic within nerve bundles will require multiple sensors and greater sensitivity to compensate for shielding by the perineurium, a tight resistive sheath enclosing the bundle ~20 microns thick with resistivity ~4000 ohm-cm.

Electric field neurosensing

The "typical" ~20 micron human neuron discharges 5-100 times per second, moving from -60 mV potential to +30 mV potential in $\sim 10^{-3}$ sec. Thus the variation in electric field at the axonal surface is ± 4500 volts/m. Since electric field sensors can detect 100 volt/m fields up to GHz frequencies, an electric sensor attached by circumaxonal cuff or pressed against the axonal surface (possibly at the node of Ranvier) should readily detect each action potential discharge. (The diameter of human nerve axons is 0.1-20 microns.) By 1998 silicon-to-neuron extracellular junctions already permitted direct stimulation of individual nerve cells *in vitro* without killing the cells, and extracellular electrodes were commonly used to detect neuronal electrical activity noninvasively, both *in vivo* and *in vitro*. Artificial electric fields may also be employed to trigger or moderate neural signals. Cell membrane capacitance is typically ~0.01 picofarads/micron^2, and varies with the state of the health of the cell.

Magnetic field neurosensing

The magnetic flux density caused by a single action potential discharge is ~0.1 microtesla at the axonal surface, which may be detected by a "compass oscillator" type magnetosensor with a ~KHz maximum sampling rate.

It might be possible for artificial magnetic fields to directly influence neural transmissions. Even a static 65 millitesla field has been shown to reduce frog skin Na^+ transport by 10-30%. Each neuronal discharge develops an electrical energy of ~20 picojoule (~10^{10} kT), far smaller than the magnetic energy stored in a B = 1.4 tesla field of a permanent micromagnet traversing an L^3 = (20 micron)^3 volume which is B^2 L^3 / 2 μ_0 ~ 6000 pJ. If properly manipulated, such a field may be sufficient to enhance, modulate, or extinguish a passing neural signal.

Neurothermal sensing

While neurons come in many shapes and sizes, our "exemplar" ~14,000-micron^3 neuron discharging ~90 mV into an input impedance of ~500 Kohms produces ~0.2 microampere current per pulse and generates a continuous (average) 100-300 pW of waste heat as measured experimentally. The discharge rate of 5-100 Hz can produce brief surges up to ~2000 pW during a high-frequency train, but the duty cycle of such trains is far less than 100%, reducing time-averaged dissipation to the observed 100-300 pW range.

Single impulses are measured experimentally to produce 2-7 microkelvin temperature spikes in cold or room-temperature mammalian non-myelinated nerve fibers and ~23 microkelvins in non-myelinated garfish olfactory nerve fibers at an energy density ranging from 270-1670 joules/m^3-impulse from 0-20°C. In non-myelinated fibers the initial heat occurs in two temperature-dependent phases: a burst of positive heat, followed by rapid heat reabsorption (called the *negative heat*). The positive heat derives from the dissipation of free energy stored in the membrane capacity, and from the decrease in entropy of the membrane dielectric with depolarization. An L ~ 20-micron neuron in good thermal contact with an aqueous heat sink at 310 K has thermal conductance L K_t ~ 10^{-5} watts/K, so trains of 5-100 Hz impulses lasting 1 second should raise cellular temperature by 10-30 microkelvins; up to ~200 microkelvin thermal spikes from such trains have been observed experimentally. These events are easily detectable by nanoscale thermal sensors capable of ~1 microkelvin sensitivity up to ~1 KHz. The ~microkelvin heat signature of individual impulses or very short pulse trains can probably be temporally resolved because the minimum pulse repetition time is ~10 millisec (at 100 Hz) which is much longer than the thermal time constant for an L ~ 20-micron neuron (thermal conductivity K_t = 0.528 watts/m-K and heat capacity C_V = 3.86 × 10^6 joules/m^3-K for brain tissue) which is t_{EQ} ~ L^2 C_V / K_t ~ 3 millisec.

Direct synaptic monitoring

The synaptic cleft between the axonal presynaptic terminal and the dendritic postsynaptic membrane is 10-20 nm in most synapses, although in the vertebrate myoneural junction it may be as large as 100 nm. Contact area per bouton is ~1 micron^2, giving a total gap volume of ~10^7-10^8 nm^3. The density of acetylcholine receptors is highest in muscles along the crests and upper thirds of the junctional folds (~10,000/micron^2), and is lowest in the extrasynaptic regions (~5/micron^2). Each action potential discharge triggers the release of ~10^4-10^5 molecules of acetylcholine into the gap volume of an active neuromuscular junction (diffusion time ~1 microsec), raising c_{ligand} from near zero to ~3×10^{-4} molecules/nm^3 (~0.0005 M) in ~1 millisec, followed by near-complete hydrolyzation by acetylcholinesterase during the 1-2 millisec refractory period. Into the gap volume may easily be inserted a ~10^5 nm^3 neurotransmitter concentration sensor able to measure ~100 acetylcholine molecules in ~1 millisec, thus detecting pulses at the fastest discharge rate. A similar device could be used to precisely regulate neuro-transmitter concentration at the junction, and hence the neural signal itself, under nanodevice control. Simple electrochemical and mechanochemical artificial synapses have been demonstrated.

Other neurosensing techniques

Many other noninvasive neurosensing techniques are readily conceivable. For instance, counting rotors or sodium magnetic resonance imaging could detect changes in the ionic composition (e.g., Na^+, K^+) of the periaxonal fluid before, during, and after discharge.

Cellular RF and Microwave Oscillations

Starting in 1968, H. Frohlich, observing that millivolt electrical potentials maintained across cell membranes ~10 nm thick give rise to huge fields ~10^7 volts/m possibly producing an electret state, theorized that membrane molecules must be highly electrically polarized and thus could interact to produce coherent surface acoustic vibrational modes in the 10-100 GHz (microwave) frequency range; the longest wavelength is about twice the membrane thickness. Interestingly, this frequency span is very close to the maximum trigger/reset frequency for bioelectronic molecules. Note that (~100 mV) (1.6×10^{-19} coul) ~ 4 kT, so a membrane molecule with a single charge on either end should be reliably reoriented by a depolarization wave, coupling pressure waves and electrostatic field fluctuations. However, the direct detection of 10-100 GHz millimeter radiation by non-nanotechnological means is experimentally difficult and controversial because the tests must be performed in vivo in close proximity to an actively metabolizing cell in water — and water strongly absorbs microwaves over macroscale ranges (e.g., ~99% absorption in 3 mm at 100 GHz).

Nevertheless, active cells have shown enhanced Raman anti-Stokes scattering, an effect ascribed to the converse of the Frohlich oscillations. In one study, the normalized growth rate of yeast cultures was enhanced or inhibited when irradiated by CW microwave fields of ~30 watts/m^2 of various frequencies; growth rate data spanning 62 separate runs revealed a repeatable frequency-dependent spectral fine structure with six distinct peaks of width ~10 MHz near 42 GHz. Investigations of related phenomena are ongoing and voluminous; the interested reader should peruse *Bioelectromagnetics*, the archival journal of this field.

100 GHz waves attenuate only ~1% after passing through ~3 microns of soft tissue. A single electron injected into an integral membrane protein could act as an oscillating dipole, making a 300 volt/m signal 10 nm from the protein antenna with an energy transfer of ~0.004 kT per cycle; ~1000 oscillating electrons could produce a measurable field. A 20-micron diameter cell modeled as a nonuniform spherical dipole layer with transmembrane dipoles located 10 nm apart and embedded in a dissipative medium could produce 10^2-10^5 volt/m microwave fields 1-10 microns from the cell surface.

Thus, a variety of rf and microwave electromagnetic emanations may in theory be detectable both within and nearby living cells which could prove diagnostic of numerous internal states. Such states may include cytoskeletal dynamics, metabolic rates, plasmon-type excitations due to the collective motion of ions freed in chemical reactions, positional, rotational or conformational changes in biological macromolecules and membranes, internal movements of organelles and nerve traffic conduction, cellular pinocytosis, cellular reproduction events, cell membrane identity (e.g., distinguishing erythrocyte, Gram-positive and Gram-negative cell coat conductivities at 10 KHz), and cell-cell interactions.

Macrosensing

Macrosensing is the detection of global somatic states (inside the human body) and extrasomatic states (sensory data originating outside the human body). While the treatment here is necessarily incomplete, the discussion nevertheless gives a good feel for the kinds of environmental variables that internally-situated nanodevices could sense. Not all capabilities outlined here need be available on every nanorobot, since injection of a cocktail of numerous distinct but mutually cooperative machine species allows designers to take full advantage of the benefits of functional specialization. In many cases, a given environmental variable can be measured by several different classes of sensor device. However, since these devices are microscopic it is in theory possible to operationalize almost all of the macrosensing capabilities described below in one patient using just a billion devices (~10 mm^{-3} whole-body deployment density), a total volume of ~1 mm^3 of nanorobots or ~0.1% of the typical ~1 cm^3 therapeutic dose.

Acoustic Macrosensing

Cyto-auscultation

Can sounds generated by a single cell be detected, and thus be useful for diagnosis? Probably not, given that low frequency acoustic radiators are notoriously inefficient. For example, mitochondrion organelles of the giant amoeba *Reticulomyxa* are shuttled back and forth by 1-4 cytoplasmic dynein motors while riding on the outside of a bundle of 1-6 microtubules. Each dynein motor generates 2.6 pN of force and drives the mitochondria at up to ~10 micron/sec, developing 0.3-1.0 × 10^{-16} watts of mechanical power within each 320-nm diameter organelle. Taking each organelle as a cylindrical acoustic radiator with a mechanical input power of $P_{in} \sim 10^{-16}$ watts at $\nu \sim 1000$ Hz, the output acoustic pressure at the organelle surface is only ~10^{-9} atm, an acoustic power intensity $I \sim 10^{-15}$ watts/m^2 which is not detectable by micron-sized nanorobots. Nevertheless, cells and intracellular elements are capable of vibrating in a dynamic manner with complex harmonics that can be altered by growth factors and by the process of carcinogenesis, so the possibility cannot be completely ruled out.

Blood pressure and pulse detection

Blood pressure ranges from 0.1-0.2 atm in the arteries to as low as 0.005 atm in the veins. The systolic/diastolic differential ranges from 0.05-0.07 atm in the aorta and 0.01-0.02 atm in the pulmonary artery, falling to 0.001-0.003 atm in the microvessels, or 0.003-0.005 atm if the precapillary sphincter is dilated. In venous vessels, pulse fluctuations are 0.002-0.010 atm in the superior vena cava, 0.004-0.006 atm in the subclavian vein, ~0.004 atm in venules generally, and ~0.0005 atm in the brachial vein. There is also a ~0.05 Hz random fluctuation in the microvessels with amplitude on the order of 0.004-0.007 atm. Both blood pressure and pulse rate can be reliably monitored

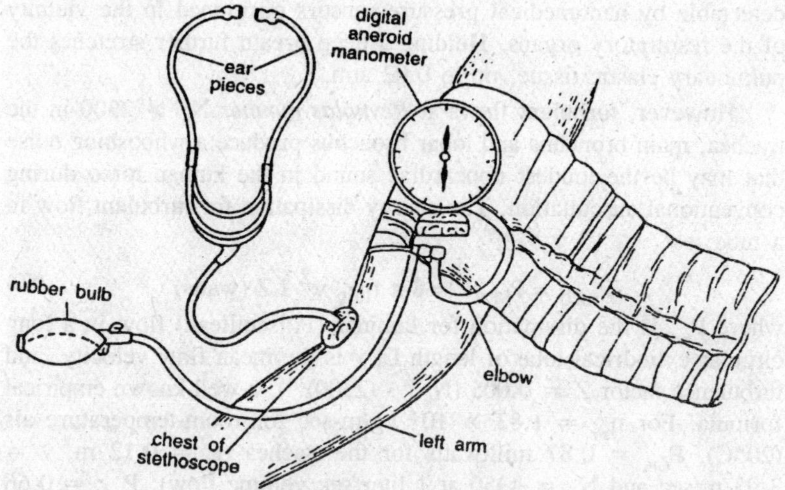

Fig. 7.1. Blood pressure and pulse detection.

by a medical nanodevice virtually anywhere in the vascular system using a (68 nm)^3 pressure sensor with ~0.001 atm sensitivity.

Pulse propagation through body tissue is somewhat muted due to absorption in compressible fatty membranes, but most cells lie within 1-3 cell-widths of a capillary so the cardiac acoustic signal should still be measurable using more sensitive detectors. The time-averaged interstitial pressure in subcutaneous tissue is 0.001-0.004 atm.

Arterial pulse waves (vascular oscillations) carry subtle messages about the health of internal organs and the arterial tree. The idea of using pulse waves for diagnosis has a long history dating back 2000 years in China. For example, in the *Book on Pulse Waves* by Wang Shu-He (201-285 AD), waves detected by manual probing are classified using such subjective and qualitative descriptors as floating, deep, hidden, rapid, slow, moderate, feeble, replete, full, thready, faint, weak, soft, slippery, hesitant, hollow, firm, long, short, swift, running, intermittent, uneven, taut, string-tight, gigantic, or tremulous. Abnormal waves were empirically related to disease states. Wave data gathered by nanodevices could make possible a theoretically sound, quantitative system of noninvasive observation, classification, and diagnosis as a supplement to other nanomedical tools.

Respiratory audition

The variation of mechanical pressure over a complete respiratory cycle is ~0.003 atm in the pleura, ~0.002 atm at the alveoli, detectable by nanomedical pressure sensors positioned in the vicinity of the respiratory organs. Holding a deep breath further stretches the pulmonary elastic tissue, up to 0.02 atm.

However, turbulent flows at *Reynolds number* $N_R > 2300$ in the trachea, main bronchus and lobar bronchus produce a whooshing noise that may be the loudest noncardiac sound in the human torso during conventional auscultation. The energy dissipation for turbulent flow in a tube is

$$P_{turb} = P_{lam} Z = 8 \pi \eta_{air} v^2 LZ \text{ (watts)} \qquad ...(50)$$

where P_{lam} is the dissipation for laminar (Poiseuillean) flow in a long circular cylindrical tube of length L, v is the mean flow velocity, and turbulence factor $Z = 0.005 (N_R^{3/4} - (2300)^{3/4})$, a well-known empirical formula. For $\eta_{air} = 1.83 \times 10^{-5}$ kg/m-sec for room-temperature air (20°C), $P_{turb} = 0.87$ milliwatts for the trachea (L = 0.12 m, v = 3.93 m/sec and $N_R = 4350$ at 1 liter/sec volume flow); $P_{turb} = 0.66$ milliwatts for the main bronchus (L = 0.167 m, v = 4.27 m/sec and $N_R = 3210$); and $P_{turb} = 0.09$ milliwatts for the lobar bronchus (L = 0.186 m, v = 4.62 m/sec and $N_R = 2390$), totalling ~1.6 milliwatts acoustic emission from a ~120 cm^3 upper tracheobroncheal volume. This is a power density of 13 watts/m^3 corresponding to a pressure of 4×10^{-5} atm assuming a 300 millisecond measurement window at the maximum respiration rate.

The amplitude of an acoustic plane wave propagating through tissue attenuates exponentially with distance due to absorption, scattering and reflection. The amplitude is given approximately by

$$A_x = A_0 \, e^{-\alpha F x} \qquad ...(51)$$

where A_0 is the initial wave amplitude in atm, A_x is the amplitude a distance x from the source, and a is the amplitude absorption coefficient. The function F expresses the frequency dependence of the attenuation. For pure liquids, $F = F_{liq} = \nu^2$ (Hz^2); for example, $\alpha_{liq} = 2.5 \times 10^{-14}$ sec^2/m for water at room temperature. However, for soft tissues, $F = F_{tiss} \sim \nu$ (Hz). Assuming <10 KHz bronchial turbulence noise of initial amplitude $A_0 = 4 \times 10^{-5}$ atm, $A_x \sim 3.7 \times 10^{5}$ atm through 1 meter of typical soft tissue; $A_x \sim 3.4 \times 10^{-5}$ atm even if 0.1 meter of bone is interposed in the acoustic path. Either A_x is reliably detected from anywhere in the body using a >(210 nm)^3 pressure sensor.

Mechanical body noises

Many other mechanical body noises should be globally audible to properly instrumented medical nanodevices. If normal chewing motions (of hard foods) release 1-10 milliwatts in a ~100 cm³ oral volume with a ~1 sec jawstroke, power density is ~10 watts/m³ or ~10^{-4} atm of tooth-crunching noise. A stomach growl registering 45 dB (vs. 30 dB whisper, 60 dB normal conversation) at 2 meters has a source power of 160 milliwatts; released from a 10 cm³ gastric sphincter volume gives a ~2×10^{-6} atm acoustic wave, detectable throughout the body. Walking and running releases 20-100 joules/footfall for a 70 kg man; assuming the energy is absorbed within a ~1 cm thickness or within ~1 second by the sole of the foot, Eqn. 52 implies an upward-moving planar compression wave of 0.4-2.0 atm, easily detectable by acoustically instrumented nanodevices body-wide. (Shoe insoles dissipate energy and alter the shock wave pulse shape.) Hand-clapping generates 0.02-0.2 atm pulses, also easily detectable.

Lesser noises including ~30 millisec hiccups at (4-60)/min, intestinal and ureteral peristalsis, sloshing of liquid stomach contents, heart murmurs, a tap on the shoulder by a friend, nasal sniffling and swallowing, clicks from picking or drumming fingernails, crepitations, manustuprations and ejaculations, the rustling noise of clothing against the skin, flapping eyelids, anal towelling, bruits (including murmurs and thrills) due to vascular lesions, dermal impact of water while showering, copulatory noises, urethral flow turbulence during urination, transmitted vibrations from musical instruments, creaking joints, and squeaking muscles can be detected locally if not globally. Implantation of significant interconnected in vivo diamondoid structures may produce increased sensitivity to internal noises, due to the extremely low acoustic absorption coefficient of diamond.

Vocalizations

Average source power for conversational speech in air is ~10 microwatts at the vocal cords (60 dB), up to ~1000 microwatts for shouting (90 dB) and as little as 0.1 microwatts (30 dB) for whispering. Vocal cord surface area ~1 cm², giving an acoustic intensity I ~ 0.001-10 watts/m². (Using the decibel notation, dB = 10 $\log_{10}$ (I/I_0), where I_0 ~ 5×10^{-13} watts/m² in air, I_0 ~ 1×10^{-16} watts/m² in water.) In a planar traveling wave, pressure amplitude A_p (N/m²) is related to power intensity I by

$$A_p = (2 \rho v_{sound} I)^{1/2} \ (N/m^2) \qquad ...(52)$$

For water at 310 K, ρ = 993.4 kg/m³ and v_{sound} = 1500 m/sec, therefore A_p = 0.0005-0.05 atm for speech, detectable by nanodevices throughout the body due to minimal attenuation at audible frequencies.

Other easily detectable vocalizations include whistling, humming, coughing, sneezing, rales, wheezing, expectorating, eructations, flatus, vomiting, hawking and noseblowing.

In the case of spherical waves diverging from an omnidirectional transmitter of radius r at a distance X from the transmitter, average intensity declines inversely as the square of the distance and so I in Eqn. 52 must be replaced by $(I\ r^2 / X^2)$.

Environmental sources

Can *in vivo* nanodevices directly detect sounds emanating from the environment outside of the body, such as other people talking in the same room or a door slamming? The waves from an external acoustic source of power P_R watts travel through the air and, upon arriving at the air/skin interface a distance x_R from the source with amplitude $A_{incident}$, are transmitted through the interface with amplitude $A_{transmit}$. For a specular reflector — interface dimensions (human body ~ 2 m) > acoustic wavelength (~0.03-3.0 m for typical audible sounds in air) — with acoustic impedance Z_1 and Z_2 on either side of the interface and perpendicular incidence,

$$A_{transmit} = A_{incident}\left[1 - \text{abs}\left(\frac{Z_1 - Z_2}{Z_1 + Z_2}\right)\right] \quad ...(53)$$

Acoustic impedance, like the speed of sound, is essentially frequency-independent over the nanomedically-relevant range of ultrasonic frequencies. For Z_{air} = 400 kg/m²-sec and assuming Z_{skin} ~ 1.6×10^6 kg/m²-sec, then $A_{transmit} = (5 \times 10^{-4})\ A_{incident}$. In other words, there is ~99.95% reflection from the air-skin interface, which is why coupling mediums like gels and oils are commonly employed in ultrasound imaging. If an immediately subdermal nanodevice can detect a minimum $A_{transmit}$ ~ 10^{-6} atm, then from Eqn. 52 and simple geometry:

$$P_R = \frac{2\pi x_R^2 A_{transmit}^2}{\rho v_{sound}} \quad ...(54)$$

For STP (1 atm, 0°C) air, ρ = 1.29 kg/m³ and v_{sound} = 331 m/sec. If the minimum detectable pressure ~10^{-6} atm ~$A_{transmit}$, then at a distance of x_R = 2 meters the acoustic source must have a power of ~2000 watts, far exceeding the ~1 milliwatt output of a person loudly shouting. To hear normal conversation at x_R = 2 m, minimum nanodevice detector sensitivity falls to 7×10^{-11} atm requiring a subdermal pressure nanosensor ~(17 micron)³ in size (Eqn. 28), roughly the dimensions of a single human cell; other methods may prove more efficient.

Of course, an *ex vivo* acoustic nanosensor may receive sound that has passed through no interface, hence may detect pressure waves ~3 orders of magnitude lower in amplitude. Assuming $A_{transmit} = A_{incident}$, $P_R \sim 600$ microwatts, so *ex vivo* nanorobots with a 0.3 micron^3 sensor could hear people shouting at $x_R = 2$ meters. To hear talking (~10 microwatt source) requires a 2.4 micron^3 ex vivo pressure sensor (limit ~10^{-7} atm), from Eqn. 28.

Optimally positioned and calibrated nanomedical pressure sensors could directly measure changes in the ambient barometric pressure to within $\pm 10^{-6}$ atm. Normal atmospheric variation due to weather ranges from 0.94-1.05 atm; such slow moving changes are readily monitored. Very near the Earth's surface, the air pressure P at altitude h above sea level is approximated by $P = e^{-k_p h}$ (atm), where $k_p = 1.16 \times 10^{-4}$ m^{-1} at 20°C; at sea level, a 10^{-6} atm change in pressure reflects a change in altitude of only ~1 cm. However, the opening or closing of a door inside a (~5 m)^3 room that displaces >125 cm^3 of air also causes a minimally detectable >10^{-6} atm pressure pulse. Other sources of environmental pressure variation such as infrasonic (~0.2 Hz) microbaroms from offshore ocean storms, wind entering through open windows, forced-air currents from central heating or A/C systems, or even the movements of nearby people and pets may be detectable and thus may further confuse the measurement, reducing absolute accuracy unless suitable corrections are made.

Proprioceptive Macrosensing

Kinesthetic macrosensing

Using a navigational transponder network, a population of ~10^{11} nanodevices each spaced an average ~100 microns apart throughout the body tissues can determine relative location to a positional accuracy of ~3 microns and an angular accuracy of ~2 milliradian, with the data updatable once every millisecond. Thus dispersed, the network can continuously monitor and record the relative positions of all limbs with worst-case (cumulative error) ~0.8 mm accuracy over a 2-meter span. These devices can prepare high-resolution dynamic maps of body position, velocity, acceleration and rotation, allowing precise real-time digital kinesthesia during sports or artistic activities such as gymnastics, pole vaulting, or ballet; during transportative activities such as driving cars around hairpin turns, roller-coaster rides, military aircraft maneuvers and space launches; during precision tool-using such as needle-threading, antique watch repair, or while using the fingers as measurement calipers; during self-defense activities requiring complex motions such as karate or judo; and during emergency situations such as automobile crashes and tumbling motions during falls from great heights. All such sensory data is readily outmessaged to the patient or user in real time.

Nanorobots could assist in path integration or the reconstruction of experienced limb trajectories. Given the ability to determine rotation rates and to measure applied forces and torques, the network should also be able to forecast the anticipated future positions of limbs. By comparing these projections to actual results, the network can then infer the viscosity of the medium in which the activity is taking place (e.g., air or water), whether the environment is stable or is translating or rotating in some direction, and whether the human user is physically supported or in free-fall. Internal nanodevices can directly measure if a body is sitting, standing, laying prone, inverted, falling, floating or diving, and then communicate that information directly to the human user. Detection of patient activity states—e.g., the patient is sitting, standing, or walking—can be used to control nanorobot behaviours, activate or deactivate outmessaging displays, and so forth. A simple macroscale wearable tactile "compass belt" that would convey desired geographical directions has already been proposed.

Orientational macrosensing

Since nanopendular sensors can reliably determine the direction of the local gravity field vector to within ~2 milliradians in ~10^{-4} sec, the navigational network can poll its members and arrive at an accurate consensus on which direction is up. Nanodevices affixed to relatively stable hard body parts will exhibit more consistent orientational readings. Measurement of the gravity vector allows the vertical orientation of the rest of the body to be precisely fixed in space, especially useful for gymnasts, trapeze artists, underwater divers in murky lakes, and vestibular-impaired individuals, to whom this information may be outmessaged directly.

The gravitational vector may also be indirectly measured, albeit more slowly and less accurately, by monitoring the body's natural reactions to changes in the axis of gravitational loading. For instance, the variation in head-to-toe hydrostatic pressure for a 1.7 m tall, 70 kg human standing in a 1-g gravity field is ~0.17 atm; the ability to measure a systematic cross-body differential of 10^{-6} atm allows the detection of a ~10^{-5} g change along the lengthwise aspect, or ~10^{-4} g change along the transverse aspect. Thus a continuously updated whole-body barostatic map allows the human body to serve as a three-dimensional gravity/orientation sensor.

Body weight measurement

A network of nanodevices can inventory the volume and density of each of the body's $\sim 10^8$ $(1 \text{ mm})^3$ voxels using a combination of acoustic ranging, somatic mapping, and flowmetry. Each voxel is identified as fat, muscle tissue, bone mass, interstitial fluid, and so forth. These measurements allow the body's weight to be precisely computed as volume times density, rather than the usual method of determining weight using gravitational force data. The existence of a natural physiological "ponderostat," a crudely analogous humoral body-mass detector, has been proposed; indeed, the insulin-leptin system has been found to serve a related purpose, and is easily eavesdropped by medical nanorobots.

Gravitational geographic macrosensing

Medical nanodevices can measure variations in the gravity field to $\sim 10^{-6}$ g's for L = 20 micron gravimeters in a measurement time t_{meas} = 2-9 millisec. This implies that *in vivo* nanodevices can take precise measurements of their latitude and altitude relative to sea level ~ 100 times every second. Gravity increases toward the poles and at lower altitudes. Specifically, using the formula of Cassinis (accounting for rotational and polar flattening effects on the Earth) with the Bouguer correction to the free air variation by altitude (assuming flat topography), measured gravity g_{meas} is given approximately by

$$g_{meas} = g_0\left[1 + k_1 \sin^2(\theta_L) - k_2 \sin^2(2\,\theta_L)\right] - k_3\, h + k_4\, h \rho_{earth} \qquad \text{...(55)}$$

where θ_L = terrestrial latitude (equator = 0°), h = height above sea level in meters, g_0 = 9.78039 m/sec^2 (equatorial sea-level value of g), $k_1 = 5.2884 \times 10^{-3}$, $k_2 = 5.9 \times 10^{-6}$, $k_3 = 3.086 \times 10^{-6}$ sec^{-2}, $k_4 = 4.185 \times 10^{-7}$, and ρ_{earth} = 5522 kg/m^3.

Since sea-level g varies from 9.78039 m/sec^2 at the equator to 9.83217 m/sec^2 at the north pole, a 20-micron gravimeter ($\Delta g = 10^{-6}$ g) detects a change in position of 1 arcmin of latitude or ~ 1900 meters north/south along the Earth's surface. Similarly, since at 45° latitude g varies from 9.806 m/sec^2 at sea level to 9.803 m/sec^2 at 1000 meters altitude, a 20-micron gravimeter detects a change in altitude of ~ 3.3 meters (e.g., upstairs vs. downstairs in a house). For comparison, in 1998 high-quality commercial gravity gradiometers measured gradients of $\sim 10^{-9}$ g/meter and allowed the compilation of micro-g ($\sim$1 milligal) resolution aerial gravity maps; atom interferometers measured the gravitational acceleration of atoms to a precision of 10^{-10}.

To achieve such phenomenal positional accuracies, the nanodevice must be able to computationally resolve several complicating factors. First, localized mass concentrations representing nonuniformities in crustal density produce residuals of up to ± 0.0006 m/sec^2, which may be removed from the data using a standard map of known terrestrial isostatic variations and anomalies. Indeed, matching observations to such a map could provide useful longitudinal information as well. Another complication is the variation in gravity due to tidal forces amounting to $\sim 3 \times 10^{-7}$ g's, twice daily, which lies at the limits of detectability for a 20-micron gravimeter. Other minor geodesic and terrain-related corrections, too complicated for discussion here, may also need to be applied in certain circumstances. Note that the presence of nearby heavy objects does not influence measurement accuracy: a 100-ton building 10 meters away adds a lateral acceleration of only 7 $\times 10^{-9}$ g's to a human body.

One final complication is that patient movements create kinematic accelerations that must be distinguished from the gravitational accelerations. Gravity readings can be corrected by taking derivatives of the signals from kinesthetic monitoring to give gravity in the reference frame of the patient's room, and many individual measurements may be averaged to improve accuracy since the gravity vector normally changes only very slowly.

Electric/Magnetic Macrosensing

Vascular-interstitial closed electric circuits

In vivo studies of the electric properties of blood vessels shows that the walls of veins and arteries present a specific electric resistance $\sim$200 times greater than the conductive medium of blood (plasma) — roughly 200 ohm-m vs. 0.7 ohm-m. Thus blood vessels are properly regarded as relatively insulated conducting cables which can electrically connect an injured tissue with surrounding noninjured tissue. Capillaries form an electric junction with the interstitial fluid, so the electric gradient can be canceled by ionic transport. Cell membranes are insulating dielectrics with resistance and capacitance, penetrated by ionic channels or gates for ionic transport. Additionally, in 1941 Szent-Gyorgyi suggested the possibility of semi-conduction in proteins, a theory which has since been elaborated by many investigators.

B. Nordenstrom has proposed that the above schema represents a system for selective electrogenous mass transport of material between blood and tissue which he calls Vascular-Interstitial Closed (electric) Circuits (VICCs) — in effect, an additional electro-circulatory system operating in parallel with the well-known diffusive, osmotic, and hydrodynamic mechanisms of regular bloodstream materials transport. As long as ions can leak through open pores and migrate through ion channels, long distance transport cannot take place and the VICC system remains primarily a local (e.g., within tissues), not global, electrical circuit. Connections also exist with conductive media including cerebrospinal fluid, bile, and urine.

Nanorobots capable of monitoring the status of local VICCs may rapidly and efficiently acquire a wealth of systemic information without the need for direct inspection of the affected tissues. For example, when a working muscle produces catabolic products such as lactic acid, an electrochemical potential gradient develops between the muscle and surrounding tissue in a known manner. Injured tissues become polarized in relation to surrounding tissue, as initial catabolic degradation products acidify the tissue. Malignant neoplasms, benign neoplasms, and internally necrotic granulomas all may polarize electrically in relation to surrounding tissue; a vascular thrombus also contains ionized material. Large deflections in the diffusion potential of blood occur if blood is deoxygenated or is infected with Gram-negative bacteria. Blood changes its electric potential from +500 mV to +1000 mV during spontaneous coagulation, and the spontaneous electric potential at the site of a crush injury of the iliac crest in rats oscillated several times between +190 mV and -60 mV over a four day experimental trial.

Nordenstrom proposes that *in vivo* electrophoresis via the VICC system may also play a role in mediating leukotaxis. When tissue is artificially polarized by electrodes, accumulation of granulocytes with margination and development of pseudopods is the result. In one experiment, exposure of a mesenterial membrane to a 1 microampere 1 volt field for 30 minutes stimulated diapedetic bleeding near the anode. At higher power, the arterio-capillaries contracted, narrowed, and emptied of blood cells, while the veno-capillaries and venules widened and filled with granulocytes. Monitoring such actions of local VICCs experiencing natural voltage fluctuations could alert medical nanorobots to the presence of local injuries, tissue changes, or pathologies that otherwise might go unnoticed for a considerable time.

Electric/magnetic geographic macrosensing

Detection of electric fields to ~100 volts/m, the normal vertical atmospheric gradient, in theory might allow determination of aboveground altitude to ~1 m accuracy if the time between recalibrations is very brief. However, a human body (which is a good conductor), standing on level ground, acquires a slight negative surface charge and becomes "part of the ground" electrically. This distorts upward the equipotentials that usually run parallel to the ground, thoroughly corrupting an altitude measurement. Nearby lightning hits and storms are detectable, which is useful.

However, medical nanodevices with appropriate magnetometers can measure variations in magnetic field to ~0.1 microtesla using a (660 nm)^3 permanent magnet sensor, in a measurement time t_{meas} ~ 0.6 millisec. Geomagnetic field maps of the Earth's surface are highly nonisotropic in both latitudinal and longitudinal directions; knowledge of the field plus the absolute direction of true north (e.g., using a nanogyroscope) gives longitude information from the separation of the planetary magnetic and spin poles. This implies that properly equipped *in vivo* nanodevices can establish their latitude and longitude on Earth's surface ~1000 times every second by this means. In particular, the horizontal component of the geomagnetic field vector ranges from 0-41 microtesla from magnetic pole to magnetic equator; the independent vertical field component ranges from 0-70 microtesla. Thus a 0.1-microtesla sensor should resolve each component to an accuracy of ~11 arcmin or ~20 km in both latitude or longitude. Terrestrial surface magnetic anomalies (e.g., iron deposits) produce local variations of 0.03-30 microtesla and are thus detectable. Additional complications to these measurements include artificial magnetic field sources, the 0.01-0.1 microtesla solar daily variation, occasional magnetic storms producing erratic geomagnetic fluctuations of 0.01-5 microtesla, sudden-commencement ionospheric electrojets up to 0.3 microtesla, and various long-term secular variations in the geomagnetic field.

Radio, television, and direct broadcast satellite signals probably are not detectable by individual nanodevices, but such detection may be indirectly achieved *in vivo* using nanorobot-accessible dedicated macroscale antennas implanted in the body.

Piezoelectric stress macrosensing

The piezoelectric effect is "the production of electrical polarization in a material by the application of mechanical stress". Many materials in the human body are piezoelectric, including tendon and elastin, dentin and bone. This polarization, or surface charge, varies as the physical stress imposed on the material changes over time. Thus, measurement of the piezoelectric effect in bone or tendon can provide information on the physical loads that are being carried by these materials. For example, a shear stress applied along the long axis of a bone alters the polarization voltage that appears on a surface at right angles to the axis. In another experiment, piezoelectric surface charges on a loaded human femur were measured to range from -131 to +207 picocoulombs/cm^2 (-8 to +13 charges/micron^2) from one end of the bone to the other end, depending upon position along the shaft. Real-time monitoring of this data would permit specific inferences as to the amount of load the bone was carrying and from what direction, including bending, shearing, and twisting forces, from which whole-body activity states could subsequently be inferred. Knowledge of these surface charge variations might also be exploited in osteographic or functional navigation.

Optical Macrosensing

In vivo medical nanodevices can gather little useful optical information from the external environment. Here's why.

In biological soft tissues, scattering dominates absorption except in the pigmented layers of the epidermis and stratum corneum. Thus the propagation of light in tissues may be regarded as occurring in two steps.

In the first step, optical photons of intensity I_0 falling perpendicularly on skin are transmitted according to Beer's law through tissue to a depth z with a transmitted intensity of

$$I_z = I_0\left(1 - r_{sp}\right)e^{-\sigma_t z}\left(\text{watts} / m^2\right) \qquad ...(56)$$

where r_{sp} = specular reflection coefficient for visible light (Fresnel reflection at the air-tissue surface) ~ 4%-7%, or 0% if the original light source lies within the body; and the transmission coefficient σ_t = σ_a (absorption coefficient ~ 300 m^{-1}) + σ_s (scattering coefficient ~30,000 m^{-1}) for various human soft tissues at optical wavelengths. (Coefficient values for the most heavily pigmented skin layers may be 5-7 times higher.) Typical values for σ_t ~ 10,000-100,000 m^{-1}, average ~30,000 m^{-1} for soft tissue, although exceptionally clear tissues with σ_t = 1000 m^{-1} have been reported. Thus the mean free path of an optical photon in human soft tissue is 10-100 microns, average ~ 30 microns (~1.5 tissue cell-widths), up to an extreme maximum of ~1 mm for the most transparent tissues known. For σ_t ~ 30,000 m^{-1}, at

z = 150 microns, $I_z / I_0 \sim 0.01$ and ~99% of all photons have been scattered at least once from their initial path.

In the second step, in an unbounded medium the patch of fully scattered photons continues to propagate through the tissue via diffusion until all photons are absorbed. The complicated governing diffusion equation has not yet been been completely solved analytically, but the asymptotic diffusive fluence is given roughly by

$$I_d \sim I_i \, e^{-\sigma_d z} \left(\text{watt} / \text{m}^2\right) \qquad \text{...(57)}$$

where σ_d is the diffusion exponent or effective attenuation coefficient (averaging ~900 m^{-1} for typical soft tissues but with an extremely wide range reported over the ultraviolet, visible, and near-infrared wavelengths, from 10-1,000,000 m^{-1}.) As a crude approximation, the initial intensity of the fully scattered photon patch $I_i \sim I_0 (1 - r_{sp}) a_{scat}$, where $a_{scat} = \sigma_s / (\sigma_s + \sigma_a) \sim 0.987$ is the albedo for single particle scattering.

Over the optical band from 400-700 nm, incident intensity I_0 = 100-400 watts/m^2 when standing in direct sunlight; artificial lighting in homes and offices is typically 0.1-10 watts/m^2; moonlight provides only $I_0 \sim 10^{-4}$ watts/m^2; and the absolute threshold for human vision is ~10^{-8} watts/m^2. For $\sigma_d \sim 900$ m^{-1}, the intensity ratio of transmitted/incident visible light falls to $I_d / I_0 = 0.1$ at z = 2.5 mm depth (~eyelids), ~10^{-4} at z = 1 cm depth, and ~10^{-11} at z = 2.8 cm — a depth at which the tissue would be completely dark to the human eye even under direct sunlight illumination at the outermost skin surface.

An optical nanosensor with N_{sensor} receiver elements, with each receiver element having an area A_e = 1 nm^2 and capable of single-photon detection, requires an integration time for reliable detection of e^{SNR} photons given by

$$I_{meas} \sim \frac{h \, \nu \, e^{SNR}}{I_d \, N_{sensor} \, A_e} \text{ (sec)} \qquad \text{...(58)}$$

where $h = 6.63 \times 10^{-34}$ joule-sec and $\nu = 4.3\text{-}7.5 \times 10^{14}$ Hz for optical photons. Let us assume that a patient is standing in $I_0 \sim 400$ watt/m^2 direct unfiltered sunlight, and that t_{meas} = 1 sec, SNR = 2, N_{sensor} = 25,000 elements giving a nanorobot eyespot of $A_e N_{sensor}$ = 0.025 micron^2, and that σ_d = 900 m^{-1}. Then from Eqns. 57 and 58 the maximum tissue depth for optical photon detection is $z_{max} \sim 17$ mm ($I_d \sim 10^{-4}$ watt/m^2) which includes a volume of tissue comprising up to ~30% of total body volume. Changes in illumination at the minimum indoor artificial level of ~0.1 watts/m² are visible to z_{max} ~ 7 mm depth. This eyespot, if exposed to air on the outermost surface of the skin, is just sensitive enough to detect full moonlight.

Our general conclusion is that variations in normal indoor lighting may be directly measurable by nanorobots stationed within the outermost ~1 cm of body tissues. Imaging, as opposed to the simple illumination detection system described here, is a much more difficult design challenge.

Neural Macrosensing

The ability to detect individual neural cell electrical discharges noninvasively in many different ways, coupled with the abilities (A) to recognize and identify specific desired target nerve cells and (B) to pool data gathered independently by spatially separated nanodevices in real time, offers the possibility of indirect neural macrosensing of complex environmental stimuli by eavesdropping on the body's own regular sensory signal traffic.

For example, facultatively mobile nanodevices may swim into the spiral artery of the ear and down through its bifurcations to reach the cochlear canal, then position themselves as neural monitors in the vicinity of the spiral nerve fibers and the nerves entering the epithelium of the organ of Corti (cochlear or auditory nerves) within the spiral ganglion. These monitors can detect, record, or rebroadcast to other nanodevices in the communications network all auditory neural traffic perceived by the human ear. Advanced speech recognition systems may permit recovery of spoken words and identification of individual

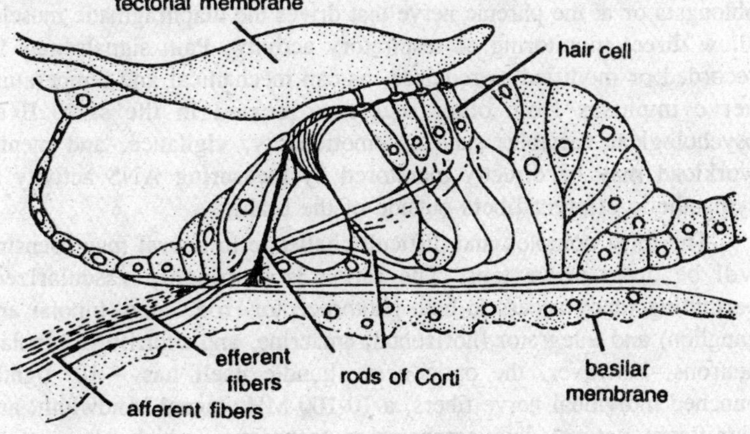

Fig. 7.2. Organs of Corti.

speakers (including vocalizations by the user), recognition of background noises, reception and validation of cues or commands spoken directly to in vivo nanosystems by authorized medical personnel, and so forth. Since properly configured monitors can also modulate or stimulate nerve impulses, these devices may add audible signals to the audio traffic, thus may be employed as hearing aids (using feedback loops), real-time language translation mechanisms, continuous vocalization/audition recorders, voice-stress analyzers, or nanodevice-user communications links.

Nanomonitors positioned at the afferent nerve endings emanating from hair cells located in the otolithic membrane of the utricle and saccule and in the cristae ampullaris of the semicircular canals allow medical nanodevices to directly record, amplify, attenuate, or modulate the body's own sensations of gravity, rotation, and acceleration, although kinesthetic sensory management might also be required for complete control. Motor neurons likewise can be monitored to keep track of limb motions and positions, or specific muscle activities, and even to exert control. Neuron-resident nanorobots may detect auditory effects induced in the brain by pulsed microwaves from external sources. Feline cochlear neurons sensitive to audible frequencies of >300 Hz respond to single microwave pulses at a threshold-specific absorption rate of 6-11 watts/kg-pulse, while in humans the threshold for effect, which depends on energy per pulse, may be as low as ~ 0.02 joule/m^2-pulse for people with low hearing threshold.

Olfactory and gustatory sensory neural traffic similarly may be eavesdropped by nanosensory instruments. Nerve taps in the medulla oblongata or at the phrenic nerve that drives the diaphragmatic muscles allow direct monitoring of respiratory activity. Pain signals may be recorded or modified as required, as can mechanical and temperature nerve impulses from other receptors located in the skin. Even psychological variables such as emotionality, vigilance, and mental workload may be directly monitored by measuring ANS activity in sympathetic efferent fibers outside of the brain.

The most complex and difficult challenge in neural macrosensing will be optic nerve taps. The retina is thoroughly vascularized, permitting ready access to both photoreceptor (rod, cone, bipolar and ganglion) and integrator (horizontal, amacrine, and centrifugal bipolar) neurons. However, the optic nerve bundle itself has $\sim 10^6$ tightly bunched individual nerve fibers, a 10-100 MHz signal bandwidth, and significant natural data compression techniques which all must be untangled in real time. Developing algorithms capable of interpreting raw optical nerve traffic, say, to recognize a specific human face or a specific scene in the vision field, would prove a significant research challenge. (Direct monitoring of photo-receptors or retinal membrane potentials and other techniques may simplify untangling of the signal compression.) Rapid visual field identification using artificial neural nets is also a subject of much current research interest. Eyeball rotations (e.g., via resident intradevice nanogyroscopes), eyelid position, pupil aperture, and lens accommodation under ciliary muscle control must be monitored to supplement the vision field analysis.

Other Macrosensing

We have only scratched the surface of the potential for nanorobot macrosensing. Medical nanodevices can quantitatively monitor variables not normally accessible to human consciousness, such as:

1. Hormone and neurotransmitter levels;
2. Gastroelectric oscillations and skin conductivity;
3. Pupil dilations;
4. Drug and alcohol (and breakdown product) concentrations in the blood;
5. Internal organ damage or malfunction;
6. Digital real-time performance data on vital organs or limbs (e.g., continuous kidney volume throughput, pancreatic insulin output, cholesterol metabolism in the liver, or lactic acid production in specific muscles during exercise); and
7. Continuous mapping of thermoregulatory isotherms from which the temperature and heat capacity of the external medium (e.g., swimming pool water, cool night air) and the presence of sunlight or shade on the skin (via dermal thermal differentials) may be inferred.

Various chemical substances cyclically increase and decrease in serum concentration with time of day, permitting crude temporal macrosensing. Other substances (e.g., aldosterone) significantly vary in serum concentration depending upon whether the patient is supine or standing, thus permitting limited biochemical postural macrosensing.

If nanorobots can leave and re-enter the body, then macrosensing may include direct sampling of the external environment upon demand and the possibilities for remote data acquisition become virtually limitless.

8

NANOSURGERY

One of the earliest and most rudimentary applications for nanotechnology may well be in what many might consider a frivolous cause—the alteration and enhancement of human appearance, otherwise known as the *beauty business*. Why? Because the technology can be relatively simple, there is a great demonstrated demand, and vast sums of money can be made. Even with the crude techniques available today, the *cosmetics industry* is thriving and lucrative. Once the means are at hand to actually perform safely, painlessly, and inexpensively the miracles that advertisements now promise, the race will be on.

MARKET POTENTIAL

How big the beauty business? That depends on how define it, but a few representative figures will provide a rough idea. According to various sources, the worldwide gross volume for toiletries in 1990 was in the range $14 to $18 billion. That's just for traditional personal hygine products such as powers, sprays, perfumes, and deodorants.

The appearance enhancement market goes far beyond that. The diet industry is said to gross $33 billion annually. A recent survey by *Glamour* magazine found representative young American women spending from $550 to $400 a year on their looks, including such items as contact lenses, gym sessions, and bicycles for exercise. And, according to a recent *Wall Street Journal* article, the total spent for breast implants, hair transplants, facelifts, tummy tucks, and other surgical repair operations comes to about one billion dollars a year, most of it for women but a growing percentage (about 30 per cent by some estimates) for men.

These expenditures are for goods and services that, by and large, perform far less miraculously than their advertisements would lead one to believe. Indeed, beauty-related goods and services are named in ways that suggest a level of performance unattainable in practice, such as "permanent" waves that last a few months at most.

Far more pervasive, and insidious, are adds that use *gorgeous models* to imply—without really saying—that using the product will make you look beautiful too. Surely no one consciously believes such advertising, but the implication is there. The subconscious naively accepts and reacts.

No one knows what people would pay for products that really did transform one's appearance. It's reasonable to suppose, however, that it would at least equal and perhaps far surpass what they now pay for cover-ups and nostrums.

And this is not the full extent of the market for *cosmetic nanosurgery*. Once this technology begins to mature, it become possible to address beauty needs that previous generations simply bore without hope or recourse. Typical examples might include completely scarless facial surgery, rebuilding poorly set bones to normal contours, and repairing nerve damage that affects facial muscles.

Applications of Cosmetic Nanosurgery

Hair Colour

To get an idea of cosmetic nanosurgery's potential, consider a hair-colour product as a simple example. This might take the form of a rudimentary, *nanomachine*—little more than a carefully designed molecule—that simply circulates in the bloodstream until it finds itself in a hair follicle in the scale. Then it inserts itself into one of the follicle's melanocytes, the cells deep within the hair-root bulb that colour the hair, where it benignly but persistently but regulates the production of delivery of the pigment melanin.

Hair colour, except in the case of *gray hair*, depends less on the amount of melanin than on its distribution. *Red hair* has regularly spaced melanin granules, whereas in brown hair the melanin granules are irregularly spaced. *Black hair* contains large melanin granules, an in *blond hair* they are relatively small.

While the effect of a single melanocyte changing to a different mode of melanin production would be detectable only with a microscope, a small syringe could easily contain more such nanomachines than there are melanocytes on your whole scalp. As a result, a brunette could bleach her hair once and never again have to worry about dark roots growing out. She would, in effect, become a blond from the inside out, requiring only an occasional booster shot to replace any nanomachines that get trapped accidentally in the hair shaft and dragged, out of the follicle.

Further, just as in normally *blond hair*, no two strands of hair would have exactly the same shade. (That's how you can tell a bleach job from the real thing). The colour of each hair shaft would depend on how completely the melanocytes had been converted to the new mode, which would naturally vary somewhat from follicle to follicle. And, since hair waviness is also controlled by what happens in the follicle, it might be subject to similar modification with a suitable nanomachine.

Such a hair-colour nanomachine could be rudimentary. It would require no on-board computer, just the ability to recognize first a scalp follicle and then a melanocyte. It could do both by binding temporarily to an appropriate surface protein on the target cell. Then it would simply bind to some structure within the melanocyte in such a way as to influence melanin formation. There would be no need for the self-replication machinery an assembler would require; in fact, self-replication would be a real disadvantage in this and many other applications.

Some sort of self-replicating machinery would be necessary to produce these *rudimentary nanomachines* in useful numbers, of course. This machinery need not be completely human constructed, however. It should be possible to engineer a number of bacterial strains to produce the various components of these simple nanomachines. These components would then self-assemble when they were mixed together, as virus components do.

A similar nanomachine could be programmed to have the opposite effect—to restart melanin production for those who wish to restore the hair colour of their youth. It would probably need to be more sophisticated than the nanomachine we just described. There are several causes of gray hair, so it would need to be "*smart*" enough to recognize and evaluate symptoms and prescribe a cure. Alternatively, one might design several different varieties of nanomachine, one for each known cause of gray hair, and inject the proper one to treat a particular patient based on the result of testing.

Suppose someone who has taken this treatment changes his or her mind. Suppose gentlemen begin to prefer brunettes, or fashion suddenly decrees that redheads deserve to have more fun. Would the new blonds be stuck with their choice?

Not at all. It should be a simple matter to design a similar hair-follicle-seeking nanomachine to simply turn off the first one. (It would take a different and more sophisticated nanomachine to turn a natural blonde into a brunette.). In fact, both nanomachines could then allow themselves to be incorporated into the growing hair shaft and thus eliminate themselves from the body. This flexibility would also be desirable from a commercial standpoint, giving changing fashions, a chance to provide a continuing market.

It has been suggested that gene therapy, already in clinical trials for certain rare diseases, could probably address thee same needs and be equally reversible. Gene therapy is also likely to be available long before functional nanomachines appear. However, gene therapy will have major political obstacles and formidable buyer resistance to overcome before it becomes accepted for taking care of anything other than life-threatening diseases.

Skin Colour

A similar set of nanomachines, perhaps even simpler because they would have no need to seek but *hair follicles*, would allow people to control their skin colour. These machines could be made to act locally—to deal with *birthmarks*, *liver spots*, and *freckles* for example—or allowed to simply enter whatever melanocyte they encounter anywhere on the body. People could get whatever skin shade—or eye colour—they choose, lighter or darker. The social consequences of such innovations are perhaps best left to the imagination. There can be little doubt, however, as to its market potential.

Baldness

Before leaving the subject of hair follicles, it seems evident that there would be stupendous market for a nanomachine that would restart hair growth in follicles that had turned off. There is probably no branch of the cosmetic industry that has garnered more business with less benefit to the customer than the potions that promise hair restoration. A product that actually worked would be an instant sensation.

Lest anyone argue that hair restoration is an impossible dream, it might be well to mention that not all hair loss is permanent, even now. Chemotherapy and various medical conditions can cause complete hair loss, which reverses itself when the therapy is discontinued or the condition is corrected. And it seems intuitively reasonable that a nanomachine working from inside the hair follicle, right where the action is, should have a better chance of success than any amount of medication trying to fight its way in from outside the scalp.

One problem in developing any such product would be to guarantee that it would function only where one wants it to. No one wants hair sprouting from their forehead or eyelids, for example. (Yes there are hair follicles on your eyelids. They just produce very short and nearly invisible hairs.) Similarly, women with thinning hair (it happens) would not be interested in a hair restorer that also grows beards. The product would have to include strong safeguards against indiscriminate action.

To accomplish this, nanomachines would need to have some way of "*knowing*" where they are in the body. However, this need not be nearly as complicated as might be imagined. One simple way might be with subcutaneous injections of a colourless chemical marker—an invisible tattoo—that would gradually seep into the surrounding tissues. This would set up a localized concentration gradient that the nanomachine could recognize as its start-up signal. Once the nanomachine had become attached to a particular follicle it would stay there, indifferent to the gradual dissipation and eventual disappearance of the chemical marker.

Unwanted Hair

Unwanted hair is an ongoing problem for many men and women: hardly anyone really enjoys shaving. A nanomachine that would enter hair follicle cells and stop unwanted hair production once and for all would be welcomed with *enthusiasm*.

An alternative and possibly simpler way to avoid the need to shave, although it might involve much more mature technology, would be with a one-time external application of a *nanomachine-laden depilatory cream*. In this case, the machines would attach themselves permanently to the stubble ends and busily convert the protein molecules of each hair stub (almost all keratin) into harmless, odorless gases such as methane, nitrogen, water vapour, and carbon dioxide, which would unobtrusively dissipate. It should be possible to get them to digest each hair stub faster than it grows out, eliminating shaving forever.

Sulfur, which constitutes four percent to six percent of hair by weight, is not really a troublemaker, even though all of its volatile compounds are smelly (mercaptans, sulfur dioxide, hydrogen sulfide) and some are skin irritants or toxins. Elemental sulfur, on the other hand, is relatively inert. It should be safe to let it accumulate in little crystals until the next washing removed it. The only potentially troublesome nonvolatile constituent of hair protein appears to be phosphorus, which is present in very minor amounts. *Phosphorus* itself can ignite on contact with air, producing skin-irritating oxides and acids. However, it is unnecessary to carry the breakdown of the hair protein that far. It should be simple to program the nanomachine to leave the phosphorus in a harmless water-soluble organic compound that would wash away in the next rinse.

As a further refinement, these nanomachines could be made to depend on a continuous supply of atmospheric oxygen. For example, the energy to do their work could come from flameless oxidation-reduction reactions the nanomachines would be promoting. This might provide a way to regulate their activity; whenever the supply of oxygen diminished, the nanomachines would slow down and wait for it to pick up again. This would be likely to happen whenever the end of the hair stub sank far below the skin surface and became covered with sebum (the natural hair lubricant secreted into each follicle by its sebaceous gland). In any case, some such regulator might be advisable to assure that no overenthusiastic nanomachines could work their way deep into a follicle and somehow damage it.

For this scheme to work, it would be important to give the nanomachines the ability to distinguish between hair keratin and the skin keratin, and keep them from harming the latter. One way to do this might e to give each nanomachine at least five other arms (or attachment sites) in addition to the one for manipulating keratin structure, and require them to link up in a way that would be easy on hair an hard on skin. For instance, we might make the manipulator arm inoperative unless the other arms were linked to two adjacent *cystine molecules*—the predominant amino acid component of keratin—perhaps by recognizing cystine's characteristic disulfide group, and also to at least three other nanomachines. Hair is almost all keratin, whereas skin keratin is mixed in with other components, making it hard for the nanomachines to fulfill the attachment criteria on skin.

Nanomachines that failed to attach themselves in this way would be neutralized with a follow-up rinse. Those already attached to keratin, either on the skin or on a hair stub, would thereby become immune to the rinse.

Any nanomachines that did attach themselves to skin keratin would find slow going in comparison with the others on the hair stubs. They would keep running into nonkeratin components that they would be unequipped to disassemble. Long before they had made much headway, the skin particle they were on would flake off. Once such a nanomachine ran would in all probability come apart before it could attach to another source of keratin.

At the same time, this multiple-attachment requirement would make it possible for each correctly attached nanomachine to stay stuck to the hair shaft even as it cuts away the ground under its own "feet." Whenever one of its attachment sites to the hair shaft became undermined, it would just grope around with the free arm until it found a fresh handhold and then keep digging on the other side.

One obvious precaution would be to prevent this nanomachine from attacking the keratin in fingernails, which appears to be chemically indistinguishable from that in hair. The easy way to do this would be to wear gloves, but that still leaves open the possibility of damage through carelessness. It would undoubtedly be safer to provide a positive deterrent.

The solution might be to make the nanomachines susceptible to attack by some simple solvent such as acetone, a common nailpolish-remover ingredient and something not normally found on the face. This would also make it possible to readjust the boundary of the area within which the nanomachines can act if one wants longer sideburns or decides to grow a mustache.

Permanent Breath Freshener

A similar topical application would address another personal hygiene concern: bad breath originating in the mouth. As most of us are aware, halitosis frequently results from decaying food particles that perfunctory brushing has failed to dislodge. A useful preparation would bond itself firmly to the tooth enamel and prevent anything else from sticking. It would also actively remove anything that became mechanically wedged.

In this case one would start with a thorough professional cleaning to remove any *accumulated plaque* and *tartar*. The dental technician would then paint on a solution teeming with nanomachines that cling tenaciously to every bit of exposed tooth enamel and link up to form a complete, invisible surface film around each tooth. Each of these nanomachines would then go to work disassembling any foreign matter that came in contact with the tooth into harmless and bacteriostatic gases or liquids, which would quickly be removed by saliva.

The same sort of idea could have many other applications such as always-clean windows, self-cleansing dishes, the permanently spotless and odor-free bathroom, and a true antifouling paint for boats. The latter would enjoy a multimillion-dollar market all by itself. For these applications the nanomachines could have much simpler attachment and attack criteria.

To return to the *fresh-breath nanomachine*, such a system would obviously need to be "*smart*" enough to recognize gum and tongue tissue and leave them alone. It might also need to be periodically reapplied, as normal chewing and brushing would tend to remove it from exposed surfaces. However, the protection would still be there on the hard-to-reach surfaces, and that's where it's most needed in any case. As an added bonus, the nanomachines could be programmed to construct oil of peppermint (or whatever aromatic essence might be desired) out of the organic molecules they have been dismantling.

Although such a rudimentary nanomachine system would certainly improve mouth cleanliness and go a long way to prevent cavities and tartar buildup, it would not by itself be enough to prevent periodontal (gum) disease if it were used as an excuse to stop gum massaging and flossing. That advance will probably have to await the development of fully capable cell-repair nanomachines, which would conquer all diseases including gum disease. Until then we will have to keep supplying manually the exercise our gums need to keep them vigorous and healthy, a discipline that is necessary because our usual bland diets of refined foods fail to provide the stimulation. One hopes that the promoters of a breath-freshener system will market it as a between -brushing safeguard instead of a dental hygiene cure-all.

Wrinkle Repair

As cosmetic nanosurgery technology matures, it should become possible to undertake more ambitious transformations such as skin rejuvenation. They saying, "*Beauty is only skin deep*," simply serves to emphasize how very important the skin is.

The skin is a very complex and little appreciated organ, exposed to a wide range of environmental assaults that would simply destroy any of our other vital organs. It resists extremes of heat and cold, dryness and abrasion, biologic, organic, and inorganic poisons, invasive organisms, the force of gravity, and continual bending and flexing. No wonder it begins to look somewhat worn after a few decades!

Total and spontaneous skin renewal will probably have to await the arrival of fully capable cell-repair nanomachines. These would be able to analyze whatever ails a particular cell or tissue component and restore it to its pristine condition. Along the road to that ideal, however, there will surely be jobs that more modest nanomachines can fill.

One of these will probably be the removal of wrinkles. *Wrinkles* arise from many causes, and it is unlikely that any single remedy will cure them all. It should be feasible to design special nanomachines to attack specific symptoms, however, and it would certainly be possible to employ a number of different kinds of nanomachines simultaneously.

Most wrinkles occur in response to *persistent folding*, as with crows feet around the eyes and smile lines, and the deep nasolabial folds that develop on either side of the mouth. Part of the problem is that the skin has lost elasticity, although there may be any number of contributing factors.

A start at treating one of the basic symptoms would be a nanomachine designed to improve the ski's elasticity, or at least to prevent further deterioration. What gives youthful skin the ability to adapt gracefully to the body s contours and movements while remaining tough enough to resist the wear and tar of everyday life is its supply of collagen fibers. Old skin is softer and floppier than young skin; it has less collagen, either because its cells make less collagen than they used to or because they produce too much collagenase, an enzyme that destroys collagen fibers. Most likely, both processes are at work.

To attack this problem, a nanomachine could be designed to seek out tne cells that produce collagenase and slow or stop them. Another and possibly simpler approach would be a nanomachine that attaches itself to collagenase in such a way as to jam its collagen-cutting mechanism. Either way, it would be necessary to ensure that the nanomachines would act only in the skin, where we want them to, and to monitor the process carefully to keep it under control.

One simple way to keep the nanomachines from attacking *collagenase* anywhere but in the skin would be to make them temperature sensitive. They could remain inert and circulating in the bloodstream until they encountered a high temperature, say 110°F (easily obtained locally at the surface with a hot pack). Then they would migrate out of whatever capillary they were in and go to work.

As with the previous products, a *feature* of such a nanomachine would be its inability to reproduce. This would permit control of the action by limiting the dose. But whereas for other applications the degree of change was a matter of vanity, in this case it would be safeguard against removing so much collagenase as to make the skin leathery. Nonreproducing nanomachines also permit an incremental approach to the treatment. Instead of trying to prescribe the exact dose needed, the therapist could schedule a series of session in which no single application would contain enough of the product to do any harm.

Slender Now

A final example deals with what appears to be the number one beauty concern of American women: obesity. Here we venture beyond mere surface appearance and into bodily restructuring. And it must be admitted at the outset that not all kinds of obesity will yield to nanosurgical control. It appears that some obesity is genetic, some psychological, and some imaginary.

That said, however, there is still a vast market for a treatment that would short-circuit the ceaseless roller-coaster ride of conventional diet-splurge cycles. The ideal weight control product would allow a person to eat more or less what one wanted, continue to live a sedentary life, and maintain a figure that reasonably approximates one's ideal proportions. It should also be capable of spot application, to reduce a person's hips or abdomen for example, or wherever the problem appears to be.

At the same time, the product should induce a feeling of well-being rewarding the user with positive sensations to make it easy to stick with the program and keep using the product. Just how that might be accomplished I leave as an exercise for the reader. As a hint, consider the role of endorphins, those natural mood-altering chemicals that your body is constantly ready to produce.

One possible approach would enlist the aid of the body's own immune system. A nanomachine could be designed to home in on a fat cell, plant a marker on its surface that falsely identifies it as a pathogen, and let the nearest phagocyte take over. An easy way to accomplish this would be to instruct the nanomachine to fabricate the marker on the spot by altering a surface protein so that the cell is no longer recognizable as "part of the family."

To make sure that such a product works only where needed (a person needs some fat storage capability, after all), it would probably be wise to make this nanomachine temperature sensitive too. Surplus fat generally lies just under the skin, and it is notoriously good insulator, so it is easy to produce local temperature gradients across it. Any surplus nanomachines could just keep circulating until they were either eliminated naturally or put to use at some new problem elsewhere.

Techniques and Strategies

Any protein product designed to circulate through the body and perform some specific task on a certain class of cells must fulfill several criteria. It must first of all conform to a rigid set of specifications imposed by the body's immune system. It must also have some way of recognizing the target cells and, in some cases, some way of "knowing" when it has arrived at an appropriate location in the body.

Foiling the Immune System

The need to get past the immune system's detectors is a daunting requirement in view of that system's perpetual vigilance and almost devilish versatility. As we have learned more about the system's functioning, however, small chinks have begun to appear in its armor. Presumably these chinks will widen with further research, of which we can expect a considerable amount in the ten to twenty years that are likely to elapse before the first nanosurgical products appear.

However, there is one surefire way to protect such a product from the immune system, even in our present state of ignorance: package it is something that your immune system instantly recognizes as completely familiar. And what could be more familiar than red blood corpuscles of your own blood type? After all, routine blood transfusions trigger no immune response.

A case might be made for using white blood cells, since these are naturally able to leave the bloodstream, crawl through tissue, and sense their environment. It might appear that hitchhiking nanomachines could simply wait, reading the macrophage's sensory input until it finds itself at its intended destination, and then slip out to do its work.

There are at least three major drawbacks to this scheme. In the first place, blood contains about 600 red cells for every white cell. Getting enough white cells to do the job would involve either taking and separating a large volume of blood or cultivating large numbers of white cells outside the body. Second, it seems quite unlikely that information about the white blood cells' environment travels to some central organ analogous to a brain, where a hitchhiking nanomachine could read it. More than likely, the cell's sensory equipment is confined to the tips of its probing tendrils and is designed to be simply oblivious to contact with "self" materials of whatever kind. In other words it reacts only when it senses something it recognizes as foreign, and then it would react in exactly the same way, wherever it happens to be in the body. Finally, it may be possible to slip a nanomachine past a macrophage's defenses and into the cell at the outset, say by inactivating the white cell with cold. It is hard, however, to imagine how an emerging nanomachine could escape instant destruction by the fully active macrophage it had been ridding. It would be much like the case of the smiling young woman from Niger, who went for a ride on a tiger.

On the other hand, slipping nanomachines into red blood cells would be easy and completely harmless, assuming the product is correctly designed to be nontoxic and to penetrate the cell membrane without damaging the cell, as viruses do now. A sample of blood of the correct blood type would be separated into its components by centrifuge or otherwise. With no white cells to interfere, the nanomachines could infiltrate the red blood cells upon contact. A rinse to remove stragglers that failed to make contact would probably be in order before injection.

Packaging the nanomachines inside red blood cells gets them past the immune system, but it isolates them from many of the clues that might be used to activate them. Any attempt to poke a molecular fragment complex enough to act as a sensor out through the cell wall would invite attack by the immune system. What to do?

One possible approach would be to take advantage of the red blood cell's internal environment. All of the nanomachines we have considered are supposed to do their work at or near the body's surface. And that is where blood temperature is lowest and where the hemoglobin in red blood cell's becomes most oxygen-depleted. The nanomachines should be able to recognize these environmental clues while still inside the red blood cells.

Once the nanomachines get the "jump signal" that tells them they are in their "*drop zone*," they emerge from the red blood cell and begin to search for their programmed target cell, which they can recognize by its surface proteins. Some nanomachines may fall prey to wandering phagocytes, but most of them, having been released close to their targets, should succeed. Those that do will be safe from phagocytes as soon as they slip inside their target cells.

Sensory Equipment

So far we have postulated nanomachine molecules able to act on environmental cues such as temperature, oxygen depletion, and chemical concentration. The question is, how do we make molecular machines with these sensitivities? The answer is, with sensing mechanisms much like those found in natural molecules.

By now we have discovered many examples of proteins (enzymes) that bind to a certain shape of molecule, often with exquisite selectivity. The common feature of all these examples is a receptor site on one molecule that closely matches a shape on the other molecule. In most such instances, once the two molecules have come together they tend to stick. In other words, it takes only one molecule to fill the site and trigger a given reaction.

To make a sensor that responds only above a certain concentration of the target chemical, we need something more. We need to give the nanomachine two weakly binding receptors linked in such a way that both must be occupied before the nanomachine responds. That way nothing happens until the concentration is high enough to provide a reasonable chance that the second site will be filled before the molecule occupying the first one shakes loose. It would be possible to control the concentration required to trigger the device by adjusting the binding energy, and thereby the residence time, in the receptor sites.

For a sensor to detect above-normal temperatures, we need something similar but opposite. In this case we start with the active site already filled with a target molecule in such a way as to jam the nanomachine's mechanism. However, we design the attraction between the two so that they break apart under molecular bombardment above a certain temperature, freeing the nanomachine.

For a below-normal temperature sensor, we start with a nanomachine consisting of target and receptor molecules already connected by a flexible hinge, but with enough mismatch between their shapes that they are only weakly attracted at normal temperatures. As the temperature decreases, the two halves of this hinged ensemble spend more and more time in contact. Eventually, they stay together long enough for a slow-acting latch to function, holding them together permanently.

When a predetermined sequence of triggers fires, the nanomachine is activated to perform its intended function. We can envision the activated nanomachine unfolding itself into the required shape like one of those "transformer" toys, an obedient robot ready to go into action.

Intercommunications

For most of these applications, each nanomachine would be on its own, carrying out its instructions on atoms and molecules within reach of its manipulator arm or within a single target cell. For control over hair waviness (which requires that the nanomachines affect the hair's cross-sectional shape), however, hundreds of nanomachines in a particular follicle must work in concert. This means that they will have to communicate, at least in a rudimentary way.

Since such traits as hair colour and hair waviness are clearly inherited, it seems reasonable that the eventual channel for this communication would be genetic—by manipulating a person's DNA. But long before we have nanomachines capable of such fine control, we should be able to find ways to coordinate less-sophisticated groups of them to override the genetic instructions.

One such crude signaling method for hair waviness could be by means of a chemical gradient. The first nanomachine to gain a foothold in a particular follicle would stimulate its cell to generate a chemical marker, thereby staking its claim as the controlling nanomachine. This chemical marker would inhibit each subsequent nanomachine from emitting the same signal, thereby establishing a chemical gradient from one side of the hair follicle to the other. From this, each nanomachine could estimate its position with respect to the "master" nanomachine and modify its cell's hair-making activity accordingly.

Licensing

It should be clear that none of the techniques so far described, with the possible exception of the depilatory and tooth-protector nanomachines, could in any way be marketed over the counter. Each of the injectable products involves at least some technical sophistication and probably would need to be administrated by medical professionals. What will be the legal ramifications?

In the first place, as injectable pharmaceuticals, each of these products would come under the licensing authority of the Food and Drug Administration. (I think we an assume the continued jurisdiction of the FDA. This agency has, on the whole, done yeoman's service in protecting the American public from hazardous nostrums, despite recent criticism of its slowness in approving untried methods of controlling AIDS. And even totally useless government agencies have amazing powers of self-preservation.)

FDA approval is a long costly process, involving extensive testing for both safety and efficacy. This process would be undertaken primarily by drug companies, not cosmetic manufacturers, since externally applied cosmetics need to be certified only for safety.

At present, the preliminary testing of any such new technology would be done on animals, although there are many opposed to this practice. And there are valid arguments against animal tests in this case. It would not be easy to prove, for example, that a nanomachine designed to work in mice or rabbits would necessarily perform properly in humans.

At the same time, the idea of testing a brand new technology on humans is far from attractive and involves serious ethical questions. A possible way out may be to develop surrogate human tissues on which to perform these preliminary tests. Tissue culture methods so far developed, however, work only on single cell types. They are incapable, without considerable development, of producing the complex structures of even such a relatively simple organ as skin.

We would have no problem, of course, if we had self-reproducing cell-repair nanomachines to work with. They could monitor the internal workings of each cell in a growing population, reproducing in synchrony with the cell they inhabit, and guide the development of its descendants along predetermined pathways to produce any desired tissue component. Just for the testing of cosmetics alone, such technology could grow acres of human skin, complete with nerve endings, capillaries, hair follicles, sweat glands, and even subcutaneous musculature and fat cells—thriving under glass in racks of oversized Petri dishes in a laboratory. (It would also be important to cultivate mucous membranes, since these are much more sensitive to many irritants than the outer skin.) Incidentally, this system could also be used to grow all sorts of replacement organs for transplantation, relieving the chronic shortage. But by the time such capabilities matured, organ transplants are likely to be unnecessary except for the treatment of severe, life-threatening injuries. In-body repair, using injected nanosurgical nanomachines, would be far preferable in every way whenever there was a time for it to work, and it could be used to treat conditions we now tolerate until they become emergencies.

Long before cell-repair nanomachines become available however, a much simpler solution will have presented itself. Recent experiments show great progress in grafting human tissue onto animals without provoking an immune reaction. Carried just a bit further, this xenograft technique should enable us to produce living surrogate organisms whose tissues mimic those of humans closely enough for reliable test purposes but with just enough of a nervous system to control such vital function as respiration, digestion and circulation. Naturally, they would have no sensory equipment that would allow them to feel pain. It's hard to see how animal-right activists could object to laboratory tests on living nonanimals such as a pillow-shaped object covered with human skin.

ECONOMICS

What will be the likely response of the cosmetic industry? Probably not much—at first. As usual, the first feeble attempts will be greeted with derision and ascribed to the work of crackpots and charlatans. As the procedures became more mature and routine, however, and more affordable, the established companies might be forced to choose between opposing the new technology or buying in to it.

Some of the largest of them will probably hedge their bets by doing both. After all, Chevrolet completes as fiercely against Pontiac as it does against Ford. It will be interesting to watch the media battles that will predictably ensue.

DISTANT PROSPECTS

I have tried so far to touch on only the easily foreseeable and immediate consequences of the simplest nanosurgical nanomachines, but it is also worthwhile to sketch more distant possibilities. Once fully capable cell-repair nanomachines hit the market, there will be little the imagination can conjure up that would be impossible. Permanent rejuvenation of face and body is only the most obvious result.

Nonsurgical Nanomachine-based Cosmetics

In the end, however, it is unlikely that cosmetic nanosurgery will completely supplant conventional cosmetics. After all, there is something to be said for being able to adjust one's makeup to one's mood or costume, and to change back and forth between different looks quickly. And it should be noted that the primary consumers of beauty preparations today are not the elderly, who might presumably need them most, but the already young and beautiful.

Instead, it is likely that nanomachine techniques will take over the task of manufacturing cosmetic, a job for which programmable, self-reproducing nanomachines are ideally suited. The ability to manipulate individual atoms and promote or inhibit specific reactions could lead to all sorts of new fragrances and materials tailored to the cosmetic industry's needs.

Some cosmetics, in fact, might consist of nanomachines. Who could resist a fingernail polish or eye shadow, for example, made of units that automatically aligned and spaced themselves to produce a diffraction grating? Rainbow colours would flash from such a product, depending on the lighting and on the angle from which it was viewed. The colours might also be made to change with skin temperature or other environmental cues.

Total Makeover

Eventually, under the care of a competent practitioner backed up by talented molecular programmers and a responsive pharmaceutical industry, it should become possible to mold the face and body to whatever shape might be desired. Each person who cared to could achieve his or her own ideal of physical perfection or, for that matter, whatever frightening or gruesome effect they wanted.

Inevitably, there will be people who don't know how to leave well enough alone. Many who never liked their own youthful appearance will opt instead to copy some popular model or other sex symbol. It could become very confusing, with dozens of pop-idol look-alikes crowding the parks and boulevards of our future metropolis. Some may not relish the prospect, but we may never see the last of the Elvis clones.

9

AGE OF NANOTECHNOLOGY

Everyday life in the *age of nanotechnology* could be wildly different from what it is now, but it doesn't have to be. In the early twentieth century, people were anxious to exchange outhouses for indoor plumbing, and oil lamps and candles for electric lights. Home-making drudgery was ameliorated by everything from stoves you didn't have to build a wood fire in to washing machines. Since then we got TVs, microwaves, and Roomba the robot vacuum cleaner.

In fact, modern middle-class family life is probably too easy in a purely physical sense. We don't get the exercise our bodies need for optimal health. What use is it to add more laboursaving devices, if it only means you have to spend more time on an exercise device?

One of the classic blunders of prediction in early science fiction was the notion of the food pill. It would be a major time saver avoiding all the folderol about cooking and serving and cleaning up—or even having to build dining rooms in the first place. Just take three square pills a day.

The problem, of course, is that most people like to eat. It's a social occasion, a significant chunk of why people enjoy life. (Early science fiction was fairly puritanical, or it might have gone on to suggest that babies should be made in factories so that people would not have to bother with sex.) Even people with busy schedules and little time to spare welcome mealtimes as an excuse to take a break.

Some people like cooking, but quite a lot of the food eaten today is processed and packaged for convenience. Many who do cook do so for economic reasons or because fresh, home-cooked food tastes better and is more nutritious than processed stuff.

Nanotech synthesizers in factories, then in restaurants, and finally in homes, will improve the situation. Modern processed foods suffer from two constraints that reduce quality. First, they are generally processed in central locations and then shipped and stored (and the same is true of a lot of produce). This means that what you eat is not fresh and has been optimized for handling and storage: picked green, inundated with preservatives, bred (or genetically engineered) for shelf life instead of flavour.

The second constraint is simply cost. The reason so much of food is sugar and starch these days is that sugar and starch are cheap compared to fats and protein. (And often instead of natural fats you'll get the cheaper, but dietetically harmful, trans fats.) A surprising percentage of what goes into processed foods is right out of a chemical reactor.

Molecular synthesis will be able to make foods that are considerably closer to natural ones than current processed products. You can eat meats no animals were killed to obtain and crops no wildlife habitat was displaced to grow. No release of genetically modified organisms into the environment will be necessary to include as much of whatever vitamins and nutrients you need into whatever you like to eat. And the foods will be synthesized fresh just before being cooked or eaten, with no need even for a refrigerator. Indeed they could simply be synthesized cooked, with no need for a stove.

The dishes could be synthesized along with the food, and then simply dropped dirty into the recycler. Of course, with nanotechnology you could make solid diamond plates that would come clean with the merest wave under the faucet, but the long-term trends are clear: manufactured items get cheaper, and space gets more expensive (on Earth, anyway). So the typical domestic arrangements will tend toward making things when and where they are needed, and recycling rather than storage.

Clothing

His grey flannel suit fitted him with a staggering perfection the whiteness of his shirt was dazzling, his tie shamed the rainbow. His soft felt hat appeared to he having its first outing since it left Bond Street. His chamois gloves were clearly being shown to the world for the first time. On his left wrist was a gold watch, and he carried a gold-mounted ebony walking-stick.

Everything, you understand, quietly but unmistakably of the very best, and worn with that unique air of careless elegance which others might attempt to emulate, but which only the Saint could achieve in all its glory.

What about clothing? The near-term products from current research include nanofibers, such as buckytubes, that are very strong. Strength of the cloth isn't enormously important in clothing, but the feel is.

Silk is an extremely strong material. It is as strong as fancy aircraft aluminum alloy, five times as strong as ordinary aluminum of the same size—or thirteen times as strong by weight. It comes in monofilaments about 5 microns in diameter, one-twentieth the width and one-four hundredth the area of a human hair. One fiber, completely invisible to the human eye, could support more than a gram (a large pill, for example).

What this strength does for clothing is that a silk thread can be quite strong enough for wearability and yet still be extremely fine. This means it can be extremely light and flexible, which are the properties of silk cloth we like. (The smoothness of feel, for example, comes from the fact that the silk threads are small and close together.) The strongest fibers we know about in theory at present are buckytube fibers, which are about 150 times as strong as silk. That means that a buckytube garment, strong enough for all normal wear and tear and to hold all your junk in its pockets, would be so diaphanous that you'd have a hard time knowing whether you had it on or not.

Clothing is one of the very few really old human technologies. For almost all of our history, we wore clothes made from animal skins and woven plant fibers. Only in the twentieth century did that change, and then only to include synthetic fibers. Most of these still do not feel as good as, say, cotton, although they are getting better fast. (It may even be the case that we have evolved to like the feel of woven plant fiber on our skins—we have been wearing clothes that long!)

The big difference with clothing will be that your closet is a synthesizer, not a storage bin. You'll have a "magic mirror" that shows what you'll look like in whatever you are thinking of wearing. Most likely you'll be able to talk to it to select and modify the style or fit. Then it's synthesized, you put it on, wear it, and toss it into the recycler. No laundry.

The major functions of clothes are for protection from the environment, modesty, and social display. Technology, nano or otherwise, has only an indirect effect on social display. First, the clothes you wear serve to identify you with a class or group, a uniform formal or informal: T-shirts and sandals for programmers, suits and ties for businessmen, clerical collars for the First Estate. Clothing makes a statement of personal style as well as a group identification. As with many matters of style, the logic seems sometimes to run backward: a business suit has its value because of what it prevents you from doing. It thus advertises that you are a person who does not need to do manual labour. (This advertisement was particularly pronounced in upper-class women's fashions of the Victorian era, for example.)

It seems quite likely, then, that there will remain a market for hand-tailored clothing of classical materials for quite some time, and that the more expensive such clothes are, the better status symbols they will make. Especially since it will be easy enough to wear a nearly unnoticeable nanosuit underneath and obtain all the benefits anyway.

Nanosuit

Imagine a fabric just 5 microns thick, the thickness of that invisible filament of silk. It covers your whole body, like an extra layer of skin. (There is a separate hole for each individual hair.) With about two square meters of skin to cover, it is the same weight as two teaspoons of water. To get an idea of how thin it is, take one drop of water and spread it over your hands as if washing them. To get an idea how heavy the suit will feel to wear, swallow two teaspoons of water and walk around naked.

The *fabric*, however, is not woven of threads. It is a machine. To get an idea how much machinery could be in it, look under the hood of your car. You will see wires and tubes of about a quarter inch diameter, some pipes one or two inches in diameter, pulleys half an inch wide, shafts the same width, subassemblies in the six-inch range. Now go to a Wal-Mart or similar giant warehouse store. Imagine that the ceiling is eight stories high, and that the building itself covers the entire continent of Asia (or North America, twice). Now pack it full of machinery like the engine compartment of your car—that's your nanosuit, scaled up by a factor of five million.

The major function of the *suit* is the same as the animal skins our ancestors wore tens of thousands of years ago—to keep warm. With the *nanosuit*, however, there are a few advances. To make the fabric flexible yet strong, it contains two or more layers of plates that can slide past each other, like scales on a fish. Each plate is hollow and contains vacuum, and has a reflective coating to prevent radiative heat transfer. Among the machinery are plenty of reversible heat pumps. When the environment is too hot, they pump the heat away from your body (requiring an energy source); if it's too cold, they act as heat engines, picking up energy from the flow of heat from your body to the outside. (We note that most environments on Earth where the outside temperature is hotter than the human body are blessed with strong sunlight, which could be used as an energy source.)

You are now set to go play tennis outdoors in Nome, Alaska, in January (assuming you have a lighted court). If you look at a globe you'll begin to see just how much difference such a relatively simple nanotech application could make to the human race. It could easily double the livable land area and substantially ameliorate one of the major energy needs.

If you like to live on mountaintops, or under the sea, extend the nanosuit to provide oxygen. Molecular sieve oxygen concentrators exist today, and using powered machinery designed at the molecular level would make them easily wearable. For gills, you'd need a forced-water feed to the filter, since dissolved oxygen in water is at a much lower concentration than in air. On the other hand, if you cared to carry a power source, a rebreather that reconstituted the oxygen from the CO_2 you exhaled would be relatively straightforward.

Computers and other high-tech gadgets as part of clothing have been in the labs for some time (notably, the wearable computers of MIT'S Media Lab). Although not integrated into clothing, the set of gadgets that a well-dressed businessperson carries has been expanding rapidly in recent decades, past the watch to the pager, cell phone, and Palm Pilot. All these functions could be built into the nanosuit, and many more besides. Although the suit will be less than a tenth the thickness of a typical human hair, you could put the equivalent of 100 billion modern-day PCs into 1% of its volume. Surely unnecessary, but it is clear that the processing power for any and all computing, communication, interface including virtual reality, and like needs would be available.

The suit could be covered with *tunable optical reflectors* and thus be a high-resolution video display, showing whatever the wearer desired. It could even have phased-array optics, that is, a *holographic video* display (there a use for a few billion processors!), and be able, among other things, to render the wearer nearly invisible. On the other hand, the sensory capacities of a suit, if desired, could be just as much advanced over natural human ones and would probably be able to unmask an "invisible" suit fairly easily. In practice, such capabilities would probably be used mostly for expressions of personal style, including full colour and motion tattoos, projection of the appearance of holes through one's body, presentation of oneself as a skeleton or a member of the opposite sex, or other personal idiosyncracies.

Little House on the Prairie

Although marks of wear and soiling may disappear in the self-cleaning and self-repairing structures of the future, perhaps their value as mementos will be replaced by video recordings, already common, and similar improvements as time goes by. Indeed it would not be difficult for a house to record everything that transpired inside. At molecular storage densities, a gigabit signal for a century could be stored in about 3 cubic millimeters, the volume of a toenail clipping.

The house itself will probably be a lot more configurable than the static structures we live in today. Objects, including furniture, will be cheap, and space will be at a premium. Alternatively, and for the same reason, houses will he put in more and more remote areas and will be responsible for the well-being of people under extreme conditions. If a vanishingly thin skinsuit can keep people warm in the Arctic, the walls of a building can easily do so. Warming and insulating houses from the cold, and cooling them in warm climates, is widespread current technology; nanotech will just make it cheaper, more effective, and less obtrusive.

Power in remote areas can be had from *solar collectors*. Sunlight is too diffuse for economical collection with current technology, but with *nanotechnology* you could do something like replace the grass on your lawn with a carpet that looked and felt like grass, never needed mowing, and acted as a solar energy collector. This would give you enough area to pick up a decent amount of power without having to build giant structures all over the place. The other current problem with solar power is storage. Nanotechnology should be able to handle that one fairly easily.

Another current technology that makes quite comfortable living is the *houseboat*. Nanotechnology could bring the cost down and improve the reliability enough for houseboats to be a viable alternative to land-based housing. Living on the high seas would be significantly more feasible with synthesizers for most everyday needs. Depending on social and political arrangements and personal preferences, there could be single-family boats and/or cruise ship-sized floating cities.

Nanotechnology could also make living underwater a viable option. Submarine dwellings would be less vulnerable to storms than surface ones, but in good weather, much more energy is available on the surface as sunlight. The ultimate trade-offs remain to be seen.

Topless Towers of Ilium

Christopher Marlowe referred to the towers of *ilium* as topless not because they did not have roofs, but because they were unexcelled in height. But a *nanoengineered tower*, with adamantine structural materials, could be so tall it would literally disappear into the sky: more than ten miles or so up, most of the blue of the sky is below you. To eyes on the ground, it would fade into invisibility; even in the absence of clouds or haze. On a clear night, of course, its lights could is seen all the way up. Towers could easily be 100 kilometers (60 miles) tall; much taller than that and they begin interfering with satellites.

In 1956, Frank Lloyd Wright designed The Illinois, a mile-high tower that would accomodate one hundred thousand people. (It was never built.) Scaled up by a factor of sixty, such a tower would have a footprint of about ten square miles and have more than thirty thousand floors, and thus three hundred thousand square miles of floor space. By scaling up Wright's calculations it should be able to house twenty billion people. That would not work in practice (nor would Wright's actual design), because you cannot simply use apartment building densities for something that has to include city-scale infra structure, services, and transportation. However, you could almost certainly house one hundred million people, maybe up to a billion. Thus you'd need somewhere between five and fifty such towers to house the Earth's current population.

After the first mile or so, the tower would need to be pressurized just as airliners are. I certainly would not want to live in such a tower, but plenty of people live in apartment buildings now; it is a lots better than nothing. It could be made much more comfortable than current apartment living with the space versus objects trade-offs discussed above, and built-in virtual reality. You'd rarely experience the box you were actually living in.

Virtual reality is a term that is used today for various advances in entertainment and user interface technology, but it's all heading for the same basic capability: to make it seem, to all your senses, that you were in a different, often an imaginary, situation than you are.

Today, an Omnimax theater does a good job with sight and sound. Full-fledged virtual reality will extend that in two ways.

As far as your senses are concerned, the other major sense (actually a cluster of related senses) is feel. However, the really hard part will be to allow you to interact with the simulated world, manipulate it, do something not in the script.

Virtual reality can be divided into two categories, depending on where the "world" originates. In pure virtual reality, it is simply made up, in some computer's imagination, like the world of a video game. In *telepresence*, it is a real place—you just don't happen to be there in actuality. You would most likely use the same interface to do both ends; in one case it would be connected to a computer (as would other people who wished to join you in the imaginary place); in the other, it would be connected to a robot that would "be you" in the distant, real, place.

Android Robots

Nanotechnology will make high-fidelity *android robots* affordable (they are close to technically possible now). A major use of such androids will be for telepresence. There will be a wide range of applications, ranging from tourism to business meetings to scientific expeditions. It will be fairly simple to make the remote android reconfigurable to reflect the appearance of the user. It will also be fairly simple to make the appearance of the android not reflect the appearance of the user.

The most unbounded form of virtual reality, including telepresence, will be by nerve taps, where synthetic signals representing the signals your sensory organs would be getting from the imaginary world are injected into the appropriate sensory nerves, and the signals going through your motor nerves are intercepted and used to play your part in the simulation.

The real danger from these possibilities is that the virtual world is so rich and full of possibilities that it may be quite difficult to get people to pay any attention to the real, physical, one.

An Interlude behind Closed Doors

One trend that was very strong in the West during the twentieth century was the loosening of moral strictures on sexual activity. It's often claimed that this is because of the contraceptive pill, hut it is quite a common phenomenon in an increasingly cosmopolitan society. Ancient Greece and Rome are examples.

As with other forms of entertainment, nanotechnology will make possible a wide variety of new sexual modes. Simplest is that nanomedicine is very likely to make contraception even easier than it is today, and for either partner. It will also substantially reduce sexually transmitted diseases. And drugs and treatments that enhance sexual performance and satisfaction are certain to appear. The immediate result could be a 1960s—like sex fad.

Another thing that nanotechnology will bring is android robots. It seems quite likely that toys very difficult to distinguish from human beings will be available, given how much people like to play games with other people. These robots would not only look and feel like humans, but would be intelligent, emotional, and come with completely individual personalities. There will be a strong debate about the morality of using these toys in ways that would be unacceptable with real humans.

Sex

Sex with real people through telepresence will be popular as well; not only will it be possible to do it in places distant from where you happen to be, but in places where, for example, it would be quite uncomfortable for a human to be naked, like a high mountain peak or at the bottom of the ocean.

The most common form of *sex*, though, would most likely be purely virtual. It highly likely that the process of dating will be as accelerated as that of business meetings. Your account on the Internet dating site would come with access to virtual restaurants, parks, entertainments, and bedrooms. Instead of typing in a chatroom, you could engage in anything from a moonlight stroll to an orgy.

Because of the cleverness of interfaces, though, some of the problems of current Internet dating will remain. Certainly there will be a temptation to improve one's appearance through the interface; it will be possible to spend the night with someone and still not be sure what the person really looks like. (On the other hand, nanotech plastic surgery will be popular as well—so maybe most people will really look like they wished they did.)

Pure virtual reality opens up virtually limitless possibilities, for games of all kinds, for assignations with partners real, imaginary, or mixed. The human participants don't even have to experience the same thing; one could be having sex and the other debating politics, if the interface program were sufficiently clever.

10

NANOTECHNOLOGY AND LIFE

Before diving into the nanotechnology of molecular manufacturing, let have an overview of where we are now. This can be divided into two major parts: biology and current lab nanoscience. Then in each case will try to characterize the differences between what's current and what coming.

NANOPARTICLES

As explained in the introduction, much of what you will see labeled as nanotechnology will he simply techniques that are extensions of current manufacturing and materials science, which happen to involve something smaller, shorter, or thinner than 100 nanometers. A classic example, and the most common one found in items on the market, is ultra fine powders.

When you stir sugar into your coffee, it dissolves. Sugar, like many substances, is made of molecules—groups of atoms held together by covalent bonds. The molecules themselves are held together by what are sometimes called *van der Waals bonds* and sometimes called *non bonded interactions*—forces much weaker than the covalent bonds. When sugar dissolves, the molecules disconnect (each one remains intact), and float free, independently, within the coffee. Now pour in some milk. Milk is called a colloidal suspension, because there are things in the milk such as globules of fat that do not dissolve in water. However, the globules are small enough that they don't just sit on top in a layer like oil would, but remain dispersed, acting a bit like supermolecules.

Particles up to the size of a bacterium or so, a billion times the volume of a sugar molecule, can form suspensions like this. Flour is ground to this size at its finest. The reason flour is ground so fine is that it's useful in many processes to have a very even mixture of some thing that doesn't technically dissolve. The finer you can get your powder, the more even your mixture and the better properties it has for many applications.

The rubber used in tennis balls, for example, is mixed with clay before being "cooked" to give it the proper weight and consistency. If the clay can be divided into nanometer-scale instead of micron-scale particles, the resulting rubber is more resistant to air leaking through it, just as sand makes a better dam than gravel.

Many ultrafine powder applications are now being marketed as nanotechnology, and typically they offer increased stain resistance, easier cleaning, and similar improvements over previous versions where the powders were "merely" microscopically fine.

Demron

A recent product advertised as nanotechnology is *Demron*, the radiation shield material. It's a polymer composite of polyurethane and polyvinyl chloride that is mixed with a variety of various ultrafine powders. The powders are of substances that include heavy atoms, and they act in a concerted way to absorb and scatter radiation of various wavelengths. The overall result is a thin, rubber like sheet that stops x—rays and the like, which no one would have believed possible a decade ago.

Demron is an ingenious invention and a true advance in technology. The size of the particles essentially allows them to be spread more evenly and to force each x-ray to contend with more different substances in a given space. The set of different substances is the key; and then the particles' size allows the material to be wearably thin.

Coatings can have different properties at the nanoscale as well. *Titanium dioxide* is a brilliantly white substance that is used as a pigment in paints. But a layer of titanium dioxide only 15 nanometers thick is transparent, and can be applied to glass to make it self-cleaning. It has a photocatalytic effect that helps sunlight break down dirt, and a hydrophilic effect that causes water to form even films that carry the dirt away. So a rain or spray from a hose will get the glass as clean as if you'd washed it with detergent.

Carbon Nanotubes

One of the structures most commonly identified with nanotechnology in the popular press is the *carbon nanotube*, or "*buckytube*." The buckytube is essentially a strip of graphite rolled up into a cylinder. Graphite is amazingly strong—it's been used for years in pricey high-performance items like fighter jets and golf club shafts—but buckytubes could be stronger because, being seamless cylinders, there's no edge where a tear can start.

Buckytubes are big news in the nanoworld because they can be made today. You can buy them in bottles. The synthesis process is still a bulk technology; you can't specify what kind of nanotube you'll get, and you typically get a mixture of different kinds. The kinds vary by diameter and chirality. If you imagine a nanotube as being formed from a strip of graphite cut out of a graphite sheet, chirality means the angle you cut the strip with respect to the sheet's grain. (Chirality has a different meaning in ordinary chemistry.)

Nanotubes, as currently made, are marvelous for research. Lab scientists don't mind picking out ones with specific properties using scanning probes, and their properties are so varied that they are a wide-open field for research. The electrical properties in particular vary widely; some nanotubes are conductors, some semiconductors. Transistors and diodes have been made from nanotubes.

The other main property of nanotubes in their natural state is that they are slippery. Often, nanotubes form in a nested configuration, sometimes called *MWCNTs* or *Multiwalled Carbon Nanotubes*. Inner and outer tubes slide and rotate easily against one another, like the wire and casing of a bicycle brake cable. This could be useful in many applications, but the slipperiness makes it difficult to use the nanotubes' strength in a composite material. However, special additives have been developed to help the matrix (the glue holding the fibers together in a composite) "grab" the tubes better.

Micro and Nanolithography

Another technology that's been called *nano* is the *lithography* used to make *electronic microchips*. The size of transistors and width of wires on a modern microprocessor is typically under 100 nanometers today. The basic process is not unlike, in principle, painting using a sprayer and a stencil. Other parts of the process are like developing a photo where light makes a chemical change to parts of the chip. The masks for these can be large and are focused down on to the chip with lenses. Once the chemical changes are made, chemical processes are used to construct or modify layers of material in the patterns thus laid out. These techniques are good for making millions of copies of a working device at these scales.

These lithography techniques can also be used to make machines with moving parts at that scale. These are larger, and more to the point, much less precise, than the nanomachines we talk about in the rest of the book. They are generally still called micromachines. Because of the process by which they are made, they tend to have a very two-dimensional character. Imagine trying to build a machine out of shapes that you cut from foam-core craft board. You can glue them together, if you like. The rub is that every one has to be lying flat.

The other rub is that they won't rub. Micromachinery has a phenomenon called "*suction*," which is a combination of being sticky and having high friction. The stickiness is due to van der Waals attraction and the friction is due to the fact that the parts are not atomically smooth. This makes it difficult to make bearings, sliders, and so forth, which severely constrains the kinds of machines you can make.

Even so, there has been quite a bit of progress in micromachines. For example, Zyvex, a Texas nanotech start-up, uses the lithographic techniques to build manipulators that can be used to pick up and assemble other micromachines.

The next step beyond microlithography is *electron-beam nanolithography*. Lithography with light is limited to micron sizes because the wavelength of light won't let you focus it any more finely. You can get some slack by using shorter and shorter wavelengths of light, up into the ultraviolet, but after that you're trying to use x-rays, which are difficult to handle. It's easier to handle electrons, which can have a nice short wavelength but still be manipulated by magnetic fields. Thus electron microscopes can see much finer details than optical ones.

Electron beams of sufficient power can carve material into shapes as well. They are used, for example, to make the stencils for ordinary lithography of microchips. Current top-of-the-line e-beam machines can cut shapes to about a ten-atom precision. The machines, of course, are expensive, and only make one part at a time, so commercial products can't be made this way. Even so, such parts are within shouting distance of the atomic scale if ways can he developed to manipulate them.

Nanomanipulation

The hump to be gotten over in manipulation is that third dimension. Manipulating atoms and molecules in two dimensions has been done routinely for more than a decade. A scanning probe microscope is essentially a hunk of quartz or other piezoelectric material, that is, one that expands or contracts slightly when a voltage is applied to it. We can carefully control voltages across our hunk and cause the far end of it to move, and we can control the movement with much better precision than the width of a single atom.

A number of techniques will produce a point that is atomically sharp. With skill, simply cutting a thin metal foil at an angle with scissors can work. Take such a sharp point and mount it on your hunk of piezoelectric material, and you have a tool that can be used to "see" or move atoms and molecules. The molecules stick to a surface by van der Waals attraction and can be pushed around with a lot of care and effort.

Imagine you have a plate full of peas that have been coated with honey. Put them on the floor, run upstairs, cut a hole in the floor above, and reach through with a fishing pole. You can now manipulate the peas about as well as we can manipulate atoms with a scanning probe. The pole is sticky, too, so you can drag stack them as well as just pushing. But it's a tricky, time-consuming process. You could readily imagine being able to make two-dimensional patterns with your peas, but building three-dimensional machines, as you can imagine, is quite difficult.

On the other hand, if you had a plate of sticky machine parts you had to assemble into a robot arm that would then do all the work for you, you might be willing to spend a great deal of effort and keep at it through numerous false starts. Especially if it were worth a billion dollars to get the first one working.

Another thing you could do with your fishing pole is attach a quill pen and put an inkwell next to the plate. Then you could write, draw pictures, and so forth. If you had thick ink that dried hard, you might even be able to build up a structure of some kind. When you do this with a scanning probe, it's called "dip-pen nanolithography." You can see that the term "lithography" is getting about as stretched as "nanotechnology" itself. It's come to mean any kind of printing or writing process at all, instead of the original meaning of a specific contact printing method.

A process that's more like the original lithography is sometimes called *soft nanolithography*. It would be more descriptive to call it "nanorubber stamps." Take a block of material with a raised design on one side, dip it in ink, and touch it to a surface. You get a print of the design. This works even if the stamp is a few nanometers in size. Molding—melting something, pouring it into a mold, and letting it harden into the shape thus formed—also works down to the nanoscale.

Molecular Beam Epitaxy

If you have a CD or DVD player, there is a good chance you own a device that is atomically precise—in one dimension, that is. Many of the solid-state lasers used to read the disks are made by molecular beam epitaxy. This is a technique that involves growing an object by exposing it to a very diffuse gas of atoms, made by vapourizing small amounts of some (usually solid) material in a vacuum. This technique can achieve remarkably fine control over the thickness of the layers thus produced, indeed controlled, to within an atom of thickness. To get a desired sequence of layers, you expose your growing object to a series of different gases. A wide variety of devices, solid state with no moving parts, like a transistor, can be made using the quantum properties of the thinness of the layers. Lasers, unusual kinds of transistors, Fabry-Perot etalons (like a transistor, but using light), and the like can be made.

Nanoelectronics

The most advanced capabilities seen in labs today are in the area of *nanoelectronics*. A wide variety of molecular-scale transistor-like switches has been discovered, and wires and memory elements fabricated using different phenomena. What is more, the path to making these into useful devices may well be shorter than one involving machines with moving parts. Electronics is the main application of *microtechnology*, for the same reason: microchips don't have to have moving parts.

The short way to *nanoelectronics* is by self-assembly. For example, you can synthesize your switches and wires chemically—after all, they're just fancy molecules—and chemically attach them to pieces of DNA. The DNA will have been carefully arranged so that when it self-assembles, it will drag the molecular electronic devices together into a working circuit. This will probably have been done in a lab by the time you read this, but it probably won't have been reduced to efficient enough practice to be commercially viable yet.

Nanomemories

Another near-term product that seems likely is one that uses scanning probe techniques to read and write data at molecular densities. If you've ever built a crystal radio (from scratch, not a kit), you could probably build a scanning probe—and you will also have some under standing of how finicky and tricky they are to use. Scanning probes were, after all, invented in IBM's labs. For almost two decades now, they have been capable of reading information that was written at *atomic resolution*. At a byte per square nanometer, this entire book would make a bathmat for a bacterium. The trick is in writing, reliably and quickly, and in reducing the scanning probe from a lab instrument for use by a skilled technician, to a completely automatic gadget you can put in a box and forget. 'That improvement will come about, in the main, with the ability to manufacture the scanning probes more precisely, and in quantity.

Nanobatteries

By far the slowest-moving part of the technology associated with computers is batteries. Batteries have improved some over the years, but everything else from processors to displays has improved so much faster that batteries seem to be moving backward. This will not change greatly until we have fuel cells of battery size and weight, but current-day *nanoscale technology* can offer a spate of modest improvements until then. For example, the amount of power a battery can produce is limited by the amount of surface interface between its electrodes and electrolytes. With a nanostructured surface, you can have a lot more area in a smaller space and with less weight.

For example, a really neat trick involves *nanograss*, a structure that consists of an array of silicon nanorods sticking straight up from a surface like grass from the ground. (They're all the same height and in regular rows and columns, so the fakir's bed of nails might be a more accurate image.) Droplets of electrolyte sit up on the tips and don't touch the sides, under normal circumstances. This cuts the actual interface area down to virtually nil, which reduces leakage and battery rundown to practically zero when it's not in use. When it is in use, a voltage or other stimulus can be used to force the droplet down between the spikes, increasing the surface area of contact enormously. And because the *nanograss battery* is silicon based, it is compatible to be fabricated as part of an electronic chip, producing self-powered chips for non power intensive applications.

Fuel cells are another area where current nanoscale technology could very well enable a significant advance. Quite a few different technologies are being looked at today. The basic idea is to find some way to do reverse electrolysis. You may have seen or done the experiment where you put two wires into water and run a current through it. *Hydrogen bubbles* up from one wire and oxygen bubbles up from the other—direct conversion from water to its constituent elements using current. The opposite is also possible: direct conversion of hydrogen and oxygen to water, producing current. This is what happens in a *proton-exchange membrane* (PEM) *fuel cell*.

Proton-exchange Membrane Fuel Cell

PEM fuel cells are a current technology; you can buy them in a variety of sizes, ranging from power backup for your house to micro-sized units for your laptop. They are costly, as yet, compared to other power sources. The quip in the industry is, "Buy a micro fuel cell, get a laptop free."

Advances in nanoscale technology, such as carbon nanotubes for electrodes, might well be involved in the continuing improvement of fuel cells. At the same time, a biotech-based project to harness the sugar-digestion machinery of the mitochondrion to run a fuel cell. Whatever works!

SELF-ASSEMBLY

Another way to put things together at the nanoscale is what called *self-assembly*, or *Brownian assembly*. This is how the molecular machines inside the cell get put together. The trick here is that the parts to be put together have to be atomically precise (or nearly so). Then you dissolve them in some fluid, often water, and let them float around. When the ones that fit properly come together by chance, the stickiness (van der Waals and some other forms) causes them to stay together. Take two bricks and break each one in half. Apply a thin layer of glue to the all the broken surfaces. You will find that two matching halves stick together much more strongly than two nonmatching halves.

How do we get atomically precise parts? Currently, we have to use chemistry or molecular biology. For example, if we can make a molecule of DNA, by whatever means, we can use some of the cell's machinery to duplicate it in great quantities. It turns out that DNA is a good molecule for doing self-assembly. Scientists, particularly Nadrian Seeman and his group at NYU, have been building some fairly complex structures, and even some moving machines out of DNA.

Their latest effort, as this book is written, is a pair of walking legs. The legs don't balance or anything like that. What the group has constructed is a pair of somewhat *floppy legs* and a *DNA footpath*. The "feet" stick to successive "footprints" in the path, unsticking from the last and going to the next by infusing the whole business with a series of different chemicals that alter the properties in just the right way. This makes for a cumbersome machine, requiring rinsing and mixing for each little controlled motion, but if you can get a machine that works, you can use it to build a faster one next.

Application of Nanotechnology

Currently, the applications of nanotechnology in products you can buy are things like the nanoscale powders that produce new (and useful) materials properties. There's some advance in catalysts and chemical analysis, mostly in specialized applications. Batteries, fuel cells, and data storage may well soon see some benefit from today's nanoscale science. But to get to the really impressive capabilities of nanomanufacturing, you have to be able to build nanomachines. Right now in the labs, a handful of experiments is addressing prototypes of individual parts, such as bearings, that will go into the nanomachines. Once we do get all the parts, we still have to figure out how to put them together.

Imagine a set of parts you have to put together to build a machine. They have to go to a specific set of positions in a specific sequence. Suppose you could have each one made with a semiflexible rod connecting it to a sort of luggage tag that had a complex surface on both sides. The bottom surface would match the top surface of the tag of the previous part, and its top would match the bottom of the next. The tags would self-assemble into a stack in only one sequence, and the rods would be such as to force the parts into the appropriate configuration, in the right order, as the tags were stacked in that sequence. You could complete the assembly process by dissolving away the rods.

The big difference between today's *protonanotechnology* and the full-fledged, *mature technology* that will forge the shape of our future is simple. Current practice involves factories that make nanoscale products. Mature nanotechnology will involve nanoscale factories: working machines, atom-for-atom precise, that make things.

Life: Nature's Nanotechnology

Of course, the world is crawling with nanomachines already assembled, working, and making things. Each living cell is an amazing factory jammed with molecular machines. Most of these machines are made of protein, that is, long chains of amino acid molecules that are tangled up into a compact shape like a bead necklace rolled into a ball.

It's the specific shape of these protein molecules that gives them their function. Often a specific machine, or enzyme, is formed of more than one actual molecule, and the shape (and distribution of electric charge) of the molecules causes them to stick together properly.

This sticking together properly is the key to how everything in the cell works. A cell has a number of different compartments to keep various kinds of molecules separate, but nowhere near the total number of kinds of molecules. So in general, to make a machine composed of molecule A and molecule B, the machinery simply makes an A and a B and lets them float around until they bump into one another and stick. This floating and sticking process is called *self-assembly*, when you are talking about putting the parts of a machine together.

Once the protein machine is put together, it performs some function, for example, as an *enzyme*. An enzyme takes one or more input molecules and causes a chemical reaction to occur, producing one or more output molecules. The protein pocket that snugly fits the input molecule(s) holds it in such a way that the reaction is favoured; that makes it a catalyst for that reaction. These reactions can be anything from building up amino acid molecules from smaller ones, to oxidizing sugar for energy, to breaking down old proteins for spare parts.

If the reaction "*runs downhill,*" producing more energy than it consumes (such as oxidizing sugar), the enzyme can do it without any help. If it runs uphill, such as in building big molecules, the enzyme needs to get energy from somewhere. Typically, the energy comes from special molecules such as ATP that serve as fuel. The reaction then is more complex, involving breaking down the ATP at the same time as doing the other, desired, reaction. Mechanisms like this allow the enzymes to add energy to their products. But where does the ATP come from?

ATP (*adenosine TRI—phosphate*) is formed from, and broken down to, ADP (*adenosine DI—phosphate*) and an extra phosphate—the parts being used over and over to transport energy from one reaction to another. The energy comes from the oxidation of sugar and fat in the mitochondria in higher forms of life. (Bacteria have a different way of getting energy from sugar without oxygen, fermenting rather than burning it.) The *mitochondrion* is a double-walled compartment in a cell. The oxidation goes on inside the inner wall, involving a complicated cycle of reactions that ultimately winds up creating a voltage difference across the inner wall. It's similar in principle, though

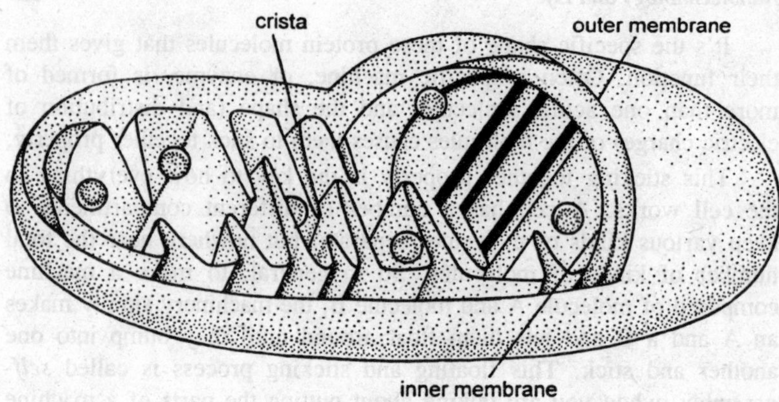

Fig. 10.1. Mitochondria

wildly different in detail, to what happens in a PEM (*proton-exchange membrane*) fuel cell.

The inner wall of the mitochondrion is covered with what looks like, in a micrograph, an aerial photo of a forest. The trees are molecular machines called ATP synthase. (Decoding the biologyspeak, that means "machine that makes ATP.") The ATP synthase machine

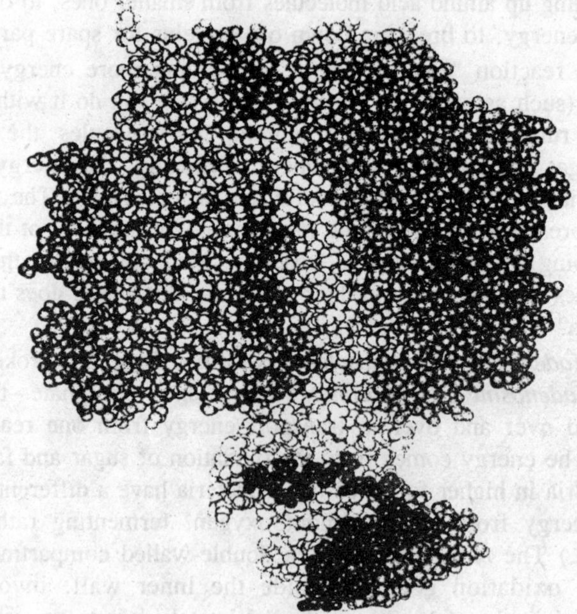

Fig. 10.2. ATP synthase.

has three main parts: an electric motor, a shaft, and a mill that takes ADP molecules and pops the extra phosphate on by mechanical force.

Most of the molecular machines in the cell are not this obviously mechanical, but ATP synthase shows that they could be. It demonstrates electric motors: quite a few motors are found in the cell, several of which are rotary. (Note that intracellular motors typically run on proton rather than electron transfer, but that's a technicality.) It shows a shaft connected to a motor, transmitting mechanical torque and power. Most important, it demonstrates a process that causes a specific "uphill" chemical reaction pushed by mechanical force.

One *ATP synthase machine* can produce something like one hundred ATP molecules per second.

Mother of All Machines

Let's look at one other machine, out of the thousands in the cell, in some detail. It is the *ribosome*, and it is the key to all the molecular machinery in the cell, because it's the one that makes proteins.

A protein is a wadded-up string of amino acid molecules. Twenty different kinds of them occur naturally and each has different properties. Some are stiff and some are flexible. Some are large and some are small. Some are charged and attracted to water; some are oily and like to bunch up away from the water. The shape of a protein molecule ultimately takes depends on the sequence of types of amino acids in the protein. Thus, to make a specific molecule, you have to specify particular sequence, and then build a string of amino acids that meet your specifications.

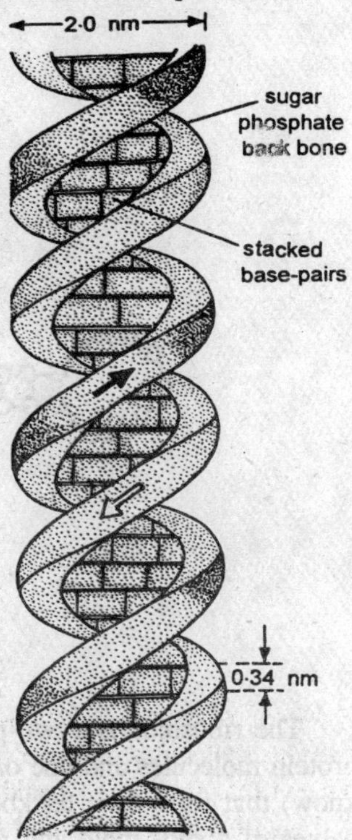

Fig. 10.3. Double-stranded DNA.

Protein sequences are stored in the DNA of the cell. DNA is double string of just four kinds of submolecules (called *nucleotides*), so it takes three of them to refer to one amino acid type. The nucleotides are commonly denoted A, C, G, and T. So if we had piece of DNA that read GCATGCAACCCA, we would parse it GCA TGC AAC CCA and it would mean (a tiny snip of) a protein consisting of alanine, cysteine, asparagine, and proline in that order. The correspondence between the nucleotide triplets and amino acids is called the genetic code.

To make a protein the information permanently stored in the DNA is copied off into a "*work order*" as a molecule of RNA, which is similar to one of the two strands of DNA. The RNA is then "read" by the ribosome to produce the protein. The work order RNA is called messenger RNA or mRNA for short. Another object made of RNA, this one called "*transfer RNA*," or tRNA, acts as an amino acid holder, and thus comes in twenty different forms. Each form corresponds to one of the amino acids, and in fact, each one gets attached to a molecule of the appropriate amino acid by a special enzyme made for that specific purpose. It has the acid (e.g., alanine) at one end and a triplet (e.g., CGT) complementary to the acid's code on the other. The tRNAs are not just amino acid molecule holders; they are labeled amino acid holders.

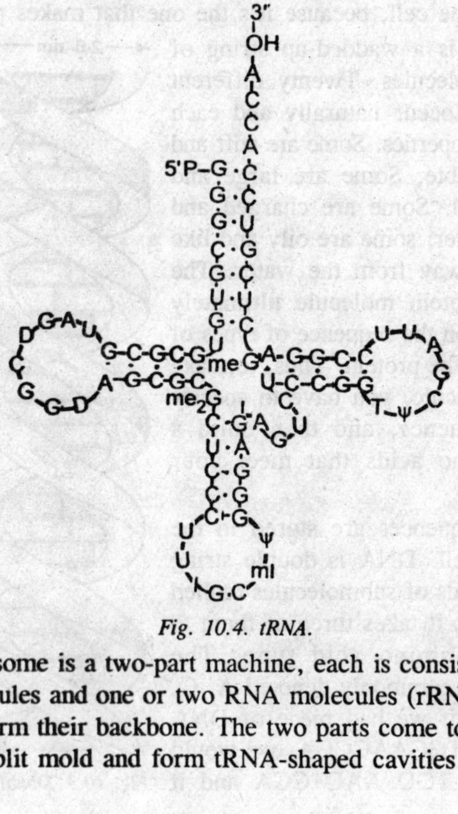

Fig. 10.4. tRNA.

The ribosome is a two-part machine, each is consisting of several protein molecules and one or two RNA molecules (rRNA, if you must know) that form their backbone. The two parts come together like the halves of a split mold and form tRNA-shaped cavities when they do.

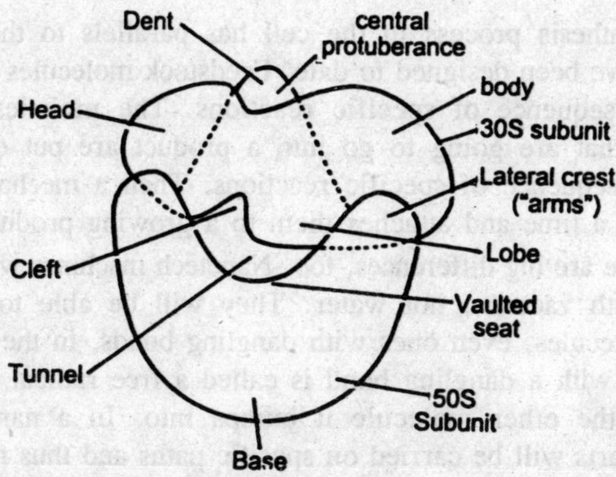

Fig. 10.5. Ribosome.

In operation, the ribosome holds a strand of mRNA along one edge, so that one triplet forms the end of a cavity. This causes the cavity to fit just the tRNA that holds the amino acid the triplet specifies. And in the cell, if there's an object, and a cavity that fits it, pretty soon the object will be found sitting snugly in the cavity. That's the magic of self-assembly.

The two parts of the ribosome reciprocate back and forth against each other, not unlike the slide and frame of an automatic pistol. Each stroke jacks another tRNA through the series of chambers, advances the mRNA "tape" one triplet, bonds the amino acid molecule of the latest tRNA to that of the previous one, and unbonds the previous one from its holder. The holders are popped out to he reloaded, and the protein "chain" grows by one amino acid "link" at each step.

Each stroke of the ribosome is powered by the breakdown of a GTP (Guanosine Triphosphate) molecule into GDP, very similar to the ATP-to-ADP process that powers most of the rest of the cell's activities.

Ribosome's in human cells can add two amino acids per second to the protein they are making, and in bacteria up to twenty.

From Biology to Nanotechnology

The molecular machines in the cell have many similarities with the ones that will form the core of nanotechnology. First and most important, they build atomically precise products by controlling specific bond-forming and bond-breaking reactions. They are—in instances like the ribosome—digitally controlled; that is, they receive instructions encoded as a sequence of discrete symbols. They can use electric motors, turning shafts and reciprocating advance mechanisms.

The synthesis process in the cell has parallels to the nanotech ones that have been designed to date. Feedstock molecules are broken down in a sequence of specific reactions. The moieties (parts of molecules) that are going to go into a product are put on holders, again by a sequence of specific reactions. Then a mechanism takes them one at a time and attaches them to a growing product.

But there are big differences, too. Nanotech machines will typically be filled with vacuum, not water. They will be able to use more reactive molecules, even ones with dangling bonds, in the processes. A molecule with a dangling bond is called a free radical and causes damage to the other molecule it bumps into. In a nanomachine, molecular parts will be carried on specific paths and thus not allowed to bump into things that they could disrupt. Moving along specific paths, on conveyor belts or in robot grippers, means that the parts will go faster and get to where they are going with more regularity. Thus, a nanotech machine performing a specific reaction on a stream of incoming molecules should be able to do millions per second, instead of ATP synthase's one hundred or so per second.

Rather than distributing energy in one-shot cartridges (the ATP molecules), nanomachine systems will tend to do it with wires, shafts, cables, and pushrods. Rather than using the intricate control processes of the cell, involving chemical messengers that are still not completely understood, nanomachines typically use cams, linkages, relays, and computers.

The parts-assembly mechanism in a nanomachine would begin making a part because a computer program told it to, not because mRNA happened to run into an idle ribosome. The mechanism itself would look more like a robot arm. The arm would take molecule holders from conveyor belts, rather than waiting for one that fits to float by.

A big difference is that the nanomachine will build a three-dimensional part in three dimensions, rather than spinning a chain of beads that fold into a shape. This means that nanomachine parts could he much more rigid than proteins, and could be made with many fewer building-block types. The robotic arm might need five or fewer incoming conveyor belts of holders, instead of twenty. It would have many more choices of where to attach the new moiety, compared to the ribosome that can only tack it to the end of the string.

Stiffer, more rigid parts not immersed in water means that nanomachines must operate in a wider range of temperatures than living cells (below freezing, for example). Operating at higher rates of speed and using higher energy reactions are just the normal differences we expect between machines and living things. With nanomachines, though, we will get to keep many of the things about life that are due to its basis in molecular machinery: complex, efficient metabolism; powerful yet silent motion; intricate structure; self-repair; growth—in short, all those qualities that make life a wonderful and seemingly mysterious thing.

11

IMPROVEMENTS

There is a broad and slippery slope between merely patching up someone who has been injured, or who suffers a genetic deficiency, and augmenting people to better capabilities. Humans vary widely in native ability and strength. Some are virtually immune to sunburn, others are debilitated after an hour outside on a bright day. Most are in between. The same is true of attributes ranging from height to intelligence. It is a cure to add respirocytes to someone's blood who is anemic. It is an augmentation to add them to a normal person. What about someone who is in the normal range, but some what lower than average? If you were to try to live at high altitudes, in the Andes or Himalayas, for example, you'd find that a good helping of respirocytes would just bring you up to the normal, evolved performance of your neighbours.

What about intelligence? Nanotechnology may well be playing second fiddle to biotech for true, *organic brain* upgrades for a while. Even so, people have been using technology to enhance their intelligence for millennia. It is amazing what just pencil and paper can do. You will find it labourious to multiply 629,904,623 by 984,571,094 with pencil and paper; but you will find it impossible to do it without them.

Of course, you'd never try nowadays. You had use a calculator or a computer. More to the point, you'd rarely have to do multiplication at all. Calculations would be made by software that interacted with you on a much higher level, and you'd never know how many multiplications were being done on your behalf. Software can extend your intellectual abilities in many ways; consider a tax-form preparation program or an engineering CAD system.

Imagine you have a computer, a sort of super-PDA, that sits in your pocket. It has the storage to hold a large library of reference material, as well as wireless access to the Internet when needed. It has the latest in voice and image processing software, and the computational horsepower to run it in real time. You wear a pair of glasses that can overlay a generated image on the real world, giving you captions, briefings, memory jogs, whatever, without your having to seek out a screen or lug a laptop.

The system listens to everything you say or that is said in your presence. It watches what you see with tiny cameras in the glasses. It transcribes everything so that you can review conversations at leisure. It remembers everyone who you are introduced to and can place a virtual nametag on them when you meet them again. Anytime a question is asked, it looks up or calculates the answer and puts it in a subtitle in your field of view. It translates from other languages, spoken and written, the same way. Or indeed, reading in your own language, if your gaze lingers on a word and you mumble "huh?" the definition would pop into view.

A system like this is possible right now, with two exceptions: the computer to handle the job would be backpack-sized instead of pocket-sized and the software is not really there, though all the basic capabilities exist at least in labs somewhere. With nanotechnology, you could put the entire system into a single contact lens. The smarter the computer gets, the smarter you act.

The next step, of course, is to put the computer inside. You can put it down at the *brain stem*, where it can listen in on all the nerve signals coming into the brain, using your eyes instead of cameras and your ears instead of microphones. Indeed, it can tap into the brain cross talk at higher levels, taking advantage of some of the preprocessing your visual system does, for example. Then it can inject the hints and answers into the appropriate sensory nerves.

No need to rely on the *biological human senses*, of course. We can extend the range of vision, from *telescopic* to *microscopic*, and into the infrared or the ultraviolet. We need not modify the eye; any patch of skin could be embedded with a grid of *microscopic sensors* and act as a phased-array optical antenna. (And look exactly like ordinary skin to normal human vision). You could know the exact chemical composition of anything you tasted or smelled.

It would be very useful to be able to look inside yourself and check out how all your organs were doing, from a surgeon's eye view to a microscopic examination of individual cells. In the opposite direction, there should be ways, mediated by information networks, of seeing the larger things you are a part of: community, company, nation, with an integrated and instinctively understandable display.

Enough has been written about virtual reality that it is probably not useful to say much more here, except to note that it would be a simple adjunct to the capabilities of this level of augmentation to inject a completely synthetic set of signals into your sensory nerves and transport you to an imaginary world, or simply one at a different place. There should be no more need for offices at all; if your stock in trade is information, you just take a walk in the park. Anything you could have done on a computer could be seen in the mind's eye; for your meeting, sit on a rock in a forest glade and your conferees appear in virtual form around you.

Just as useful, and probably a lot of fun, too, would be imaginary friends. With nanotechnology you can have enormous computer power stuffed into odd spots in your body, powered by all those desserts you don't have time to exercise off. This can be used to simulate a number of companions, who can work on intellectual stuff while you are doing something else. You can "*hear*" and "*talk*" to them subvocally, and they can give you presentations in your field of view. The closer they get to real AI, the closer you are to being a large, well-integrated team instead of just one person when you tackle a problem.

Elbow Grease

Current medical prostheses are mostly physical: arms, legs, and so forth. More sophisticated ones are appearing, starting with cardiac pacemakers and ranging to cochlear implants, as the technology improves to be able to simulate, to some useful level, the more complex functions of the body.

With full Stage V nanotechnology virtually the entire body could be replaced and have its capabilities enhanced. The reason to do so would not primarily be to extend strength or endurance—we would have squads of robots to accomplish whatever we needed physically—but survivability and recuperation in the case of injury.

The major watershed would be whether to retain the ability to look and feel just like a standard biological body. Given the penchant for body modification among the upcoming generations, this may not be as important a shibboleth as it is to us older folks. One physical ability I would be willing to undertake drastic modifications for would be to be able to fly. *Biological human bodies*, even with the most extreme imaginable biotech modifications, are very unlikely to be able to handle it. Human flight is a nanotech-only proposition.

A much closer and more likely motive is aging. Biotech has a number of approaches that seem to be promising, but in the long run, it may just be simpler to replace the evolved kludge with a new model that doesn't have the biological clocks that need resetting all the time. You can be immune not just to aging but to all known human diseases, eat anything (including wood, coal, candle wax, or gasoline), and live comfortably anywhere on the surface of any planet in the solar system, or in empty space.

All of this is simple and easy to take, however, compared to replacing the brain. Again we face a slippery slope. The modifications discussed above as the "*super implanted PDA*" appeared to enhance intelligence, but did not change the brain; they only talked to it. We could imagine, in the not too distant future, having a better under standing of how the brain works internally (this is an area of science that is moving at least as fast as nanotechnology, by the way). Then we could inject signals, not simply into the incoming and outgoing nerves, but inside the brain itself. Instead of seeing the answer to be read in your visual field, you had "remember" it. Your imaginary assistants would "read your mind" and not need to be told what to do. You could look at a pen and know that it was exactly 152.4 millimeters long the way you look at it now and think it is about 5 or 6 inches.

Moving in a different direction for a moment, we could take the brain and replace it with a machine that duplicated the functions of the neurons, together with other circuitry that simulated the effects endocrine balances and other chemical signaling pathways. The machine could be smaller, lighter, and much more durable than the actual neurons of the brain. More important, it could be a lot faster. In the same volume as the human brain, you could put somewhere between a thousand and a million times the raw computational power. Now your imaginary friends are not so much a family as a city—all still intimately aware of everything you want, but in a complex organ organization with highly specialized members who can work at a level of detail that you had never have time for. And it is all inside your head.

You might now be fit, for example, to live in a tribal everybody-knows-everybody community of two hundred thousand people, instead of just two hundred. You might be able to assess the risks, gauge the advantages, and intelligently design and implement complex technologies and social systems. With a little extra effort, you could probably set the clock on your VCR.

The use of biotech to augment humans, is called *Enough: Staying Human in an Engineered Age*. McKibben takes this view of improving intelligence: Of course, the problem with arms races is that you never really get anywhere. If everyone's adding 30 IQ points, then having an IQ of 150 won't get you any closer to Stanford than you were at the outset.

Getting into Stanford is far from the best reason to want to change your IQ, or that of your children. I would be much happier in a world where everyone else's IQ were thirty points higher, even if mine remained the same. Consider that crime, which is strongly correlated with low IQ, would virtually vanish. For the average person, thirty IQ points would move them from the category where they need training for a new job to the one where they can pick up a book and do it themselves, where they are much more likely to understand the consequences of various life choices, and thereby would be better off in the long run. Public discourse might be raised from its current level of least-common-denominator banal, stupid, emotion-laden sound bites to slightly more informed discussion. Not, of course, because the politicians had gotten any closer to being moral animals, but simply because the least common denominator had become more intelligent. I might find Bach instead of Beastie Boys playing on diner jukeboxes. I might be able to talk to my barber about thermodynamics instead of the temperature.

McKibben does not really find too many specifics to argue with in the prospect of augmenting humans. This is surprising, given the general tone of the book. In many cases he simply displays the ideas of the transhumanists (proponents of technological self-improvement) often quoting people I know and/or whose opinions I respect, leaving them hanging there as if they were horrific enough all by themselves. Perhaps they are, to him. A couple of problems that he identifies do deserve some thought, though.

First, as a *marathon runner*, he bemoans possible physical enhancements that would make marathon running easy. Where would be the joy of accomplishment, the feeling of having measured yourself against a challenge, if there were no challenge?

Marathon running is a game. It is not a true challenge, but one that is made up for people to measure themselves against. A true challenge, for a true human, is one where success or failure makes a difference. Enhanced humans can play games, too, and they will be just as -meaningful as today's games are. You will be measuring yourself against someone else who is enhanced. Car racing is a widely appreciated sport. So are women's tennis, soccer, and basket ball, as is Little League baseball and the Special Olympics. The fact that there are different classes of competition does not make the game less rewarding.

On the other hand, true satisfaction when you look back over your life does not come from having played games. It comes from having done things that mattered. Yes, we can meet with triumph and disaster, and treat those two imposters just the same; but when life is on the line, triumph is better. McKibben seems not to realize that there are greater challenges, worthy of *nanotech* (or biotech) *capabilities*.

Oddly enough, one of these challenges squarely answers the concerns of McKibben's earlier book, *The End of Nature*. There he worried that the advance of technology was making the Earth into a man aged park, even if we do decide to take care of it. No more truly wild lands, nothing beyond the reach of grasping bureaucrats or greedy developers.

The universe at large is wild, natural, and forever (yes, forever) beyond the ability of humans to fence in, domesticate, and tame. Of course, from where we sit on Earth today, it does not seem like a challenge; it seems impossible. For it to be a decent challenge, we have got to get out there and measure ourselves against it. As nonaugmented, biological humans, we are fish looking up at dry land. But our forefathers, the actual fish, managed to conquer the actual land. We can do as well.

The other concern in *Enough* worth considering is the notion that parents might program their children off in different directions and humanity will lose a common basis. For example, a musician might enhance the musical skills of her child but also strengthen the connections from music to feelings so much as to alter the child's basic out look. Meanwhile a cleric might program his son for religious faith. Ultimately we had get people who simply couldn't understand one another. Sort of like the English and the IRA, the Israelis and the Palestinians, or Bill McKibben and yours truly.

Consider: You have the new Humanity 1.5 Control Panel installed. You can pop up the panel, a virtual reality gadget, into your field of view. It has a bunch of gauges, buttons, sliders, and so forth familiar from software applications. But these say things like Height, Weight,

Hair Colour, a selector menu for Racial Type, a slider that varies between Fine Motor Control and Gross Motor Control as your optimum coordination regime. Scroll down some and you find Intelligence, Personality Type, Interpersonal Sensitivity Irritability, Ambition, Gregariousness, and so forth.

Twiddle the dials some. You get taller or shorter, fatter of thinner. Except for these gross physical attributes, though, you would not notice much difference until you interact with people or tackle some problem. I don't see any major objection to letting people have this kind of synthetic self-control, in general, provided some common sense is used.

OK, fine. Now you get panels for your kids. Who gets to set them? The kids? You? Schoolteachers, guidance counselors, bureaucrats? Who gets to play with yours? Policemen? Judges? Anybody who wins a lawsuit against you?

Programmable Animal

The human mind, if it is to be the physical organ of human reason, simply cannot be seen as bound and restricted by the biological skinbag. In fact, it has never been thus restricted and bound, at least not since the first meaningful words were uttered on some ancestral plain. But this ancient seepage has been gathering momentum with the advent of texts, PCs, coevolving software agents, and user-adaptive home and office devices. The mind is less and less in the head. The problem is that McKibben is a couple of million years too late. Starting (very roughly) two million years ago, we begin finding in the fossil record an artifact that archeoanthropologists call the Acheulian handaxe. In fact, it is neither particularly Acheulian nor a handaxe, but the name has stuck anyway. What it actually is, is the granddaddy of all technology.

The Acheulian handaxe was made and used by our ancestors for something like 1.8 million years. No other piece of technology is even close. When our ancestors began using it, they were *Homo erectus*; by the time people had quit making it, they were *Homo sapiens*. We shaped the stone, but the stone also shaped us.

The handaxe puzzled anthropologists for many years. It is a piece of carefully chipped rock, flat, with a sharp edge, that has an outline like a pointed egg. The trouble is that it is sharpened all the way around; if you used it for a handaxe, you had cut your hand. And there is no need to go to all the work to make a handaxe symmetrical.

Thanks to some inspired detective work by William Calvin, we finally have a good idea how the handaxe was used. It was a killer

Frisbee, to be thrown at animal herds when they were tightly packed at water holes. It works by a complex interaction of phenomena, including aerodynamics of the particular shape, the nature of animals' reflexes, and herd reactions to hunters. In particular, it works, often enough, when you are not nearly a good enough rock thrower to hit a specific animal in a vulnerable spot, but can hit the herd somewhere. Essentially the handaxe, however thrown, will flip vertical in the air, the sharp edge will hit the animal, and the point will catch instead of rolling off. The animal will flinch and trip as the herd stampedes, leaving it knocked down, trampled, and dazed enough that the *Homo erectuses* could run up and grab it

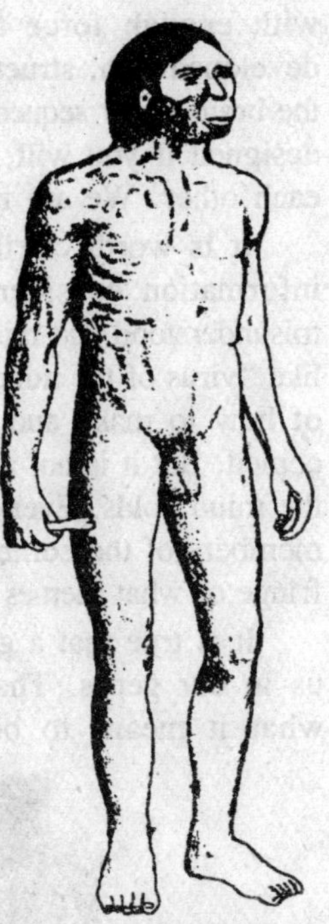

Fig. 11.1. Homo erectus.

This took some discovering. However, it did not get rediscovered again by each new hunter over the course of nearly two million years; it got copied. Indeed, the handaxe itself almost certainly evolved over many generations from hunters seeing that some stones worked better than others, and trying to chip bad stones till they looked like the good ones. So while *Homo erectus* was evolving hand-eye coordination and becoming a better stone chipper and thrower, he was also evolving the ability to imitate—and becoming a substrate.

Fig. 11.2. Homo sapiens.

In other words, the human brain became not a machine for controlling a body, as in an animal, but a machine for running programs, some of which controlled the body, but which could also speak and understand and invent and experiment and teach and learn.

By the time we became *Homo sapiens*, we had developed our coordination, which included our brains, to the point that an experienced hunter could throw a spear thirty yards and hit an animal in a particular spot with enough force to deliver a debilitating wound. We had also developed rich, structured, symbolic language, using the same part of the brain as for sequencing complex hand motions. We are now evolved, designed if you will, to copy ideas, techniques, and knowledge from each other. We are meme machines.

It is worth clarifying at this point that a meme is a pattern of information considered as a replicator. Memes have been wildly misunderstood and misrepresented by a number of authors, called things like "virus of the mind," and so forth. Memes, including the knowledge of how to make and use an Acheulian handaxe, do spread by being copied. But it is not as if that were different from any other knowledge the mind holds. Everything in there got copied from parents and other members of the community. Being an informational virus is the merest fringe of what memes are. They are the software of the mind. All of it.

It is true that a good part of what it means to be human is given us in our genes. That is also, to a 99% overlap, the same part of what it means to be a chimpanzee. That tiny difference is the

Fig. 11.3. Chimpanzee.

supercharged idea-copying machinery (Chimps copy techniques, but with about the same facility that humans mentally multiply nine-digit numbers. It can take a chimp seven years to learn to crack nuts between rocks by watching other chimps do it. We are good at copying.) All our basic emotions, patterns of behaviour toward others, even politics, are basically chimp like.

That one difference is huge, though. People are an idea substrate. We are built by taking a young, blank organism, and filling it with ideas, techniques, attitudes, concepts, knowledge. We are just as much formed by our cultural heritage as by our genetic one. Indeed it is not going too far to say that in a very real sense, we are our ideas. If you have an internal conversation with yourself, conscious creature that you are using concepts and abilities that were learned as part of your culture. A feral human cannot use language and may not even be conscious in the way we are.

What price modifying humanity if "what we are" is really "the software"? A man who is been genetically programmed to be pious, unable to doubt his faith, has been reduced to a subhuman state. His hard ware has been crippled so that it cannot run the full range of software. But so has the child raised by radical environmentalists and trained to hate the idea of technological progress. Both are reversible, by the same genre of techniques with which they were done; but in neither case would the individual want to be deprogrammed.

What is more, our genetic component is not a specification for a brain, anyway. It is a specification of an algorithm for building a brain that is heavily dependent on the environment the brain develops in. It is been shown recently that watching television as a baby can greatly increase one's chance of an attention deficit disorder. It is not hard to see why: watch a TV for a few minutes and count the number of scene shifts. Especially in commercials, it's rare for the camera to point at the same thing for much more than a second. When a brain forms under such a torrent of continual change, it will tend to be better at quickly parsing newly flashed information, but worse at deeply analyzing something that remains the same for long periods. TV is packed, intentionally, with as many attention-getting phenomena as possible. To a mind formed on TV, the attention-getting signals in the real world are sparse and weak.

In other words, we are already rewiring our kid's brains simply by the constructed environments we live in, as well as programming them with everything we teach and everything they watch us do.

In my own concept of what it means to be human, copying the software to a new substrate can preserve the essence of being human, if the new substrate is able to run the software the way the old one would. This point is very important, more so than realized by some transhumanists, I think. At the current state of knowledge in cognitive psychology and neurophysiology there is lots of stuff we do not have a clue about. We do know that, for example, emotions and the sensory feedback from the body are important; the human mind is first and foremost a body controller, and second and almost as important, an interacting unit in social groups.

So it will not be an easy or a simple job to build a compatible substrate. It will involve more than simply simulating the neurons; some kind of (at least virtual) embodiment appears necessary. However, it is by no means impossible. We have a lot to learn, but we are learning fast in the areas we need.

Now we can give some considerably more enlightened guidelines for staying human in an engineered brain. The prime notion is from computer science: "*upward compatibility*." In your new hardware, you must be able to think all the same thoughts, have all the same feelings, experience all the same stimuli as in the old. Only then can you talk about expanding the repertoire. One obvious way to go is to be able to think more of the thoughts, know more of the knowledge, and have more of the experiences that humans have always had. More memory, more integrating circuitry of the same type, faster processing. Simply being able to see things from more points of view, to use more of the knowledge available to make decisions, could make us better people—wiser as well as more capable.

Another really major improvement that would go a long way toward making people a lot happier is to add cognitive abilities that allow us to make sense of the modern, formalistic, mechanical, engineered world with the same instinctive ease that we evolved to handle the wild savanna. You can pick out a tiger slinking behind a screen of brush with a single effortless glance. The data processing involved in putting together the hints of light and shadow, patterns of motion in the visual field, and comparing the result with an enormous stored database of possibilities, is still beyond our best computer systems. It is considerably more computation than would he involved, for example, in filing the most complex income tax return. But we live in a world of taxes, not of tigers. There is no reason, in principle, why the relatively simple information-processing tasks we face today cannot humans effortless as the considerably more complex ones we already do.

Obviously more efficient, you may say, but happier? Leave your office, the boxes of steel and concrete, and stand in a park, a field, a forest. Watch the clouds, deer wandering by, or a sunset. Look up and see the scattered diamond dust of the stars in the night sky. These things are beautiful and restful to us because we are evolved to appreciate them. The good things in our created world should evoke the same sense of rightness and satisfaction.

But probably the most important class of improvements will be an ability we might call *telemagination*. Consider having the reverse of vision: the ability to create pictures instantly of whatever it is we are thinking of. Our computers already do this. Then transmit the picture, along with imagined sounds and other sensory impressions, to someone. In real time, it is a movie, produced as effortlessly as talking is today. The really revolutionary aspect of such a skill would be to open up the bandwidth of person-to-person communication. It might be possible—this remains to be seen—to compact some communication into a more condensed semantic form; but failing that, movies are pretty good.

To see the difference between people today and people with telemagination, think of the difference between a PC and a PC connected to the Internet. Or the difference between *Home erectus*, who did not have language, and *Homo sapiens*.

Adding a few new cognitive abilities, and a lot more horsepower, to the old noggin will, yes, make us into something not human as we conceive it: today. So did using the handaxe change our ancestors. We still have the *Homo erectus* mind, and indeed the chimpanzee one, as a vital part of our basis; but we have added to it, in what I believe is a good way. So will our descendants, and possibly ourselves, add to what we are today. Surely it is the most arrogant conceit to think that we, now, are the apotheosis of evolution, the best that can be, intellectually or morally. Only a fool believes himself unimprovable.

We have improved ourselves for millions of years by adding ideas to the bundle of information we consist of. Our genetic heritage of information, encoded in DNA, is given almost entirely to each of us, indeed ten trillion copies. Each cell has the whole thing and can use whatever parts of it are appropriate to the limits of its abilities. It is a good model.

We have gotten to the point where our culture, the informational heritage of humankind, is a staggering overload for any person to learn. It is not that our minds could not handle it—after all, it was all thought of by people in the first place. Our ape brains just do not have the horsepower. Our informational selves have outgrown the substrate, and we are in serious danger of falling behind our own machines. It is time for an upgrade.

We are like the bound feet of Chinese women in historical times, living in a tourniquet, unable to grow. Our minds could grow naturally to follow our culture, art, and science if only we did not have to cram it all into a few pounds of gray matter evolved to steer a hunter/gatherer around the plains of Africa. If we can only learn to stop identifying ourselves—the patterns of our thoughts and feelings—with the lump of matter those patterns are formed in, we can move into more spacious accommodations and have room to grow. Then we can emerge from our cocoon and stand on our own feet among the thinking machines of the future.

INDEX

A
Accumulated plaque, 330
Adenosine DI—phosphate, 359
Adenosine TRI—phosphate, 359
Age of nanotechnology, 341
Agoraphobia, 141
Akamba, 47
Akathisia, 140
Alcmaeon of Croton, 55
Alcoholism, 139
Alexander, 58
Alexandria, 58, 59
Allotropic, 175
Alzheimer's disease, 140
Ambroise Pare, 63
Amoeba proteus, 117
Anaphylaxis, 139
Android robots, 348
Androphobia, 141
Angioedema, 139
Anthracosis, 52
Antibiotics, 15, 85
Antiport mechanisms, 227
Antiseptic, 78
Arachnophobia, 141
Aristotelian, 58
Aristotle, 58
Artificial Cells, 125
Asclepiades, 59
Asclepius, 54
Astounding Science Fiction, 122
Astrology, 138
Atomic force microscope, 6
Atomic resolution, 356
ATP synthase machine, 361
Autocatalysis, 168
Automated engineering, 31

B
Bacteria, 13
Barbiturates, 85
Beauty business, 324
Belonephobia, 141
Benzodiazepines, 85
Big thing, 36
Bioelectromagnetics, 307
Biological human bodies, 368
Biological human senses, 367
Biomedical research tools, 1
Biosensors, 251
Biostasis, 28
Biotechnology, 42
Birthmarks, 327

Black hair. 325
Blond hair, 325, 326
Boltzmann constant, 146, 227
Book of wounds, 53
Bottom-up, 143
Brain stem, 367
Brownian assembly, 357
Brownian motion, 206
Buckytube, 351
Bulimia, 139

C

Cancers, 20
Carbon nanotube, 7, 351
Cecil Textbook of Medicine, 87
Cell biology, 23
Cell herding, 21
Cell herding machines, 21
Cell surgery, 24
Cells, 11
Charles Turner Thackrah, 66
Chemosensors, 251
Chirurgia, 57
Chlorpromazine, 85
Chocoholics, 138
Christian, 61
Chromophobia, 141
Clathratus, 167
Coffee, 138
Cola drinkers, 138
Collagenase, 332
Computer, 35
Corpus, 56
Cosmetic nanosurgery, 325
Cosmetics, 17
Cosmetics industry, 324
Cro-Magnons, 46
Cystine molecules, 329

D

De novo, 157
Demron, 351
Dendrimer molecule, 7
Dendrimers, 8, 168
Dentistry, 23
Diabetes, 26
Diagnostic procedures, 15
Dipsomania, 139
Disease idealism, 93
DNA, 5, 23
DNA footpath, 357
Domination, 138
Drop zone, 335
Drug therapies, 15
Drug-delivery vehicles, 1
Dystrophin, 13

E

E. coli, 97, 210, 258
Ecosystem protectors, 17
Electron-beam nanolithography, 353
Electronic microchips, 352
Electrostatic rotor, 40
Empedocles, 55
Engines of Creation, 113, 125
Enthusiasm, 328
Enzyme, 10, 359
Erasistratus, 58
Evipan, 85
Ex vivo, 49, 295, 313

F

Fabric, 344
Fantastic Voyage, 1, 9
Faraday constant, 227
Feature, 332
Feynman Grand Prize, 200
Fistula clubs, 66
Floppy legs, 357
Folliculalgia, 139
Freckles, 327
Fresh-breath nanomachine, 331
Friedrich Engels, 66

Fuel cell, 356, 357
Fuel cells, 356
Fullerene Science and Technology, 175

G

Galen, 60
Galen of Pergamum, 59
Galenic medicine, 61
Genetic material, 5
Glamour, 324
Gluttony, 138
Gorgeous models, 325
Grand challenges, 9
Gravidophilia, 138
Gray hair, 325

H

Hair follicles, 327
Hear, 368
Helicobacter pylori, 140
Hemoglobin, 10
Hemophobia, 141
Herophilus, 58
Hertfordshire, 64
Hippocrates, 55
Hippocratic, 61
Hippocratic corpus, 55
Hippocratic medicine, 55
Holism, 102
Holographic video, 345
Homo erectus, 372, 373, 377
Homo sapiens, 46, 372, 373, 377
Houseboat, 346
Humors, 56
Hydrogen bubbles, 356
Hypochondria, 138

I

Ilium, 347
Immune machines, 19, 20
Immune machines, 18
In cyto, 96, 101
In situ, 122
In vacuo, 146, 179, 194, 231, 262, 290
In vitro, 87, 202, 304
In vivo, 49, 96, 99, 120, 130, 132, 136, 210, 220, 238, 247, 256, 260, 265, 271, 272, 282, 294, 295, 301, 304, 312, 315, 316, 317, 318, 319

J

Jewish, 61
Journal of Micromechanics and Microengineering, 149

K

Karl Marx's, 66
King Asa, 61
Kleptomania, 138
Knowing, 328
Knowledge space, 34

L

Librium, 85
Listeria monocytogenes, 132
Lithium carbonate, 85
Lithography, 352
Liver spots, 327

M

Magnetic-sensing, 2
Magnetizations, 3
Magnetotactic bacterium, 3
Malacosteus niger, 293
Marathon runner, 370
Martial, 59
Masai, 47
Mature nanotechnology-based medicine, 17
Mature technology, 358
Medical nanotechnology, 27, 29
Medical sensors, 16
Micro-motor, 40

Microelectronics, 41
Micromechanics, 40
Micromechanics to nanotechnology, 41
Microscopic, 367
Microscopic sensors, 367
Microtechnology, 355
Miniature compass, 3
Mining, 36
Mitochondrion, 359
Mobilophilia, 138
Modeling, 36
Molecular chemists, 165
Molecular machines, 10, 12, 25
Molecular manufacturing, 10
Molecular medicine, 26
Molecular repairs, 25
Molecular surgery, 25
Molecules, 10
Multiwalled Carbon Nanotubes, 352
Muscular dystrophy, 13
MWCNTs, 352

N

Nano, 352
Nanocomputer, 18
Nanocrystal, 3, 4
Nanoelectronics, 355
Nanoengineered tower, 347
Nanograss, 356
Nanograss battery, 356
Nanoinstruments, 21
Nanomachine, 17, 325
Nanomachine-laden depilatory cream, 328
Nanomanufactured medical devices, 27
Nanomedicine, 17, 48, 49, 125
Nanoparts, 188
Nanoscale, 1, 24
Nanoscale building blocks, 2
Nanoscale engineering, 3
Nanoscale technology, 356
Nanoscale tools, 23
Nanoshell, 8, 9
Nanosuit, 344
Nanosystems, 203
Nanotech, 371
Nanotech synthesizers, 342
Nanotechnology, 1, 2, 10, 16, 346
Nanotubes, 352
Napoleon, 68
Neanderthal, 46
Negative heat, 305
Noctiphobia, 141
Non bonded interactions, 350
Noninvasive imaging, 1
Nucleotides, 361
Numerically Controlled, 43
Nymphomania, 138

O

On Medicine, 60
Ophidiophobia, 141
Organ Cryopreservation Project, 24
Organic brain, 366
Organic dendrimers, 7
Ovonics, 36

P

Painlessly, 27
Paraplegics, 13
Parthenophobia, 141
Pathological liars, 138
Percivall Pott, 66
Persistent folding, 332
Pharmaceutical industry, 43
Pharmacologia, 105
Phengophobia, 141
Phobophobia, 141
Phosphorus, 329
Piezoelectric objects, 40

Plague, 29, 62
Polygamy, 138
Polymer capsules, 8
Polymerase chain reaction, 43
Praxagoras, 58
Prismanes, 169
Prophylaxis, 102
Proton-exchange membrane, 357, 360
Protonanotechnology, 358
Pulmonary silicosis, 52

Q

Quadriplegics, 13
Quantum dots, 4, 5, 6

R

Red hair, 325
Respirocyte, 196, 197
Reticulomyxa, 308
Reynolds number, 310
Ribosome, 361
Richard Wiseman, 64
Robotics, 124
Rome, 59
Rudimentary nanomachines, 326
Runner's high, 138
Runs downhill, 359

S

Salmonella typhimurium, 120
Satyriasis, 138
Schistosomiasis, 52
Schizophrenia, 139
Scotophobia, 141
Self-assembled monolayers, 167
Self-assembly, 357, 359
Self-replicating machine systems, 188
Self-replication, 168, 197
Sex, 349
Sickle cell anemia, 13
Siderophobia, 141
Skin, 17
Smart, 326, 331
Smart materials, 31
Smokers, 138
Soft nanolithography, 354
Solar collectors, 346
Spendoholics, 138
Speophyes lucidulus, 287
Staphylococcus epidermidis, 97
Stephen Bradwell, 64
Stone Age, 49
Streptococcus, 92
Suction, 353
Suit, 344
Sulfur, 328
Sulphonamides, 85
Super implanted PDA, 369
Supramolecular chemistry, 166
Surgery, 15
Sweet tooth, 138
Symport mechanisms, 227
Syphilis, 17

T

Talk, 368
Target molecule, 246
Tartar, 330
Telemagination, 377
Telepresence, 348
Telescopic, 367
The Apologie and Treatise, 64
The End of Nature, 371
The Future and Its Enemies, 137
The Society of Mind, 110
The Temperature in Diseases, 69
Thomas Young, 67
Thorazine, 85
Thrillseeking, 138
Titanium dioxide, 351
Transfer RNA, 362
Transistors, 39

Triskaidekaphobia, 141
Trypanosoma cruzi, 132
Tunable optical reflectors, 345

U

Unbounding the Future, 112
Unwanted hair, 328
Upward compatibility, 376
Urticaria, 139

V

Valium, 85
Van der Waals bonds, 350
Venesection, 64
Virtual reality, 347
Viruses, 13
Vis medicatrix naturae, 56

W

Wagerphilia, 138
Wall Street Journal, 324
William Cheselden, 65
Work order, 362
Workaholics, 138
Wrinkles, 332

X

Xenophobia, 141

Z

Zoophobia, 141